STUDENT'S SOLUTIONS MANUAL

BEVERLY FUSFIELD

CALCULUS & ITS APPLICATIONS

TWELFTH EDITION

AND BRIEF CALCULUS & ITS APPLICATIONS

TWELFTH EDITION

Larry J. Goldstein
Goldstein Educational Technologies

David C. Lay
University of Maryland

David I. Schneider
University of Maryland

Nakhlé H. Asmar
University of Missouri

Prentice Hall
is an imprint of

Reproduced by Pearson Prentice Hall from electronic files supplied by the author.

Publishing as Prentice Hall, Upper Saddle River, NJ 07458.

ISBN-13: 978-0-321-59901-8
ISBN-10: 0-321-59901-2

2 3 4 5 6 BB 12 11 10 09

Prentice Hall
is an imprint of

www.pearsonhighered.com

Table of Contents

How To Use This Manual ..... *v*

Chapter 0 Functions 1

Chapter 1 The Derivative 17

Chapter 2 Applications of the Derivative 44

Chapter 3 Techniques of Differentiation 74

Chapter 4 The Exponential and Natural Logarithm Functions 90

Chapter 5 Applications of the Exponential and Natural Logarithm Functions 109

Chapter 6 The Definite Integral 121

Chapter 7 Functions of Several Variables 139

Chapter 8 The Trigonometric Functions 156

Chapter 9 Techniques of Integration 169

Chapter 10 Differential Equations 190

Chapter 11 Taylor Polynomials and Infinite Series 221

Chapter 12 Probability and Calculus 235

Visual Calculus Software VC1

How To Use This Manual

Proper use of this manual can help to improve your performance in your calculus course. There are detailed explanations here to guide you carefully and patiently through hundreds of exercises. And there are helpful hints and strategies for studying that several thousand of our students have used successfully over the past twenty years. Students who apply the techniques in this manual consistently score from five to fifteen percentage points higher on campus-wide final exams than students who have no systematic strategy for mastering the material in this course.

The manual is organized by the sections in *Calculus and Its Applications, Twelfth Edition,* by Goldstein, Lay, Schneider, and Asmar. Most sections begin with some remarks about the section, its general purpose and layout, or its relation to other material. Following that, there are complete solutions to every sixth problem. Essentially all of the basic skills are covered in these problems. The solutions are written in the style of the examples in the text, but they tend to contain more warnings and side remarks than in the text. At the end of many chapters you will find a brief review to help you prepare for an exam.

The way you use this manual should be determined by an overall strategy for success in your calculus course. We have listed below five things you can do to improve your performance and make the course an enjoyable experience. Hardly a semester passes without several of our students writing on their (anonymous) course evaluations that they never thought they would enjoy a math course, and yet they found themselves liking this course more than most of the courses in their major. If you have the proper background, there is no reason why you cannot do well in this course, and enjoy it at the same time!

Secrets for Success in Calculus

1. **Work on exercises nearly every day, keeping up with the lectures in your class.** You will minimize the time you need to learn calculus if you heed this advice. Concepts in calculus are usually introduced and explained in terms of concepts you have studied earlier. When you fall behind, the class explanations will be harder to understand and you will spend more time on your own trying to catch up. In this sense, studying calculus is like learning a foreign language, where each lesson builds on the previous lessons, and if you miss one lesson you may not understand a thing that is said in class.

2. **Read the text and examples carefully before you attempt the exercises.** Then try the exercises. If you get "stuck" on a problem, look for a relevant example, but don't reread the whole example. Read just enough to get you started on the exercise. If you studied the example earlier, parts of the explanation may still be floating in your mind, and the quick glance at the example may be all you need. If the problem is complex and you get stuck again, take another peek at the example. This process of trying to recall an example that you have already studied is the key to learning how to work the problem yourself.

Many students read the examples after they have tried (unsuccessfully) to work an exercise. When a problem seems difficult, a student may look for a similar example and "copy" the example with the numbers changed. In this approach very little learning takes place. *Being able to read* an *explanation and understand each step is entirely different from really knowing how to work the problem by yourself.* We cannot emphasize this too strongly. You must first attempt a problem by yourself, with no text or manual to help you. When you need help, accept only enough assistance to get started on the problem.

3. **Treat this manual as a tutor.** One function of a tutor is to point out where you are making mistakes. After you work a problem whose solution is in the manual, compare your solution to the one printed here. If your final answer is correct, have you included all the important steps? (We have included nearly all algebraic steps. You may feel comfortable with fewer steps, but be careful. Check with your instructor if you're not sure about how much detail is desired.) If your final answer is incorrect, find your first mistake and correct it. If the solution is long, then you may uncover other mistakes by reworking the part that followed your first mistake.

   Another function of a tutor is to help you on a problem when you cannot proceed further. If you get stuck on a problem that has a solution in this manual, and if the examples in the text are not sufficient, then read the first part of the solution. But only read enough to get started. *Don't sit back and watch your "tutor" work the entire problem.* You must have the practice of working it yourself. Return to the solution later for more help, if necessary.

4. **Keep a list of places where errors are likely to occur.** The text warns of some potential errors. This manual will point out many more. Identify your weaknesses. As you work exercises, make a note of the most common mistakes you make. Do the same with any quizzes you take. Then review the list before each exam.

5. **Develop a strategy for taking each test.** You should be able to walk into an exam knowing what most of the questions will concern and knowing where your own strengths and weaknesses lie.
   a. Read over the entire exam for a few minutes instead of beginning work immediately on the first problem. Then as you work one problem, your subconscious mind can be thinking about related problems that appear elsewhere on the exam. (This strategy works for math exams just as well as it does for essay exams, once you learn how to do it.)
   b. Begin the exam by working on the problem in which you have the most confidence. Successful completion of one or two problems will help to calm the "test anxiety" (or panic!) that most people feel to some extent.
   c. Move from easier to harder problems. Avoid getting bogged down in a problem where you spend a lot of time and yet have the potential to earn only a few points.

   There are other suggestions we could give, but that's probably enough for now. We'll be adding to this list at strategic points in the manual.

A Warning

Although this manual has the potential to help you tremendously, it also has the power to undermine your chances of success. Because the manual contains complete solutions to many homework problems, you will be tempted to read the explanations here instead of working the problems yourself. While this may shorten the time you spend on homework, it will have a *disastrous* effect on your exam performance! In the long run, a proper use of the manual *will* save you time and it will make the time you spend more productive, but you must use the manual wisely.

Chapter 0
Functions

0.1 Functions and Their Graphs

You should read Sections 0.1 and 0.2 even if you don't plan to work the exercises. The concept of a function is a fundamental idea, and the notation for functions is used in nearly every section of the text.

1. The notation [–1, 4] is equivalent to $-1 \le x \le 4$. Both –1 and 4 belong to the interval.

−1 0 4

7. The inequality $2 \le x < 3$ describe the half-open interval [2, 3). The right parenthesis indicates that 3 is not in the set.

13. If $f(x) = x^2 - 3x$, then

$$\begin{aligned} f(0) &= (0)^2 - 3(0) = 0, \\ f(5) &= (5)^2 - 3(5) = 25 - 15 = 10, \\ f(3) &= (3)^2 - 3(3) = 9 - 9 = 0, \\ f(-7) &= (-7)^2 - 3(-7) = 49 + 21 = 70. \end{aligned}$$

19. If $f(x) = x^2 - 2x$, then

$$\begin{aligned} f(a+1) &= (a+1)^2 - 2(a+1) = (a+1)(a+1) - 2a - 2 \\ &= (a^2 + 2a + 1) - 2a - 2 = a^2 - 1. \\ f(a+2) &= (a+2)^2 - 2(a+2) = (a+2)(a+2) - 2a - 4 \\ &= (a^2 + 4a + 4) - 2a - 4 = a^2 + 2a. \end{aligned}$$

25. The domain of $g(x) = \dfrac{1}{\sqrt{3-x}}$ consists of those x for which $x < 3$, since division by zero is not permissible and since square roots of negative numbers are not defined.

31. This is not the graph of a function by the vertical line test.

37. By referring to the graph, we find that $f(4)$ is positive since the graph of the function is above the x-axis when $x = 4$.

43. The concentration of the drug when $t = 1$ is 0.03 units because the graph tells us that $f(1) = 0.03$.

49. If the point $\left(\frac{1}{2}, \frac{2}{5}\right)$ is on the graph of the function $g(x) = \frac{3x-1}{x^2+1}$, then $g\left(\frac{1}{2}\right)$ must equal $\frac{2}{5}$. This is the case since

$$g\left(\frac{1}{2}\right) = \frac{3\left(\frac{1}{2}\right)-1}{\left(\frac{1}{2}\right)^2+1} = \frac{\frac{3}{2}-1}{\frac{1}{4}+1} = \frac{\frac{1}{2}}{\frac{5}{4}} = \frac{1}{2}\cdot\frac{4}{5} = \frac{2}{5}$$

so $\left(\frac{1}{2}, \frac{2}{5}\right)$ is on the graph.

55. Since $f(x) = \pi x^2$ for $x < 2$, $f(1) = \pi(1)^2 = \pi$.
Since $f(x) = 1 + x$ for $2 \le x \le 2.5$, $f(2) = 1 + 2 = 3$.
Since $f(x) = 4x$ for $2.5 < x$, $f(3) = 4(3) = 12$.

61. Entering **Y₁ = X ^ 3 / 4** will graph the function $f(x) = \frac{x^3}{4}$. In order to graph the function $y = x^{3/4}$, you need to include parentheses in the exponent: **Y₁ = X ^ (3/4)**.

Help for Technology Exercises: The examples in the text use the family of TI-83/84 calculators. Helpful information about the use of calculators appears at the end of most sections in subsections titled **Incorporating Technology**. You might wish to place a paper or plastic tab in your text, to help you reference the appropriate information.

0.2 Some Important Functions

The most important functions here are the linear functions, the quadratic functions, and the power functions. Absolute values appear only briefly in Sections 4.5 and 6.1, and then more frequently in Chapters 9 and 10.

1. The function $f(x) = 2x - 1$ is linear. Since a line is determined by any two of its points, we may choose any two points on the graph of $f(x)$ and draw the line through them. For instance, we find

$$f(0) = 2(0) - 1 = 0 - 1 = -1 \quad \text{and} \quad f(1) = 2(1) - 1 = 2 - 1 = 1.$$

So (0, –1) and (1, 1) are two points on the graph of $f(x)$.

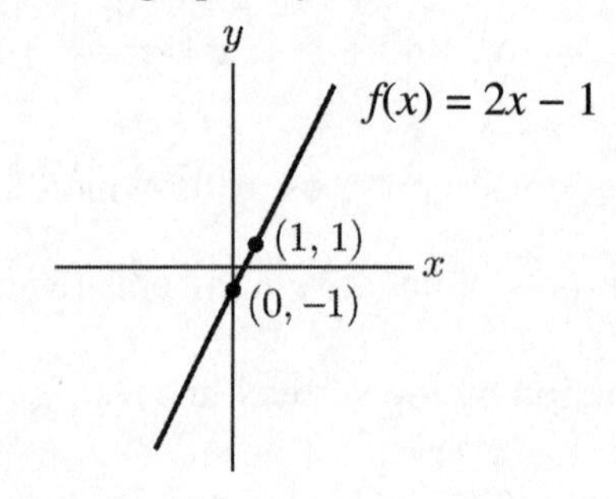

7. To find the y-intercept, evaluate $f(x)$ at $x=0$.

$$f(0)=9(0)+3=0+3=3.$$

So the y-intercept is (0, 3). To find the x-intercept, set $f(x)=0$ and solve for x.

$$9x+3=0,$$
$$9x=-3,$$
$$x=-\frac{3}{9}=-\frac{1}{3}.$$

So the x-intercept is $\left(-\frac{1}{3},0\right)$.

13. **(a)** To write $f(x)=0.2x+50$ in the form $f(x)=\left(\frac{K}{V}\right)+\frac{1}{V}$, set

$$\frac{K}{V}=0.2 \quad \text{and} \quad \frac{1}{V}=50.$$

If $\frac{1}{V}=50$, then $V=\frac{1}{50}$. Substitute $\frac{1}{50}$ for V in $\frac{K}{V}=0.2$ to get

$$\frac{K}{\left(\frac{1}{50}\right)}=0.2, \quad \text{and hence} \quad K=0.2\left(\frac{1}{50}\right)=\frac{2}{500}=\frac{1}{250}.$$

(b) To find the y-intercept evaluate the function at $x=0$.

$$y=\left(\frac{K}{V}\right)(0)+\frac{1}{V}=\frac{1}{V}.$$

The y-intercept is $\left(0,\frac{1}{V}\right)$. To find the x-intercept set $y=0$ and solve for x:

$$\left(\frac{K}{V}\right)x+\frac{1}{V}=0, \quad \left(\frac{K}{V}\right)x=-\frac{1}{V}, \quad \text{and} \quad x=\left(\frac{V}{K}\right)\left(-\frac{1}{V}\right)=-\frac{1}{K}.$$

The x-intercept is $\left(-\frac{1}{K},0\right)$.

19. $f(x)=\dfrac{50x}{105-x}, \quad 0\le x\le 100,$

$f(70)=100$ (million dollars) from Example 5,

$f(75)=\dfrac{50(75)}{105-75}=\dfrac{50(75)}{30}=125$ (million dollars).

Thus, the added cost to remove another 5% is:

$$f(75)-f(70)=125-100 \text{ (million dollars)}$$
$$=25 \text{ million dollars.}$$

From Example 5, the cost of removing the final 5% of the pollutant is 525 million dollars. This is twenty-one times the cost of removing an extra 5% of pollutant after 70% has been removed.

25. Write $y = 1 - x^2$ in the form $y = ax^2 + bx + c$. That is, $y = (-1)x^2 + (0)x + 1$, so $a = -1, b = 0$, and $c = 1$.

31. This function is defined by three distinct linear functions. For $0 \le x < 2, f(x) = 4 - x$. This graph is determined by two points, say at $x = 0$ and $x = 2$.

$$f(0) = 4 - 0 = 4, \qquad f(2) = 4 - 2 = 2.$$

So (0, 4) and (2, 2) determine this part of the graph. Draw a line segment between these points. Note that (0, 4) is part of the graph but (2, 2) cannot yet be counted as part of the graph since $x < 2$. For $2 \le x < 3$, we have $f(x) = 2x - 2$. This graph is also determined by two points, say at $x = 2$ and $x = 3$.

$$f(2) = 2(2) - 2 = 2, \qquad f(3) = 2(3) - 2 = 4.$$

So (2, 2) and (3, 4) determine this part of the graph. Draw a line segment between these points. Now (2, 2) is part of the graph, but (3, 4) cannot yet be counted as part of the graph since $x < 3$. For $x \ge 3$ we have $f(x) = x + 1$. This graph is also determined by two points, say at $x = 3$ and $x = 4$:

$$f(3) = 3 + 1 = 4, \qquad f(4) = 4 + 1 = 5.$$

So (3, 4) and (4, 5) determine this part of the graph. Draw a line segment through these points and extend it to the right as x tends to infinity.

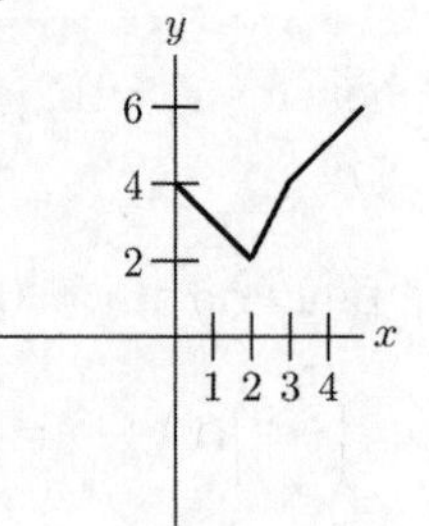

37. $f(x) = |x|$, where

$$|x| = \begin{cases} x & \text{if } x \text{ is positive or zero,} \\ -x & \text{if } x \text{ is negative.} \end{cases}$$

Here $x = -2.5$, which is a negative number, so $|x| = -(-2.5) = 2.5$. This gives $f(x) = 2.5$ when $x = -2.5$.

0.3 The Algebra of Functions

The algebraic skills in this section will be used frequently throughout the course. You should review the material here on composition of functions before you read Section 3.2.

1. $f(x) + g(x) = (x^2 + 1) + 9x = x^2 + 9x + 1.$

7. $f(x)+g(x)=\dfrac{2}{x-3}+\dfrac{1}{x+2}$.

In order to add two fractions, the denominators must be the same. A common denominator for

$$\frac{2}{x-3} \quad \text{and} \quad \frac{1}{x+2}$$

is $(x-3)(x+2)$. If we multiply

$$\frac{2}{x-3} \quad \text{by} \quad \frac{x+2}{x+2}$$

we obtain an equivalent expression whose denominator is $(x-3)(x+2)$. Similarly, if we multiply

$$\frac{1}{x+2} \quad \text{by} \quad \frac{x-3}{x-3}$$

we obtain an equivalent expression whose denominator is $(x-3)(x+2)$. Thus

$$\begin{aligned} f(x)+g(x) &= \frac{2}{x-3}+\frac{1}{x+2} \\ &= \frac{2}{x-3}\cdot\frac{x+2}{x+2}+\frac{1}{x+2}\cdot\frac{x-3}{x-3} \\ &= \frac{2(x+2)}{(x-3)(x+2)}+\frac{x-3}{(x-3)(x+2)} \\ &= \frac{2x+4+x-3}{(x-3)(x+2)}=\frac{3x+1}{(x-3)(x+2)}=\frac{3x+1}{x^2-x-6}. \end{aligned}$$

13. $f(x)-g(x)=\dfrac{x}{x-2}-\dfrac{5-x}{5+x}$. A common denominator for

$$\frac{x}{x-2} \quad \text{and} \quad \frac{5-x}{5+x}$$

is $(x-2)(5+x)$.

Proceeding as in Exercise 7:

$$\begin{aligned} f(x)-g(x) &= \frac{x}{x-2}\cdot\frac{5+x}{5+x}-\frac{5-x}{5+x}\cdot\frac{x-2}{x-2} \\ &= \frac{x(5+x)}{(x-2)(5+x)}-\frac{(5-x)(x-2)}{(x-2)(5+x)} \\ &= \frac{5x+x^2-(-x^2+7x-10)}{(x-2)(5+x)} \\ &= \frac{2x^2-2x+10}{x^2+3x-10}. \end{aligned}$$

19. $f(x+1)g(x+1) = \left[\dfrac{(x+1)}{(x+1)-2}\right]\left[\dfrac{5-(x+1)}{5+(x+1)}\right]$

$$= \left[\frac{x+1}{x-1}\right]\left[\frac{4-x}{6+x}\right] = \frac{(x+1)(4-x)}{(x-1)(6+x)}$$

$$= \frac{-x^2+3x+4}{x^2+5x-6}.$$

25. To find $f(g(x))$, substitute $g(x)$ in place of each x in $f(x) = x^6$. Thus

$$f(x) = (g(x))^6 = \left(\frac{x}{1-x}\right)^6.$$

31. If $f(x) = x^2$ then,

$$\begin{aligned} f(x+h) - f(x) &= (x+h)^2 - x^2 \\ &= x^2 + 2hx + h^2 - x^2 \\ &= 2hx + h^2. \end{aligned}$$

37. $f(x) = \frac{1}{8}x$, $g(x) = 8x+1$. Substitute $g(x)$ in place of each x in $f(x)$. Thus

$$\begin{aligned} h(x) = f(g(x)) &= \frac{1}{8}g(x) \\ &= \frac{1}{8}[8x+1] \\ &= x + \frac{1}{8}. \end{aligned}$$

$g(x)$ converts British sizes to French sizes, while $f(x)$ converts French sizes to U.S. sizes. Thus, $h(x) = f(g(x))$ converts British sizes to U.S. sizes. For example, consider the British hat size of $6\frac{3}{4}$. Since $h(x) = x + \frac{1}{8}$, the U.S. size is $h\left(6\frac{3}{4}\right) = 6\frac{3}{4} + \frac{1}{8} = 6\frac{7}{8}$, which agrees with the table.

43. On a TI-83, if $Y_1 = X/(X-1)$, set $Y_2 = Y_1(Y_1)$. Before graphing Y_2, you should "deselect" Y_1, so its graph will not appear also. Use the TRACE feature to get an impression of what the formula for Y_2 might be.

To actually determine the formula for $f(f(x))$, substitute $x/(x-1)$ for each occurrence of x in the formula for $f(x)$. (Note that $f(x)$ is not defined for $x = 1$.)

$$f(f(x)) = \frac{x/(x-1)}{[x/(x-1)]-1}, \quad x \neq 1.$$

The denominator can be simplified:

$$\frac{x}{x-1} - 1 = \frac{x}{x-1} - \frac{x-1}{x-1} = \frac{x-(x-1)}{x-1} = \frac{1}{x-1}.$$

Multiply numerator and denominator of the formula for $f(f(x))$ by $x-1$ and obtain

$$f(f(x)) = \frac{x/(x-1)}{1/(x-1)} = \frac{x}{1} = x, \quad x \neq 1.$$

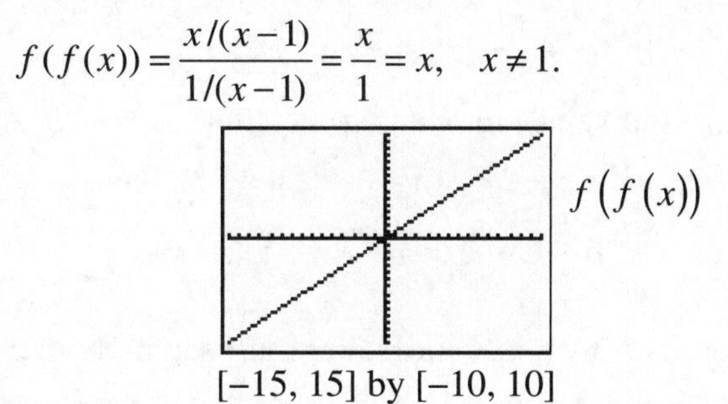

[−15, 15] by [−10, 10]

0.4 Zeros of Functions—The Quadratic Formula and Factoring

The material in this section will be used routinely throughout the text, beginning in Section 2.3. (Factoring skill is also needed for a few exercises in Section 1.4.) By Section 2.3 you must have mastered the ability to factor quadratic polynomials and to solve problems such as Exercises 33–38 (in Section 0.4). You should also be able to factor cubic polynomials where every term contains a power of x. (See (b) and (c) of Example 7.) Problems such as those in Exercises 25–32 will appear in Section 6.4 and in later sections. The quadratic formula should be memorized although you will not need to use it as often as factoring skills.

1. In the equation $2x^2 - 7x + 6 = 0$, we have $a = 2$, $b = -7$, and $c = 6$. Substitute these values into the quadratic formula and obtain

$$x = \frac{-(-7) \pm \sqrt{(-7)^2 - 4(2)(6)}}{2(2)} = \frac{7 \pm \sqrt{49-48}}{4} = \frac{7 \pm 1}{4}.$$

Thus, $$x = \frac{7+1}{4} = 2 \quad \text{or} \quad x = \frac{7-1}{4} = \frac{3}{2}.$$

So the zeros of the function $2x^2 - 7x + 6$ are 2 and $\frac{3}{2}$.

7. In the equation $5x^2 - 4x - 1 = 0$, we have $a = 5$, $b = -4$, and $c = -1$. By the quadratic formula,

$$\begin{aligned} x &= \frac{-(-4) \pm \sqrt{(-4)^2 - 4(5)(-1)}}{2(5)} \\ &= \frac{4 \pm \sqrt{16+20}}{10} = \frac{4 \pm \sqrt{36}}{10} \\ &= \frac{4 \pm 6}{10}. \end{aligned}$$

So the solutions of the equation $5x^2 - 4x - 1 = 0$ are

$$x = \frac{4+6}{10} = \frac{10}{10} = 1, \quad \text{and} \quad x = \frac{4-6}{10} = \frac{-2}{10} = -\frac{1}{5}.$$

13. To factor $x^2 + 8x + 15$, find values for c and d such that $cd = 15$ and $c + d = 8$. The solution is $c = 3, d = 5$, and

$$x^2 + 8x + 15 = (x+3)(x+5).$$

19. To factor $30-4x-2x^2$, first factor out the coefficient –2 in front of x^2. That is,

$$30-4x-2x^2=-2(x^2+2x-15).$$

Then to factor $x^2+2x-15$, find values for c and d such that $cd=-15$ and $c+d=2$. The solution is $c=5, d=-3$, and $x^2+2x-15=(x+5)(x-3)$. Thus

$$30-4x-2x^2=-2(x+5)(x-3).$$

25. If a point (x, y) is on both graphs, then its coordinates must satisfy both equations. That is, x and y must satisfy $y=2x^2-5x-6$ and $y=3x+4$. Equate the two expressions for y:

$$2x^2-5x-6=3x+4.$$

To use the quadratic formula, rewrite the equation in the form

$$2x^2-8x-10=0,$$

and divide both sides by 2:

$$x^2-4x-5=0.$$

By the quadratic formula,

$$x=\frac{-(-4)\pm\sqrt{(-4)^2-4(1)(-5)}}{2(1)}=\frac{4\pm\sqrt{16+20}}{2}$$
$$=\frac{4\pm\sqrt{36}}{2}=\frac{4\pm6}{2}.$$

So $x=\frac{4+6}{2}=5$ or $x=\frac{4-6}{2}=-1$. Thus the x-coordinates of the points of intersection are 5 and –1. To find the y-coordinates, substitute these values of x into either equation, $y=2x^2-5x-6$, or $y = 3x + 4$. Since $y=3x+4$ is simpler, use this equation to find that at $x = 5$, $y = 3(5) + 4 = 15 + 4 = 19$. Also, at $x=-1$, $y=3(-1)+4=-3+4=1$. Thus the points of intersection are (5, 19) and (–1, 1).

31. As in Exercise 25, equate the two expressions for y to get

$$\frac{1}{2}x^3+x^2+5=3x^2-\frac{1}{2}x+5.$$

Then rewrite this as

$$\frac{1}{2}x^3-2x^2+\frac{1}{2}x=0,$$

and factor out a common factor of x to get

$$x\left(\frac{1}{2}x^2-2x+\frac{1}{2}\right)=0.$$

This shows that one of the points of intersection has x-coordinate 0. To find the x-coordinates of the remaining points of intersection, apply the quadratic formula to $\frac{1}{2}x^2-2x+\frac{1}{2}=0$:

$$\begin{aligned} x &= \frac{-(-2) \pm \sqrt{(-2)^2 - 4\left(\frac{1}{2}\right)\left(\frac{1}{2}\right)}}{2\left(\frac{1}{2}\right)} \\ &= \frac{2 \pm \sqrt{4-1}}{1} \\ &= 2 \pm \sqrt{3}. \end{aligned}$$

Substitute these values into $y = 3x^2 - \frac{1}{2}x + 5$. At $x = 0$, $y = 3(0)^2 - \frac{1}{2}(0) + 5 = 5$.

At $x = 2 + \sqrt{3}$,

$$\begin{aligned} y &= 3(2+\sqrt{3})^2 - \frac{1}{2}(2+\sqrt{3}) + 5 \\ &= 3(7+4\sqrt{3}) - \frac{1}{2}(2+\sqrt{3}) + 5 \\ &= 21 + 12\sqrt{3} - 1 - \frac{1}{2}\sqrt{3} + 5 \\ &= 25 + \frac{23}{2}\sqrt{3}. \end{aligned}$$

At $x = 2 - \sqrt{3}$,

$$\begin{aligned} y &= 3(2-\sqrt{3})^2 - \frac{1}{2}(2-\sqrt{3}) + 5 \\ &= 3(7-4\sqrt{3}) - 1 + \frac{1}{2}\sqrt{3} + 5 \\ &= 21 - 12\sqrt{3} + 4 + \frac{1}{2}\sqrt{3} \\ &= 25 - \frac{23}{2}\sqrt{3}. \end{aligned}$$

Thus the points of intersection are:

$$(0, 5), \quad \left(2+\sqrt{3}, 25+\frac{23}{2}\sqrt{3}\right), \text{ and } \left(2-\sqrt{3}, 25-\frac{23}{2}\sqrt{3}\right).$$

37. A rational function will be zero only if the numerator is zero. Solve

$$\begin{aligned} x^2 + 14x + 49 &= 0, \\ (x+7)^2 &= 0, \\ x + 7 &= 0. \end{aligned}$$

That is, $x = -7$. Since the denominator is not 0 at $x = -7$, the solution is $x = -7$.

43. You can graph the function and use **TRACE** to estimate the zero as approximately 4.6. (Your estimate is influenced by your choice of viewing window.) Another procedure, which is usually more accurate, is described on page 32 in the text.

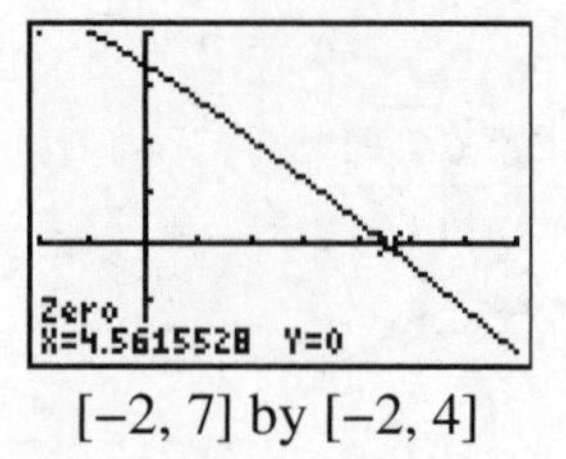

$[-2, 7]$ by $[-2, 4]$

Helpful Hint: Exercises 49–52 are somewhat difficult because you do not yet have the tools of calculus available. To find an interval that contains the zeros of $f(x)$, use the comment that follows Figure 7 in the text. In Exercise 51, for example, write

$$f(x) = 3\left(x^3 + \frac{52}{3}x^2 - 4x - 4\right)$$

and let M be the number that is one more than the largest magnitude of the coefficients 52/3, –4, and –4. That is, let $M = 55/3$. Then the zeros of $f(x)$ lie between $-M$ and M. For a first graphing attempt, use x values in the interval $-55/3 \le x \le 55/3$, or perhaps $-19 \le x \le 19$.

49. Since $f(x) = x^3 - 22x^2 + 17x + 19$, and the coefficient of the highest power of x is 1, you can let M be the number that is one more that the largest magnitude of the coefficients –22, 17, and 19. That is, let $M = 23$. Then the zeros of the polynomial lie between $-M$ and M, in the interval $-23 \le x \le 23$. Sketch the graph in the window $[-23, 23]$ *by* $[-10, 10]$

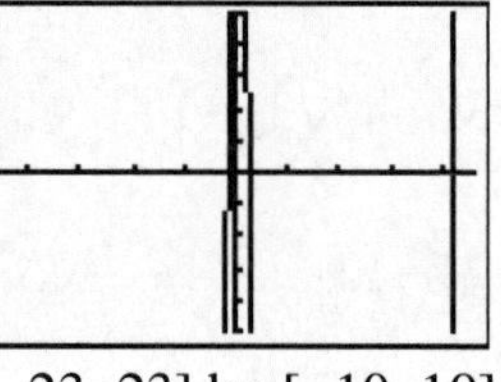
$[-23, 23]$ by $[-10, 10]$

In this first window, you see three nearly vertical lines. The "lines" connect somewhere off the screen. So enlarge the range of y-values, say, $-100 \le y \le 100$. Also, since two of the "lines" in the graph are close to $x = 0$ and one is close to $x = 23$, you can chop off most of the negative x-axis. Try the window $[-5, 25]$ by $[-100, 100]$.

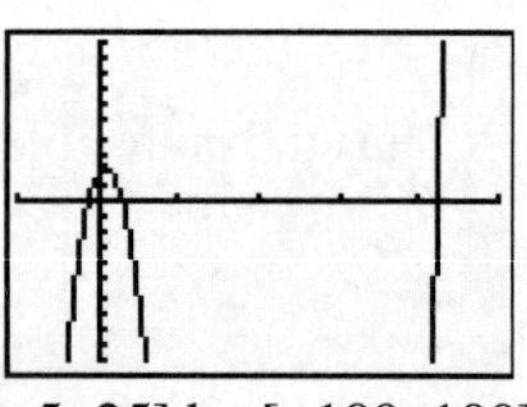
$[-5, 25]$ by $[-100, 100]$

This time, you should observe that the y-values seem to get very negative. The **TRACE** command might be helpful now, to find out about how negative the y-values become. Move the cursor along the curve until it disappears at the bottom of the screen. The coordinates of points on the graph will continue to appear on the screen even when the cursor is not visible. Watching the y-coordinates, you should see them go down to about -1300. So, try the window $[-5, 25]$ *by* $[-1500, 100]$. (When you use such a large y-range, set the y-scale to, say, 100.)

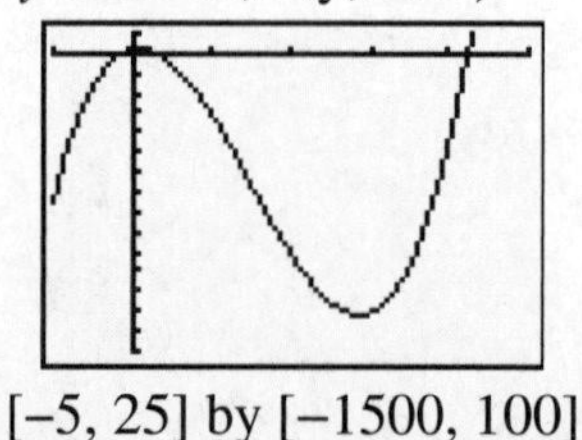
$[-5, 25]$ by $[-1500, 100]$

0.5 Exponents and Power Functions

We have found that operations with exponents cause our students more difficulty than any other algebraic skill, so we have included lots of drill exercises. Try some of each group of problems. If you cannot work them accurately with relative ease and confidence, keep working more problems. If you need more practice, get a college algebra text or Schaum's *Outline of College Algebra.* Do this immediately, because operations with exponents are used in Section 1.3 and in most sections thereafter.

Helpful Hint: We are all so accustomed to reading formulas from left to right that it is more difficult to use a formula such as $b^r b^s = b^{r+s}$ in "reverse," that is, in the form $b^{r+s} = b^r b^s$. Yet the laws of exponents are sometimes needed in reverse from the way they are written in the text. So *memorize* the following list, which incorporates the list on Page 34 of the text.

Laws of Exponents (continued)

1. $b^r \cdot b^s = b^{r+s}$

1′. $b^{r+s} = b^r \cdot b^s$

2. $b^{-r} = \dfrac{1}{b^r}$

2′. $\dfrac{1}{b^r} = b^{-r}$

3. $\dfrac{b^r}{b^s} = b^r \cdot b^{-s} = b^{r-s}$

3′. $b^{r-s} = b^r \cdot b^{-s} = \dfrac{b^r}{b^s}$

4. $(b^r)^s = b^{rs}$

4′. $b^{rs} = (b^r)^s$

5. $(ab)^r = a^r b^r$

5′. $a^r b^r = (ab)^r$

6. $\left(\dfrac{a}{b}\right)^r = \dfrac{a^r}{b^r}$

6′. $\dfrac{a^r}{b^r} = \left(\dfrac{a}{b}\right)^r$

A common use of Law 4′ is in the form $b^{m/n} = (b^{1/n})^m$. For instance, $9^{3/2} = (9^{1/2})^3 = (3)^3 = 27$, and $27^{4/3} = (27^{1/3})^4 = (3)^4 = 81$. Most instructors will assume that you can compute the square roots of the following numbers without using a calculator:

$$4, 9, 16, 25, 36, 49, 64, 81, 100.$$

You should also know the following cube roots:

$$8^{1/3} = 2, \quad 27^{1/3} = 3, \quad 64^{1/3} = 4, \quad \text{and possibly} \quad 125^{1/3} = 5.$$

It wouldn't hurt also to learn the following fourth roots:

$$16^{1/4} = 2,$$
$$81^{1/4} = 3.$$

1. $3^3 = 3 \cdot (3 \cdot 3) = 3 \cdot 9 = 27.$

7. $-4^2 = -16$. Note that the exponent 2 does not act on the negative sign. That is, -4^2 is not the same as $(-4)^2 = (-4) \cdot (-4) = +16$.

13. $6^{-1} = \dfrac{1}{6}$ (Law 2)

19. $(25)^{3/2} = (25^{1/2})^3$ (Law 4′)
$= 5^3$
$= 125$

25. $4^{-1/2} = \dfrac{1}{4^{1/2}}$ (Law 2)
$= \dfrac{1}{2}$

31. $6^{1/3} \cdot 6^{2/3} = 6^{1/3+2/3}$ (Law 1)
$= 6^1$
$= 6$

37. $\left(\dfrac{8}{27}\right)^{2/3} = \dfrac{8^{2/3}}{27^{2/3}}$ (Law 6)
$= \dfrac{(8^{1/3})^2}{(27^{1/3})^2}$ (Law 4′)
$= \dfrac{2^2}{3^2} = \dfrac{4}{9}$

43. $\dfrac{x^4 \cdot y^5}{xy^2} = \left(\dfrac{x^4}{x}\right)\left(\dfrac{y^5}{y^2}\right)$
$= (x^{4-1})(y^{5-2})$ (Law 3)
$= x^3y^3$

49. $(x^3y^5)^4 = (x^3)^4(y^5)^4$ (Law 5)
$= x^{12}y^{20}$ (Law 4)

55. $\dfrac{-x^3y}{-xy} = \dfrac{x^3y}{xy} = \dfrac{x^3}{x} \cdot \dfrac{y}{y}$
$= x^{3-1}y^{1-1}$ (Law 3)
$= x^2y^0 = x^2$

61. $\left(\dfrac{3x^2}{2y}\right)^3 = \dfrac{(3x^2)^3}{(2y)^3}$ (Law 6) [Don’t forget parentheses]
$= \dfrac{3^3(x^2)^3}{2^3y^3}$ (Law 5) [Careful: This step is where mistakes usually happen.]
$= \dfrac{27(x^2)^3}{8y^3} = \dfrac{27x^6}{8y^3}$ (Law 4)

67. $\sqrt{x}\left(\frac{1}{4x}\right)^{5/2} = x^{1/2}\left(\frac{1}{4x}\right)^{5/2}$

$= x^{1/2}\cdot\frac{1^{5/2}}{(4x)^{5/2}}$ (Law 6) [Careful: Don't forget parentheses around $4x$]

$= x^{1/2}\cdot\frac{1}{4^{5/2}x^{5/2}}$ (Law 5)

$= \frac{x^{1/2}}{(4^{1/2})^5 x^{5/2}}$ (Law 4)

$= \frac{x^{1/2}}{2^5 x^{5/2}} = \frac{1}{32}\cdot x^{1/2-5/2}$ (Law 3)

$= \frac{1}{32}\cdot x^{-2} = \frac{1}{32x^2}$ (Law 2)

73. $x^{-1/4} + 6x^{1/4} = x^{-1/4}\left(1+6x^{1/2}\right)$ or $x^{-1/4}\left(1+6\sqrt{x}\right)$

79. $f(4) = (4)^{-1} = \frac{1}{4}$

85. $A = P\left(1+\frac{r}{m}\right)^{mt}$, where $P = 500,\ r = 0.06,\ m = 1,\ t = 6$

$= 500\left(1+\frac{0.06}{1}\right)^{1\cdot 6}$

$= 500(1.06)^6$

$\approx \$709.26$

91. $A = P\left(1+\frac{r}{m}\right)^{mt}$, where $P = 1500,\ r = .06,\ m = 360,\ t = 1$

$= 1500\left(1+\frac{0.06}{360}\right)^{360\cdot 1}$

$= 1500(1.0001667)^{360}$

$\approx \$1592.75$

97. [new stopping distance] $= \frac{1}{20}(2x)^2$

$= \frac{1}{20}\cdot 4x^2$

$= 4\left(\frac{1}{20}x^2\right)$

$= 4\cdot$[old stopping distance]

0.6 Functions and Graphs in Applications

The crucial step in many of these problems is to express the function in one variable using the information given. Exercise 7 is a typical example. Both the height and width are expressed in terms of x. The function, here perimeter, is now in terms of x.

1. If $x = \text{width}$, then the height is $3(\text{width}) = 3x$.

7.
$$\begin{aligned}\text{Perimeter} &= 2\cdot\text{height} + 2\cdot\text{width}\\ &= 2(3x) + 2(x) = 6x + 2x = 8x.\\ \text{Area} &= \text{height}\times\text{width}\\ &= (3x)(x) = 3x^2.\end{aligned}$$

But Area = 25 square feet, so $3x^2 = 25$.

13. Volume $= \pi r^2 h$, where r is the radius of the circular ends and h is the height of the cylinder. Since the volume is supposed to be 100 cubic inches,

$$\pi r^2 h = 100.$$

Using the solutions to Practice Problems 0.6,

Area of the left end: πr^2,

Area of the right end: πr^2,

Area of the side (a "rolled-up rectangle"): $2\pi rh$.

Cost for the left end:	$5(\pi r^2) = 5\pi r^2$,
Cost for the right end:	$6(\pi r^2) = 6\pi r^2$,
Cost for the side:	$7(2\pi rh) = 14\pi rh$.

So the total cost is $11\pi r^2 + 14\pi rh$ (dollars).

19. From Exercise 7, perimeter $= 8x$ (for the rectangle of Exercise 1), hence

$$\begin{aligned}8x &= 40\\ x &= 5.\end{aligned}$$

Also from Exercise 7, the area is $3x^2$ (for the rectangle of Exercise 1). Therefore, since $x = 5$,

$$\text{Area} = 3(5)^2 = 75 \text{ cm}^2.$$

25. **(a)**
$$\begin{aligned}P(x) &= R(x) - C(x)\\ &= 21x - (9x + 800)\\ &= (12x - 800) \text{ dollars.}\end{aligned}$$

(b) x = number of sales, and here $x = 120$, so

$$\begin{aligned}P(x) &= P(120)\\ &= 12(120) - 800\\ &= 640 \text{ dollars in profit.}\end{aligned}$$

(c) The weekly profit function is $P(x) = 12x - 800$, from (a). A weekly profit of \$1000 yields

$$12x - 800 = 1000,$$
$$12x = 1800,$$
$$x = 150.$$

The weekly revenue is $R(x) = 21x$, so revenue $= 21(150) = \$3150$.

31. (3, 162), (6, 270) are points on the graph $y = f(r)$. The cost of constructing a cylinder of radius 3 inches is 162 cents, and similarly, for a radius of 6 inches the cost is 270 cents. Thus the additional cost of increasing the radius from 3 inches to 6 inches is $270 - 162 = 108$ cents = \$1.08.

37. $C(1000) = 4000$.

43. "Solve $P(x) = 30{,}000$" translates to "find the x-coordinates of the points on the graph whose y-coordinate is 30,000."

49. The phrase "determine when" means "find the time or times." In this case, find the values of t that make $h(t) = 100$ feet. Graphically, the task is to find the t-coordinates of the points on the graph of $h(t)$ whose y-coordinate is 100. See Example 7(d).

Chapter 0: Supplementary Exercises

Study the Review of Fundamental Concepts. Write out your own answers. Can you handle the algebra of functions as in Section 0.3? Can you factor quadratic (and simple cubic) polynomials? Have you memorized the quadratic formula? Have you thoroughly (and successfully) practiced using the laws of exponents? If you cannot answer yes to all these questions by the time you finish Chapter 1, you are not seriously interested in succeeding in this calculus course, or you need to take a refresher course in college algebra before you study calculus.

The supplementary exercises provide a brief review of the main skills of the chapter.

1. If $f(x) = x^3 + \frac{1}{x}$, then

$$f(1) = 1^3 + \frac{1}{1} = 2,$$
$$f(3) = 3^3 + \frac{1}{3} = 27 + \frac{1}{3} = 27\frac{1}{3},$$
$$f(-1) = (-1)^3 + \frac{1}{-1} = -1 - 1 = -2,$$
$$f\left(-\frac{1}{2}\right) = \left(-\frac{1}{2}\right)^3 + \frac{1}{-\frac{1}{2}} = -\frac{1}{8} - 2 = -2\frac{1}{8},$$
$$f(\sqrt{2}) = (\sqrt{2})^3 + \frac{1}{\sqrt{2}} = 2\sqrt{2} + \frac{1}{\sqrt{2}} = 2\sqrt{2} + \frac{\sqrt{2}}{2} = \frac{5\sqrt{2}}{2}.$$

7. The domain of $f(x) = \sqrt{x^2 + 1}$ consists of all values of x since $x^2 + 1$ is never negative.

13. To factor $18+3x-x^2$, first factor out the coefficient -1 of x^2 to get $-1(x^2-3x-18)$. To factor $x^2-3x-18$, you need to find c and d such that $cd=-18$ and $c+d=-3$. The solution is $c=-6$ and $d=3$, so

$$x^2-3x-18=(x-6)(x+3).$$

Thus,

$$18+3x-x^2=-1(x^2-3x-18)=-1(x-6)(x+3).$$

19. $f(x)+g(x)=(x^2-2x)+(3x-1)=x^2+x-1.$

25. $f(x)-g(x)=\dfrac{x}{(x^2-1)}-\dfrac{(1-x)}{(1+x)}.$

In order to add two fractions, their denominators must be the same. Observe that $x^2-1=(x+1)(x-1)$. Thus a common denominator for

$$\frac{x}{(x^2-1)} \quad \text{and} \quad \frac{(1-x)}{(1+x)} \quad \text{is} \quad x^2-1.$$

If we multiply by

$$\frac{1-x}{1+x} \quad \text{by} \quad \frac{x-1}{x-1},$$

we get an equivalent expression whose denominator is x^2-1. Thus

$$\begin{aligned}\frac{x}{x^2-1}-\frac{1-x}{1+x}&=\frac{x}{x^2-1}-\frac{1-x}{1+x}\cdot\frac{x-1}{x-1}\\&=\frac{x}{x^2-1}-\frac{(1-x)(x-1)}{x^2-1}\\&=\frac{x-(-x^2+2x-1)}{x^2-1}=\frac{x^2-x+1}{x^2-1}.\end{aligned}$$

31. Substitute $g(x)$ for each occurrence of x in $f(x)$ to obtain $f(g(x))$. Thus

$$f(g(x))=\left(\frac{1}{x^2}\right)^2-2\left(\frac{1}{x^2}\right)+4=\frac{1}{x^4}-\frac{2}{x^2}+4.$$

37. $(81)^{3/4}=(81^{1/4})^3=3^3=27,$ (Law 4′)

$8^{5/3}=(8^{1/3})^5=2^5=32,$ (Law 4′)

$(.25)^{-1}=\left(\dfrac{1}{4}\right)^{-1}=4$

43. $\dfrac{x^{3/2}}{\sqrt{x}}=\dfrac{x^{3/2}}{x^{1/2}}=x^{(3/2-1/2)}=x^1=x$ (Law 3)

Chapter 1
The Derivative

1.1 The Slope of a Straight Line

Slope Property 1 will help you understand the concept of the slope of a line. Property 2 is needed when you have to find the slope of a line between two points. Property 3 is the most useful, and you must memorize the point-slope form of the equation of a line. Exercises 7-10 are simple but very important and will give you practice using the point-slope form. Check with your instructor to see how much attention you should give to Slope Properties 4 and 5. They are seldom needed later in the text.

1. Write the equation $y = 3 - 7x$ in the form $y = mx + b$. That is,

$$y = -7x + 3.$$

Hence the slope is –7, and the y-intercept is (0, 3).

7. Let $(x_1, y_1) = (7, 1)$ and $m = -1$, then use Slope Property 3. The equation of the line is $y - 1 = -1(x - 7)$, or, equivalently, $y = -x + 8$.

13. By Slope Property 2, the slope of the line is

$$\frac{0-0}{1-0} = 0.$$

Since (0, 0) is on the line, the point-slope equation of the line is $y - 0 = 0(x - 0)$, or $y = 0$.

19. The x-intercept of the line is –2, so the point $(-2, 0)$ is on the line. Let $(x_1, y_1) = (-2, 0)$ and $m = -2$, then use Slope Property 3. The equation of the line is $y - 0 = -2(x - (-2))$, or, equivalently, $y = -2x - 4$.

25. The line whose equation is sought is perpendicular to $x + y = 0$, or $y = -x$. The slope of this line is –1, so by Slope Property 5 the slope m of the line whose equation is sought satisfies the equation

$$-1 \cdot m = -1$$

or m = 1. Now let $(x_1, y_1) = (2, 0)$ and m = 1, then use Slope Property 3. The equation of the line is $y - 0 = 1(x - 2)$, or, equivalently, $y = x - 2$.

31. No units are shown on the graphs in Figure 10, but you can see whether a graph has positive or negative slope, and you can see whether the y-intercept is on the positive or negative y-axis.

(a) Rewrite the equation as $y = 1 - x = (-1)x + 1$. The slope is negative and the y-intercept (0, 1) is on the positive y-axis. The only graph with these two properties is (C).

(b) Rewrite the equation as $y = x - 1$. The slope is positive and the y-intercept (0, –1) is on the negative y-axis. The graph is (B).

(c) Rewrite the equation as $y=(-1)x-1$. The slope is negative, the y-intercept is on the negative y-axis, and so the graph is (D).

(d) This must be (A) because the other graphs are already chosen. The fact that (A) works is also easy to see from the equation $y=x+1$ (positive slope and positive y-intercept).

37. Since the slope of this line is 2, if we start at a point on the line and move 1 unit to the right and then 2 units up (in the positive y-direction) we will reach another point on the line. If we start at (1, 3) and move 1 unit to the right and 2 units up, we arrive at (2, 5). So (2, 5) is on the line. A similar move from (2, 5) takes us to (3, 7), so (3, 7) is on the line. Finally, suppose that (0, y) is on the line. This point is one unit to the left of (1, 3) in the x-direction. Starting at (0, y) and moving one unit to the right and 2 units up, we arrive at $(1, y+2)$. This point is on the line and so is (1, 3). Hence we must have $y+2=3$, and $y=1$. Thus (0, 1) is on the line.

Alternatively, we can determine the equation of the line using the point-slope form:

$$y-3=2(x-1) \text{ or } y=2x+1.$$

If $x = 2$, then $y=2(2)+1=5$. If $x = 3$, then $y=2(3)+1=7$. If $x = 0$, then $y=2(0)+1=1$. Thus, the points (2, 5), (3, 7), and (0, 1) lie on the line.

43. For slope –2 and y-intercept (0, –1), the slope-intercept equation $(y=mx+b)$ is $y=-2x+(-1)$, or

$$y=-2x-1.$$

To graph this line, first plot the y-intercept (0, –1). Then since the slope is –2, start at (0, –1) and move one unit right, and then two units in the negative y-direction. The new point (1, –3) is on the line. Draw the straight line through these two points.

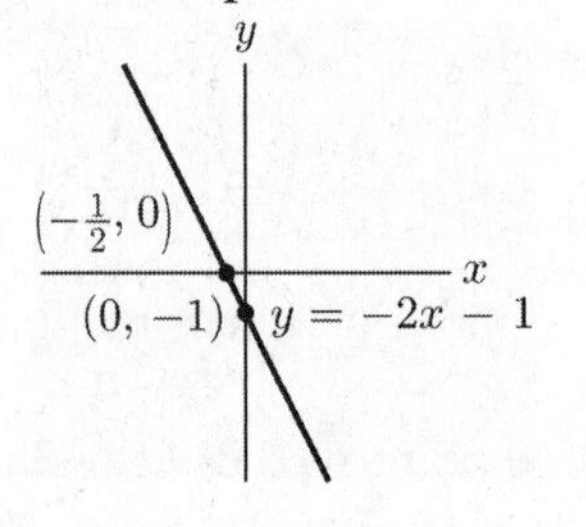

49. Let x be the number of months since January 1, 2009, and $P(x)$ be the price of gasoline per gallon x months since January 1, 2009. Since the price of gasoline on January 1, 2009 was \$4.89/gallon, the point (0, 4.89) is on our line. Also, since the price is rising at a rate of 6 cents per gallon, the slope of our line in .06. By Slope Property 3, the equation of the line through point (0, 4.89) with slope .06 is

$$y-4.89=.06(x-0)$$
$$y-4.89=.06x$$
$$y=.06x+4.89.$$

Therefore, the price of gasoline per gallon x months after January 1, 2009 is

$$P(x)=.06x+4.89.$$

April 2009 is 3 months after Jan. 1, 2009, thus the price for a gallon of gasoline on April 1, 2009 is

$$P(3)=.06(3)+4.89=\$5.07/\text{gallon}$$

and 15 gallons of gasoline on April 1, 2009 would cost $15\times\$5.07/\text{gallon}=\$80.55$.

Sept. 2009 is 8 months after Jan. 1, 2009, thus the price for a gallon of gasoline on Sept. 1, 2009 is

$$P(9)=.06(8)+4.89=\$5.37/\text{gallon}$$

and 15 gallons of gasoline on Sept. 1, 2009 would cost $15\times\$5.37/\text{gallon}=\$80.55$.

55. Assuming the total cost $C(x)$ is linearly related to the daily production level x, total cost can be expressed as

$$C(x) = mx + b$$

(a) Fixed costs are \$1500, therefore $b = 1500$. Also, we know the total cost is \$2200 when 100 rods are produced per day, thus

$$C(100) = m(100) + 1500 = 2200.$$

Solving for the slope m, we find $m = 7$. Therefore, the total cost can be expressed as a function of daily production levels as follows:

$$C(x) = 7x + 1500.$$

(b) In Example 1, we saw that the marginal cost which is the additional cost incurred when the production level is increased by 1 unit, is the same as the slope of the line. Thus, the marginal cost at $x = 100$ is \$7 per rod.

(c) The additional cost of raising the daily production level from 100 to 101 rods is the marginal cost, \$7 per rod. Also, the additional cost of raising production from 100 to 101 rods can be expressed by

$$\begin{aligned} C(101) - C(100) &= 7(101) + 1500 - (7(100) + 1500) \\ &= 707 + 1500 - 700 - 1500 \\ &= \$7 \text{ per rod.} \end{aligned}$$

61. Let $y = mx + b$ and $y = m'x + b'$ be two distinct lines. We show that these lines are parallel if and only if $m = m'$. Since two lines are parallel if and only if they have no points in common, it suffices to show that $m = m'$ if and only if the equation $mx + b = m'x + b'$. Suppose $m = m'$. Then $mx + b = m'x + b'$ implies $b = b'$; but since the lines are distinct, $b \neq b'$. Thus if $m = m'$, $mx + b = m'x + b'$ has no solution. If $m \neq m'$, then $x = \dfrac{b' - b}{m - m'}$ is a solution to $mx + b = m'x + b'$. Thus, $mx + b = m'x + b'$ has no solution in x if and only if $m = m'$, and it follows that two distinct lines are parallel if and only if they have the same slope.

Help for Technology Exercises: You may wish to review the **Incorporating Technology** section in chapter 0 to help you with the technology exercises in chapter 1.

1.2 The Slope of a Curve at a Point

This brief section should be read carefully. Example 2 and Exercises 23, 24, and 32 are very important, and so we have included a solution of Exercise 23. Of the students who do *not* use this *Manual,* at least 20% will miss an exam problem on the equation of a tangent line. If you carefully study the solution of Exercise 23 here and the solution of Exercise 43 in Section 1.6, you should have no difficulty on an exam.

1.

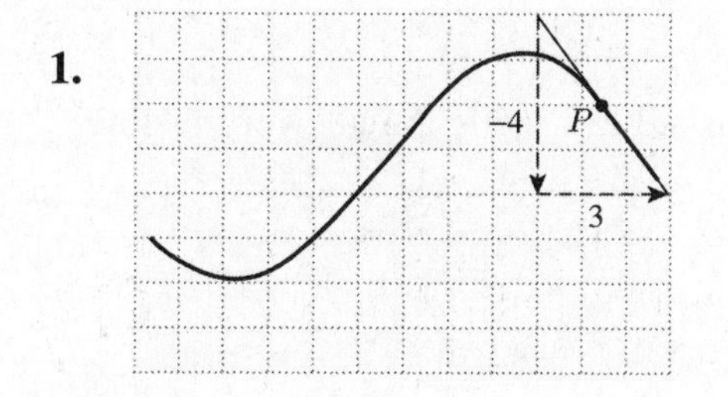

The slope of the curve at the point P is, by definition, the slope of the tangent line at P. If you move four units in the negative y-direction from P, you can return to the line by moving three units in the positive x-direction. Therefore the slope is $-\frac{4}{3}$.

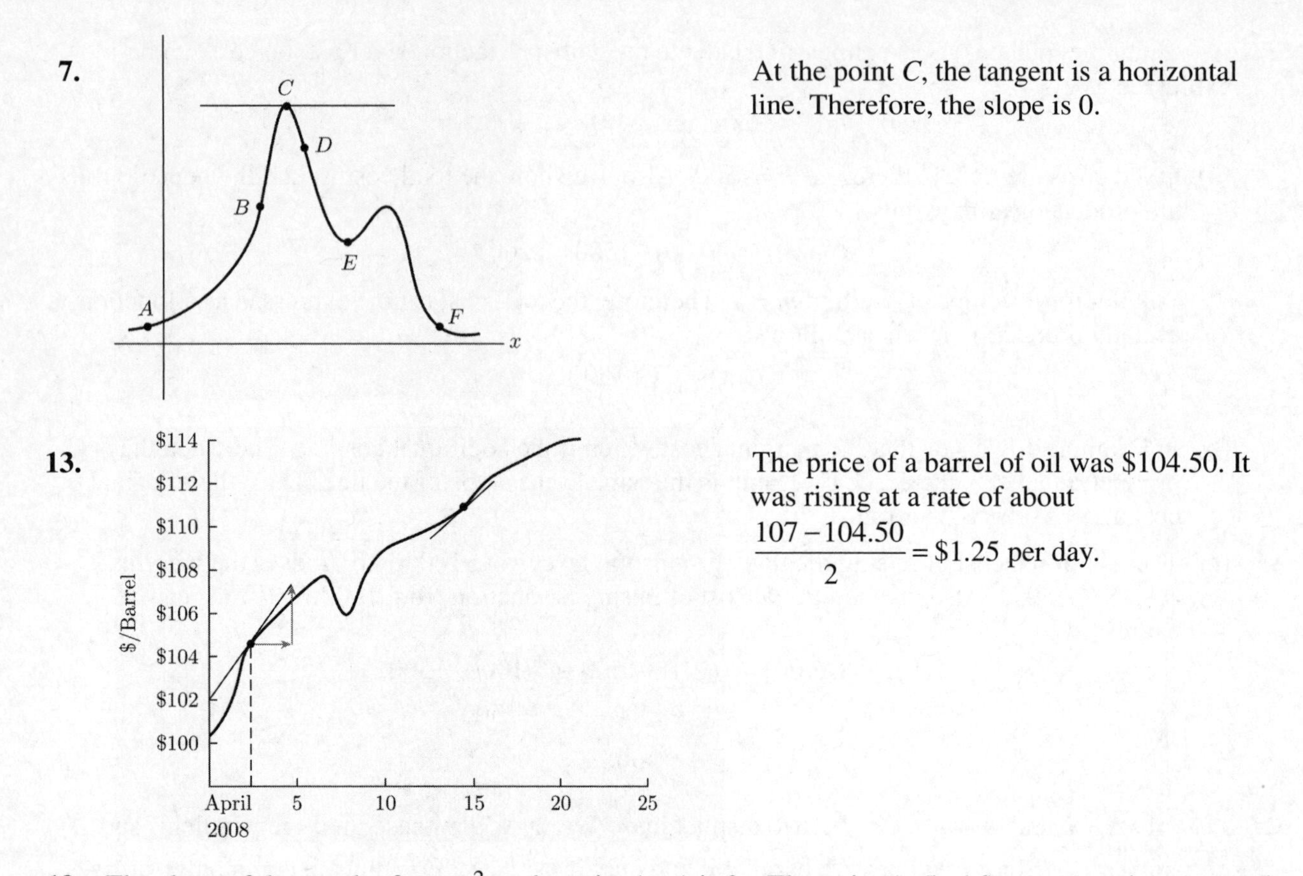

7. At the point C, the tangent is a horizontal line. Therefore, the slope is 0.

13. The price of a barrel of oil was \$104.50. It was rising at a rate of about

$$\frac{107 - 104.50}{2} = \$1.25 \text{ per day.}$$

19. The slope of the graph of $y = x^2$ at the point (x, y) is $2x$. The point $(-.5, .16)$ corresponds to $x = -.5$, so the slope at this point is $2x = 2(-.5) = -1$. Now, let $(x_1, y_1) = (-.5, .16)$ and $m = -1$ and use Slope Property 3. The equation of the tangent line through $(-.5, .16)$ is:

$$y - .16 = -1(x - (-.5)), \quad \text{or} \quad y = -x - .34.$$

23. Use the point-slope equation of the tangent line to the graph of $y = x^2$ at $x = 2.5$. Find a point (x_1, y_1) and a slope m. There are two basic principles to keep in mind.

(a) Use the *original equation* $y = x^2$ to find a *point* on the graph. For $x = 2.5$, compute $y = (2.5)^2 = 6.25$. Thus $(2.5, 6.25)$ is on the graph of $y = x^2$.

(b) Use the *slope formula* $2x$ to find the *slope* of the graph at a point. When $x = 2.5$, the slope of the graph is $2(2.5) = 5$. By definition, this slope is the slope of the tangent line to the graph at the point where $x = 2.5$.

The desired tangent line equation has the form $y - y_1 = m(x - x_1)$, where $x_1 = 2.5$, $y_1 = 6.25$, and $m = 5$. That is,

$$y - 6.25 = 5(x - 2.5).$$

Leave the answer in this form unless you are specifically asked to rewrite the answer in some equivalent form such as the slope-intercept form.

Warning: A common mistake in Exercise 23 is to replace m in the equation $y - y_1 = m(x - x_1)$ by the slope formula $2x$ instead of a *specific value* of m. But the equation $y - 6.25 = 2x(x - 2.5)$ is *not* the equation of a line. (In fact, this equation simplifies to $y - 6.25 = 5x^2 - 5x$, or $y = 5x^2 - 5x + 6.25$, which is a quadratic equation.) The equation of a tangent *line* must involve a specific *number m* that gives the slope of the desired tangent line.

25. The slope of the graph of $y = x^2$ at the point (x, y) is $2x$. Hence, if the slope is $\frac{7}{2}$, then

$$2x = \frac{7}{2} \Rightarrow x = \frac{7}{2} \cdot \frac{1}{2} = \frac{7}{4}.$$

To find the y-coordinate of the point where $x = \frac{7}{4}$, use the original equation $y = x^2$.

$$y = \left(\frac{7}{4}\right)^2 = \frac{49}{16}.$$

Hence, $\left(\frac{7}{4}, \frac{49}{16}\right)$ is the desired point on the graph.

31. The slope formula for the curve $y = x^3$ is given by the formula $3x^2$. The point $\left(-\frac{1}{2}, -\frac{1}{8}\right)$ corresponds to $x = -\frac{1}{2}$, so the slope at $\left(-\frac{1}{2}, -\frac{1}{8}\right)$ is $3x^2 = 3\left(-\frac{1}{2}\right)^2 = 3\left(\frac{1}{4}\right) = \frac{3}{4}$.

37. **(a)** By Slope Property 2, the slope of the line l is

$$\frac{13 - 4}{5 - 2} = \frac{9}{3} = 3.$$

The length of line segment d is the absolute value of the difference between the y-coordinates of the Points P and Q, i.e.,

$$d = |13 - 4| = 9.$$

(b) As the Point Q is moved along the curve toward P, the slope of the line l through the Points P and Q increases. For example, if the Point Q' lies on the curve midway between P and Q, then the slope of the line through P and Q' is greater than the slope of the line l through P and Q.

39. Review the material in the **Incorporating Technology** section on page 69 of the text to learn how to zoom in on a graph.

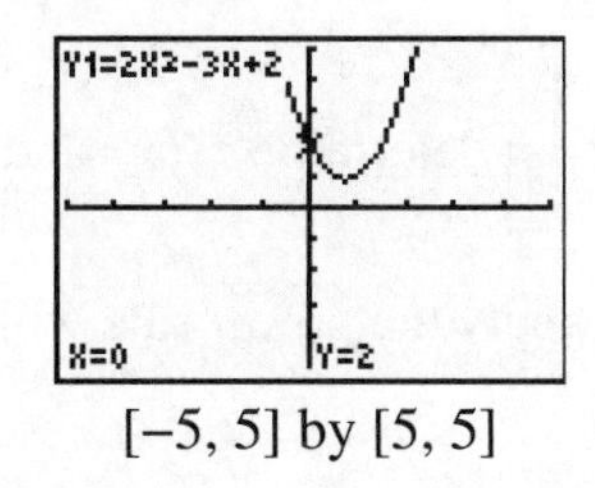

[−5, 5] by [5, 5]

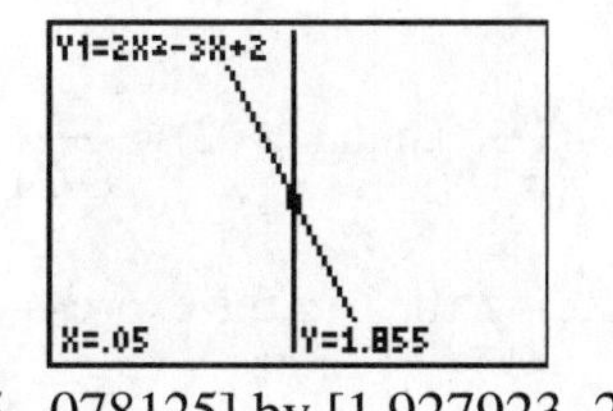

[−.078125, .078125] by [1.927923, 2.084173]

When $x = 0$, $y = 2$. Find a second point on the line using **value**: $x = .05$, $y = 1.855$

$$m = \frac{1.855 - 2}{.05 - 0} = -2.9$$

The actual value of $f'(0)$ is -3.

1.3 The Derivative

Your short-term goal for this section should be to learn the derivative formulas that appear in boxes on page 74–75 and to become familiar with the notation introduced on Page 77. This will enable you to do the homework for this section. Your long-term goal should be to have some understanding of the secant-line calculation of the derivative, as described on Pages 78–79. This material is not grasped easily. Reading it *out loud* will help you to go over the ideas slowly and carefully. Plan to review this section before your first exam and again when you reach Chapter 3.

1. If $f(x) = mx + b$, then $f'(x) = m$. Hence if $f(x) = 3x + 7$, then $f'(x) = 3$.

7. Apply the power rule with $r = \frac{2}{3}$:

$$f(x) = x^{2/3},$$
$$f'(x) = \frac{2}{3}x^{(2/3)-1} = \frac{2}{3}x^{-1/3}$$
$$= \frac{2}{3}\left(\frac{1}{x^{1/3}}\right) = \frac{2}{3\sqrt[3]{x}}$$

13. Rewrite $f(x) = \frac{1}{x^{-2}}$ as $f(x) = x^2$. Then apply the power rule with $r = 2$.

$$f'(x) = 2x.$$

19. If $f(x) = \frac{1}{x}$, then $f'(x) = -\frac{1}{x^2} (x \neq 0)$, by formula (5) on Page 75. Hence at $x = \frac{2}{3}$,

$$f'\left(\frac{2}{3}\right) = -\frac{1}{(2/3)^2} = -\frac{1}{4/9} = -\frac{9}{4}.$$

Helpful Hint: Formula (5) for the derivative of is used so frequently that you should memorize it even though it is only a special case of the power rule.

25. Use the power rule with $r = 4$:

$$\frac{d}{dx}(x^4) = 4x^{4-1} = 4x^3.$$

Setting $x = 2$ in this derivative, we find that the slope of the curve $y = x^4$ at $x = 2$ is $4(2)^3 = 4 \cdot 8 = 32$.

31. If $f(x) = \frac{1}{x^5}$, then $f(-2) = \frac{1}{(-2)^5} = -\frac{1}{32}$. To compute $f'(2)$, first determine $f'(x)$ and then substitute −2 for x in the expression for $f'(x)$. Since $\frac{1}{x^5} = x^{-5}$, apply the power rule with $r = -5$.

$$f(x) = x^{-5},$$
$$f'(x) = -5x^{-5-1} = -5x^{-6} = -\frac{5}{x^6},$$
$$f'(2) = -\frac{5}{2^6} = -\frac{5}{64}.$$

37. If $f(x)=\sqrt{x}=x^{1/2}$, then $f\left(\frac{1}{9}\right)=\frac{1}{3}$. Applying the power rule with $r=\frac{1}{2}$, $f'(x)=\frac{1}{2}x^{-1/2}=\frac{1}{2\sqrt{x}}$.

Then the slope of the line tangent to $f(x)$ at $x=\frac{1}{9}$ is $f'\left(\frac{1}{9}\right)=\frac{1}{2}\left(\frac{1}{9}\right)^{-1/2}=\frac{1}{2}(9)^{1/2}=\frac{3}{2}$. Since the tangent line passes through the point $\left(\frac{1}{9},\frac{1}{3}\right)$, its equation is

$$y-\frac{1}{3}=\frac{3}{2}\left(x-\frac{1}{9}\right).$$

43. The slope of the graph $y=\sqrt{x}$ at the point $x=a$ is found by using y' evaluated at $x=a$.

$$y'=\frac{1}{2}x^{-1/2}=\frac{1}{2\sqrt{x}}.$$

At $x=a$,
$$y'=\frac{1}{2\sqrt{a}}.$$

The slope of the tangent line $y=2x+b$ at the point $(a,\sqrt{a})$ is $m=2$. So we have

$$2=\frac{1}{2\sqrt{a}}$$
$$\sqrt{a}=\frac{1}{4}$$
$$(\sqrt{a})^2=\left(\frac{1}{4}\right)^2\Rightarrow a=\frac{1}{16}.$$

To find P: at $a=\frac{1}{16}$, $y=\sqrt{\frac{1}{16}}=\frac{1}{4}$. Therefore, $P=\left(\frac{1}{16},\frac{1}{4}\right)$. To find b:

$$y=2x+b$$
$$\frac{1}{4}=2\left(\frac{1}{16}\right)+b$$
$$b=\frac{1}{4}-2\left(\frac{1}{16}\right)$$
$$b=\frac{1}{8}.$$

49. Use the power rule with $r=8$:

$$\frac{d}{dx}(x^8)=8x^{8-1}=8x^7.$$

55. Apply the power rule with $r = \frac{1}{5}$:

$$\begin{aligned} y &= x^{1/5}, \\ \frac{dy}{dx} &= \frac{1}{5}x^{(1/5)-1} \\ &= \frac{1}{5}x^{-4/5}. \end{aligned}$$

61.

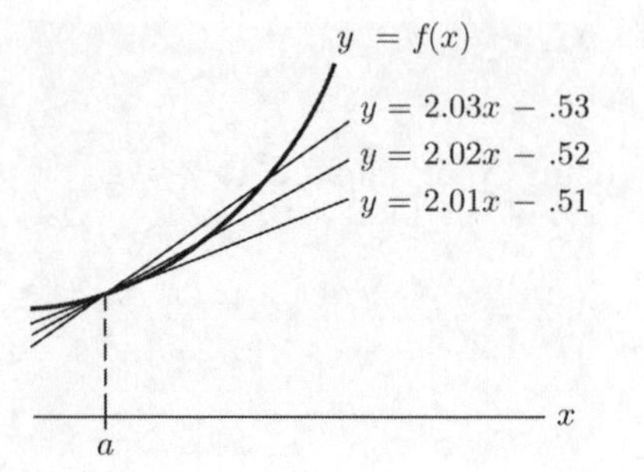

The three lines and $y = f(x)$ intersect at the point $(a, f(a))$. Take the line $y = 2.01x - .51$ at this point. Then

$$f(a) = 2.01a - .51.$$

Similarly using $y = 2.02x - .52$,

$$f(a) = 2.02a - .52.$$

So,

$$\begin{aligned} 2.01a - .51 &= 2.02a - .52, \\ 0.01 &= 0.01a, \\ a &= 1, \\ f(a) &= 2.02a - .52 = 2.02 - .52 = 1.5. \end{aligned}$$

To estimate $f'(a)$ use the secant-line calculation. The secant line $y = 2.01x - .51$ is "nearly" a tangent line, therefore the slope of this line is nearly $f'(a)$. The slope of the tangent line is slightly less than the secant line's slope of 2.01. Hence

$$f'(a) \approx 2.$$

67. $f(x) = -x^2 + 2x \Rightarrow$

$$\begin{aligned} \frac{f(x+h) - f(x)}{h} &= \frac{\left[-(x+h)^2 + 2(x+h)\right] - \left(-x^2 + 2x\right)}{h} \\ &= \frac{-x^2 - 2xh - h^2 + 2x + 2h + x^2 - 2x}{h} \\ &= \frac{h(-2x + 2 - h)}{h} \\ &= -2x + 2 - h \end{aligned}$$

73. We must apply the three-step method to find the derivative of $f(x)=7x^2+x-1$.

Step 1:

$$\begin{aligned}\frac{f(x+h)-f(x)}{h}&=\frac{\left[7(x+h)^2+(x+h)-1\right]-\left(7x^2+x-1\right)}{h}\\&=\frac{7x^2+14xh+7h^2+x+h-1-7x^2-x+1}{h}\\&=\frac{14xh+7h^2+h}{h}\\&=\frac{h(14x+7h+1)}{h}\\&=14x+1+7h\end{aligned}$$

Steps 2 and 3: As h approaches 0, the quantity $14x + 1 + 7h$ approaches $14x + 1$. Thus,

$$f'(x)=14x+1.$$

For exercises 79–91, refer to the **Incorporating Technology** section on pages 79–80.

79. $f'(0)$, where $f(x)=2^x$

```
nDeriv(2^X,X,0)
       .6931472361
```

85.

```
Plot1 Plot2 Plot3
\Y1■3X²-5
\Y2■nDeriv(Y1,X,
X)
\Y3=
\Y4=
\Y5=
\Y6=
```

```
Y2=nDeriv(Y1,X,X)
X=2        Y=12
```

[0, 4] by [–5, 40]

Using **TRACE**, we find that the value of the derivative of Y_1 at $x = 2$ is 12.

91. $f(x)=\dfrac{5}{x}$, $g(x) = 5 - 1.25x$

To solve graphically, graph the functions and use the **intersect** command to find where $g(x)$ is tangent to $f(x)$.

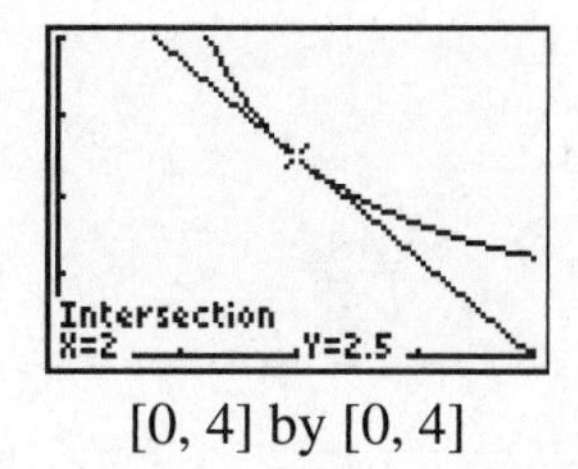

[0, 4] by [0, 4]

Thus, $a = 2$.

Alternatively, we can solve $\frac{5}{x} = 5 - 1.25x$ to find the x-value of the intersection of $f(x)$ and $g(x)$.

$$\begin{aligned} \frac{5}{x} &= 5 - 1.25x \\ 5 &= 5x - 1.25x^2 \\ 1.25x^2 - 5x + 5 &= 0 \\ 1.25(x^2 - 4x + 4) &= 0 \\ 1.25(x-2)^2 &= 0 \\ x &= 2 \end{aligned}$$

1.4 Limits and the Derivative

Instructors differ widely about how much theoretical material to include in a course using our text. We recommend that at a minimum, you read Pages 83–84, even if no exercises are assigned. The skills practice in this section are not needed for later work, but the exercises will give you valuable experience with simple limits and will improve your understanding of derivatives.

1. $\lim_{x \to 3} g(x)$ does not exist. As x approaches 3 from the right, $g(x)$ approaches 2, but as x approaches 3 from the left, the values for $g(x)$ do not approach 2. In order for a limit to exist, the values of the function must approach the same number as x approaches 2 from each direction.

7. Since $1 - 6x$ is a polynomial, the limit exists and

$$\begin{aligned} \lim_{x \to 1} (1 - 6x) &= 1 - 6(1) \\ &= 1 - 6 = -5. \end{aligned}$$

13. Using the Limit Theorems

$$\lim_{x \to 7} (x + \sqrt{x-6})(x^2 - 2x + 1)$$

$$= \left[\lim_{x \to 7} (x + \sqrt{x-6})\right]\left[\lim_{x \to 7} (x^2 - 2x + 1)\right] \quad \text{(Thm. V)}$$

$$= \left[\lim_{x \to 7} x + \lim_{x \to 7} (x-6)^{1/2}\right]\left[\lim_{x \to 7} (x^2 - 2x + 1)\right] \quad \text{(Thm. III)}$$

$$= \left[\lim_{x \to 7} x + \left(\lim_{x \to 7} (x-6)\right)^{1/2}\right]\left[\lim_{x \to 7} (x^2 - 2x + 1)\right] \quad \text{(Thm. II)}$$

$$= [7 + (1)^{1/2}][49 - 14 + 1] \quad \text{(Limits of polynomial functions)}$$

$$= (8)(36) = 288.$$

19. Since $\frac{-2x^2 + 4x}{x - 2} = \frac{-2x(x-2)}{x-2} = -2x$ for $x \neq 2$,

$$\begin{aligned} \lim_{x \to 2} \frac{-2x^2 + 4x}{x - 2} &= \lim_{x \to 2} -2x \\ &= -2(2) = -4. \end{aligned}$$

25. No limit exists. Observe that

$$\lim_{x\to 8} x^2+64=128 \quad \text{and} \quad \lim_{x\to 8} x-8=0,$$

as x approaches 8. So the denominator gets very small and the numerator approaches 128. For example, if $x=8.00001$, then the numerator is 128.00016 and the denominator is .00001. The quotient is 12,800,016. As x approaches 8 even more closely, the quotient gets arbitrarily large and cannot possibly approach a limit.

31. Since

$$f'(a)=\lim_{h\to 0}\frac{f(a+h)-f(a)}{h},$$

we must calculate

$$\lim_{h\to 0}\frac{f(0+h)-f(0)}{h}.$$

If $f(x)=x^3+3x+1$, then

$$\begin{aligned}\frac{f(0+h)-f(0)}{h} &= \frac{[(0+h)^3+3(0+h)+1]-[0^3+3(0)+1]}{h}\\ &= \frac{h^3+3h+1-1}{h}\\ &= \frac{h^3+3h}{h}\\ &= h^2+3.\end{aligned}$$

Therefore, $f'(0)=\lim_{h\to 0}(h^2+3)=0^2+3=3.$

37. We must calculate

$$f'(x)=\lim_{h\to 0}\frac{f(x+h)-f(x)}{h}.$$

If $f(x)=3x+1$, then

$$\begin{aligned}\lim_{h\to 0}\frac{f(x+h)-f(x)}{h} &= \lim_{h\to 0}\frac{3(x+h)+1-(3x+1)}{h}\\ &= \lim_{h\to 0}\frac{3x+3h+1-3x-1}{h}\\ &= \lim_{h\to 0}\frac{3h}{h}\\ &= \lim_{h\to 0} 3\\ &= 3.\end{aligned}$$

We conclude $f'(x)=3$.

43. We must calculate

$$f'(x)=\lim_{h\to 0}\frac{f(x+h)-f(x)}{h}.$$

If $f(x)=\dfrac{1}{x^2+1}$, then

$$\begin{aligned}\lim_{h\to 0}\frac{f(x+h)-f(x)}{h}&=\lim_{h\to 0}\frac{\frac{1}{(x+h)^2+1}-\frac{1}{x^2+1}}{h}\\&=\lim_{h\to 0}\frac{\frac{(x^2+1)-((x+h)^2+1)}{((x+h)^2+1)(x^2+1)}}{h}=\lim_{h\to 0}\frac{\frac{x^2+1-x^2-2xh-h^2-1}{((x+h)^2+1)(x^2+1)}}{h}\\&=\lim_{h\to 0}\frac{\frac{-h(2x+h)}{((x+h)^2+1)(x^2+1)}}{h}=\lim_{h\to 0}\frac{-h(2x+h)}{((x+h)^2+1)(x^2+1)}\left(\frac{1}{h}\right)\\&=\lim_{h\to 0}\frac{-(2x+h)}{((x+h)^2+1)(x^2+1)}=\frac{-2x}{(x^2+1)(x^2+1)}\\&=\frac{-2x}{(x^2+1)^2}.\end{aligned}$$

We conclude $f'(x)=\dfrac{-2x}{\left(x^2+1\right)^2}$.

49. We want to find $f(x)$ so that $\lim_{h\to 0}\dfrac{(1+h)^2-1}{h}$ has the same form as

$$f'(a)=\lim_{h\to 0}\frac{f(a+h)-f(a)}{h}.$$

So we want $f(a+h)=(1+h)^2$; and $f(a)=1$. From this we see that $f(x)=x^2$ and $a=1$. Note that $f(a+h)=f(1+h)=(1+h)^2$ as needed.

55. We want to find $f(x)$ so that $\lim_{h\to 0}\dfrac{(2+h)^2-4}{h}$ has the same form as

$$f'(a)=\lim_{h\to 0}\frac{f(a+h)-f(a)}{h}.$$

So we want $f(a+h)=(2+h)^2$; and $f(a)=4$. From this we see that $f(x)=x^2$ and $a=2$. Therefore, the given limit is $f'(2)$ where $f(x)=x^2$. Also, since we know $f'(x)=2x\Rightarrow f'(2)=2(2)=4$, we conclude

$$f'(2)=\lim_{h\to 0}\frac{(2+h)^2-4}{h}=4$$

61. As x increases without bound so does x^2. Therefore, $\dfrac{1}{x^2}$ approaches zero as x approaches ∞. That is

$$\lim_{x\to\infty}\frac{1}{x^2}=0.$$

67. Referring to the figure in the text, we see that as x approaches 0 from the left, $f(x)$ approaches $\frac{3}{4}$. Likewise, as x approaches 0 from the right, $f(x)$ approaches $\frac{3}{4}$. That is, $f(x)$ approaches $\frac{3}{4}$ from both the left and right as x approaches 0, so we have established that

$$\lim_{x \to 0} f(x) = \frac{3}{4}.$$

73. Examining the graph of the function $f(x) = \sqrt{25+x} - \sqrt{x}$, it appears as if the values of the function approach 0 as $x \to \infty$.

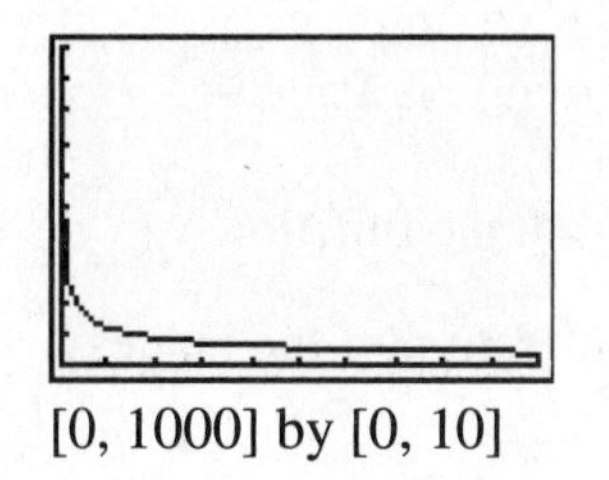

[0, 1000] by [0, 10]

To verify this, lets examine $f(x)$ for large values of x:

x	$f(x) = \sqrt{25+x} - \sqrt{x}$
10,000	$f(10{,}000) = \sqrt{25+10{,}000} - \sqrt{10{,}000} \approx .12492197250$
100,000	$f(100{,}000) = \sqrt{25+100{,}000} - \sqrt{100{,}000} \approx .0395260005$
1,000,000	$f(1{,}000{,}000) = \sqrt{25+1{,}000{,}000} - \sqrt{1{,}000{,}000} \approx .0124999219$
10,000,000	$f(10{,}000{,}000) = \sqrt{25+10{,}000{,}000} - \sqrt{10{,}000{,}000} \approx .0039528446$
100,000,000	$f(100{,}000{,}000) = \sqrt{25+100{,}000{,}000} - \sqrt{100{,}000{,}000} \approx .00125$
1,000,000,000	$f(1{,}000{,}000{,}000) = \sqrt{25+1{,}000{,}000{,}000} - \sqrt{1{,}000{,}000{,}000} \approx .00039528$

It also appears from the table that the function approaches 0 as x approaches ∞. We conclude

$$\lim_{x \to \infty} f(x) = 0.$$

1.5 Differentiability and Continuity

We want you to be aware that real applications sometimes involve functions that may not be differentiable at one or more points in their domains. So this section gives you rare opportunity to see functions whose graphs are not as "nice" as the ones we usually consider.

Exercises 1 and 7 refer to the following figure.

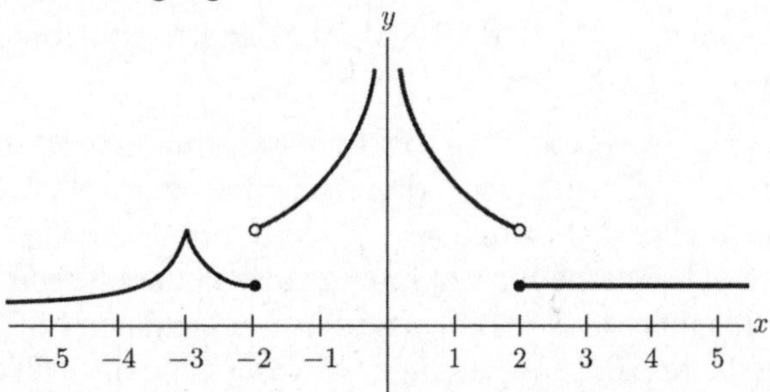

1. No, the graph in the figure is not continuous at $x = 0$ since the limit as x approaches zero does not exist.

7. No, the graph in the figure is not differentiable at $x = 0$ because the function is not even *defined* at $x = 0$. Even if the function were given some value at $x = 0$, say $f(0) = 1$, the graph would have a vertical tangent line at $x = 0$.

13. Since $f(x) = x^2$, the power rule gives a slope-formula, namely, $2x$, that is valid for all x. This slope-formula was verified in Section 1.3 using the limit of the slopes of secant lines. Thus $f(x)$ has a derivative (in the formal sense of Section 1.4) for all x. In particular, $f(x)$ is differentiable at $x = 1$. By Theorem 1 on Page 95, $f(x)$ is necessarily continuous at $x = 1$.

19. At $x = 1$ the function $f(x)$ is defined, namely $f(1) = 0$. When computing $\lim_{x \to 1} f(x)$, we exclude consideration of the value $x = 1$; therefore,

$$\lim_{x \to 1} f(x) = \lim_{x \to 1} \frac{1}{x - 1}$$

which does not exist. Hence $\lim_{x \to 1} f(x) \neq f(1)$, so $f(x)$ is not continuous at $x = 1$. By Theorem 1, since $f(x)$ is not continuous at $x = 1$, it cannot be differentiable at $x = 1$.

25. $f(x)$ is continuous at $x = a$, if $\lim_{x \to a} f(x) = f(a)$. Here

$$f(x) = \frac{(6 + x)^2 - 36}{x}, \quad x \neq 0.$$

Where $a = 0$, we have:

$$\lim_{x \to 0} \frac{(6 + x)^2 - 36}{x} = \lim_{x \to 0} \frac{36 + 12x + x^2 - 36}{x}$$
$$= \lim_{x \to 0} (12 + x) = 12.$$

So define $f(0) = 12$, and this definition will make $f(x)$ continuous for all x.

31. **(a)** Referring to the graph in the text, we see that at 8 A.M., a total of \$4000 in sales have already been made. The graph also shows that by 10 A.M., \$10,000 worth of sales have been made. Thus the rate of sales during the period between 8 A.M. and 10 A.M. is given by the slope formula given in Slope Property 2:

$$\frac{y_2 - y_1}{x_2 - x_1} = \frac{10{,}000 - 4{,}000}{10 - 8} = \frac{6{,}000}{2} = 3{,}000.$$

In the time period between 8 A.M. and 10 A.M., the department store is selling goods at an average rate of \$3000 per hour.

(b) Each portion of the graph corresponding to a two hour time period starting at an even hour is a straight line segment. Visually inspecting these line segments reveals that the segment corresponding to the time period from 8 A.M. to 10 A.M. has the largest slope (it's the steepest). Thus, the 2-hour interval with the highest rate of sales is the interval between 8 A.M. and 10 A.M. The rate of sales in this interval is \$3000 per hour, as calculated in Part (a). Note that it is possible to compute the rate of sales for each two hour time interval starting at an even hour as in Part (a), but this is not necessary for this problem.

1.6 Some Rules for Differentiation

The rules in this section are mastered by working lots of problems. Make sure you learn to distinguish between a constant that is *added* to a function and a constant that *multiplies* a function. For instance, x^3+5 is the sum of the cube function $f(x)=x^3$ and the constant function $g(x)=5$. Thus

$$\frac{d}{dx}(x^3+5)=\frac{d}{dx}(x^3)+\frac{d}{dx}5=3x^2+0, \qquad \text{(Sum rule)}$$

because the derivative of a constant function is zero. However, the constant 5 in the formula $5x^3$ *multiplies* the function x^3. Hence, by the constant-multiple rule,

$$\begin{aligned}\frac{d}{dx}5x^3 &= 5\cdot\frac{d}{dx}(x^3)\\ &= 5\cdot 3x^2\\ &= 15x^2.\end{aligned}$$

Exercise 43 is very important. But before you look at the solution here, go back and read the solution to Exercise 23 in Section 1.2. Then try Exercise 43 in Section 1.6 by yourself. Peak at the solution only if you are stuck.

1. To differentiate $y=x^3+x^2$, let $f(x)=x^3$ and $g(x)=x^2$ and apply the sum rule.

$$\begin{aligned}\frac{dy}{dx} &= \frac{d}{dx}(x^3+x^2)\\ &= \frac{d}{dx}(x^3)+\frac{d}{dx}(x^2)\\ &= 3x^2+2x.\end{aligned}$$

7. $$\begin{aligned}\frac{d}{dx}(x^4+x^3+x) &= \frac{d}{dx}(x^4)+\frac{d}{dx}(x^3+x) && \text{(Sum rule)}\\ &= \frac{d}{dx}(x^4)+\frac{d}{dx}(x^3)+\frac{d}{dx}(x) && \text{(Sum rule)}\\ &= 4x^3+3x^2+1.\end{aligned}$$

13. Write $\frac{4}{x^2}$ in the form $4\cdot x^{-2}$. Then

$$\begin{aligned}\frac{dy}{dx} &= \frac{d}{dx}(4\cdot x^{-2})\\ &= 4\cdot\frac{d}{dx}(x^{-2}) && \text{(Constant-multiple rule)}\\ &= 4(-2)x^{-3} && \text{(Power rule)}\\ &= -8x^{-3}, \quad \text{or} \quad -\frac{8}{x^3}.\end{aligned}$$

19. Since $\frac{1}{5x^5} = \frac{1}{5} \cdot \frac{1}{x^5} = \left(\frac{1}{5}\right) \cdot x^{-5}$, use the constant-multiple rule and the power rule.

$$\frac{dy}{dx} = \frac{d}{dx}\left(\frac{1}{5} \cdot x^{-5}\right) = \frac{1}{5} \cdot \frac{d}{dx}(x^{-5}) = \frac{1}{5}(-5)x^{-6} = -x^{-6}, \quad \text{or} \quad -\frac{1}{x^6}.$$

Warning: Problems like those in Exercises 13 and 19 tend to be missed on exams by many students. The difficulty lies in the first step—recognizing how to write the function as a constant times a power of *x*. Be sure to review this before the exam.

25. $\frac{d}{dx}5\sqrt{3x^3 + x} = \frac{d}{dx}5(3x^3 + x)^{1/2}$

$$= \frac{5}{2}(3x^3 + x)^{-1/2} \cdot \frac{d}{dx}(3x^2 + x) \qquad \text{(General power rule)}$$

$$= \frac{5}{2}(3x^3 + x)^{-1/2}\left[\frac{d}{dx}(3x^3) + \frac{d}{dx}(x)\right] \qquad \text{(Sum rule)}$$

$$= \frac{5}{2}(3x^3 + x)^{-1/2}\left[3 \cdot \frac{d}{dx}(x^3) + \frac{d}{dx}(x)\right] \qquad \text{(Constant-multiple rule)}$$

$$= \frac{5}{2}(3x^3 + x)^{-1/2}[3(3x^2) + 1] = \frac{45x^2 + 5}{2\sqrt{3x^3 + x}}.$$

Warning: Don't forget to use parentheses (or brackets) when appropriate in the general power rule. The next-to-last line in the solution above is incorrect if written in the form:

$$= \frac{5}{2}(3x^3 + x)^{-1/2} \cdot \underbrace{3(3x^2) + 1}_{\text{missing brackets}}.$$

31. Note that

$$y = \frac{2}{1 - 5x} = 2 \cdot \frac{1}{1 - 5x} = 2 \cdot (1 - 5x)^{-1}.$$

Hence

$$\frac{dy}{dx} = \frac{d}{dx}(2 \cdot (1 - 5x)^{-1})$$

$$= 2 \cdot \frac{d}{dx}(1 - 5x)^{-1} \qquad \text{(Constant-multiple rule)}$$

$$= 2(-1)(1 - 5x)^{-2} \cdot \frac{d}{dx}(1 - 5x) \qquad \text{(General power rule)}$$

$$= -2(1 - 5x)^{-2}(-5) = 10(1 - 5x)^{-2}.$$

Warning: Forgetting the (–5) in the last line is a common mistake. Another common error is to carelessly forget the parentheses and write the derivative as

$$-2(1 - 5x)^{-2} - 5.$$

This is definitely incorrect, and many instructors will give no part credit for such an answer. Your ability to find a derivative is of no value to anyone if your algebra is careless and incorrect.

In this *Manual*, we'll point where algebra errors are likely to occur so you can guard against them. One key to avoiding such errors on exams is to practice working carefully on your homework.

37. If $f(x)=\left(\dfrac{\sqrt{x}}{2}+1\right)^{3/2}$ then

$$f'(x)=\frac{d}{dx}\left(\frac{\sqrt{x}}{2}+1\right)^{3/2}$$

$$=\frac{3}{2}\left(\frac{\sqrt{x}}{2}+1\right)^{1/2}\cdot\frac{d}{dx}\left(\frac{\sqrt{x}}{2}+1\right) \qquad \text{(General power rule)}$$

$$=\frac{3}{2}\left(\frac{\sqrt{x}}{2}+1\right)^{1/2}\left[\frac{d}{dx}\left(\frac{\sqrt{x}}{2}\right)+\frac{d}{dx}(1)\right] \qquad \text{(Sum rule)}$$

$$=\frac{3}{2}\left(\frac{\sqrt{x}}{2}+1\right)^{1/2}\left[\frac{1}{2}\cdot\frac{d}{dx}(x^{1/2})+\frac{d}{dx}(1)\right] \qquad \text{(Constant-multiple rule)}$$

$$=\frac{3}{2}\left(\frac{\sqrt{x}}{2}+1\right)^{1/2}\left[\frac{1}{2}\left(\frac{1}{2}\right)x^{-1/2}+0\right]$$

$$=\frac{3}{2}\left(\frac{\sqrt{x}}{2}+1\right)^{1/2}\left(\frac{1}{4}x^{-1/2}\right).$$

43. The general slope-formula for the curve $y=(x^2-15)^6$ is given by the derivative.

$$\frac{dy}{dx}=6(x^2-15)^5\cdot\frac{d}{dx}(x^2-15) \qquad \text{(General power rule)}$$

$$=6(x^2-15)^5(2x)$$

$$=12x(x^2-15)^5.$$

To find the *slope m* of the tangent line at the particular point where $x=4$, substitute 4 for x in the *derivative* formula to get

$$m=12(4)[(4)^2-15]^5=48(1)^5=48.$$

For the equation of the tangent line at $x=4$, use the point-slope form with slope $m=48$. To find a *point* (x_1, y_1) on the curve when $x_1=4$, substitute 4 for x in the original equation to get

$$y_1=[(4)^2-15]^6=(1)^6=1.$$

The equation of the tangent line is

$$y-1=48(x-4).$$

Warning: Exercise 43 helped you to get the equation of the tangent line by asking you first for the slope of the line. Exam questions tend to be more like Exercise 44. You are expected to know that you need to find the slope of the line and a point on the line. [Try Exercise 44; the answer is $y-1=-\frac{5}{8}(x-2)$.] Also, see the solution to Exercise 23 in Section 1.2.

49. $f(5)=2, g(5)=4,$ and $f'(5)=3, g'(5)=1.$ Then

$$\begin{aligned} h(x) &= 3f(x)+2g(x), \\ h(5) &= 3f(5)+2g(5)=3(2)+2(4)=14, \\ h'(x) &= 3f'(x)+2g'(x), \\ h'(5) &= 3f'(5)+2g'(5)=3(3)+2(1)=11. \end{aligned}$$

55. The point of intersection of the tangent line and $f(x)$ is (4, 5), which alternatively can be stated as $f(4)=5$. The tangent line goes through (0, 3) and (4, 5), which means its slope is $m=\frac{5-3}{4-0}=\frac{2}{4}=\frac{1}{2}$. That is, $f'(4)=\frac{1}{2}$.

61. **(a)** When $8000 (8 thousand) was spent on advertising, 1200 (12 hundred) computers were sold. $A(x)$ is the amount of computers sold in hundreds when x is the amount spent on advertising, in thousands. Here we have $x=8$ and $A(8)=12$. The rate of increase of sales is 50 more computers (.5 hundred) for each 1000 more spent on advertising (1 thousand). Therefore, $A'(8)=.5$.

(b) If $9000 is spent on advertising, we can estimate the amount of computers sold by:

$$\begin{aligned} A(9) &\approx A(8)+A'(8) \\ &= 12+.5 \text{ (hundred)} \\ &= 1250 \text{ computers} \end{aligned}$$

1.7 More About Derivatives

Your ability to differentiate functions easily and correctly depends upon how much you practice. Use the exercises in this section to gain the experience you need. Don't be annoyed that you have to learn so much notation. We want you to be able to read technical articles in your field and be familiar with some of the notation you will find there.

1. $f(t)=(t^2+1)^5$

$$\begin{aligned} f'(t) &= 5(t^2+1)^4\cdot\frac{d}{dt}(t^2+1) \\ &= 5(t^2+1)^4 2t = 10t(t^2+1)^4. \end{aligned}$$

7.
$$\begin{aligned} \frac{d}{dP}\left(3P^2-\frac{1}{2}P+1\right) &= \frac{d}{dP}(3P^2)+\frac{d}{dP}\left(-\frac{1}{2}P\right)+\frac{d}{dP}(1) \\ &= 3\cdot\frac{d}{dP}(P^2)+-\frac{1}{2}\cdot\frac{d}{dP}(P)+\frac{d}{dP}(1) \\ &= 3(2P)-\frac{1}{2}(1)+0 \\ &= 6P-\frac{1}{2}. \end{aligned}$$

Helpful Hint: The notation $\frac{d}{dt}$ in Exercise 9 indicates that t is the independent variable when you differentiate. Any other letters appearing in the function represent constants, even though the values of these constants are not specified.

13. $y = \sqrt{x} = x^{1/2}, \quad \frac{dy}{dx} = \frac{1}{2}x^{-1/2}, \quad$ and

$$\frac{d^2y}{dx^2} = \frac{d}{dx}\left(\frac{dy}{dx}\right) = \frac{d}{dx}\left(\frac{1}{2}x^{-1/2}\right) = \frac{1}{2}\left(-\frac{1}{2}\right)x^{-3/2} = -\frac{1}{4}x^{-3/2}.$$

Helpful Hint: Problems involving radical signs (mainly square roots) are handled more easily when you switch to exponential notation, as in the solution to Exercise 13, above. Your instructor will probably permit you to write your answers in this form, too.

19. $f(P) = (3P+1)^5,$

$$f'(P) = 5(3P+1)^4 \cdot \frac{d}{dP}(3P+1) = 5(3P+1)^4 \cdot 3$$

$$= 15(3P+1)^4. \qquad \text{[Don't forget to multiply by 3 above.]}$$

$$f''(P) = \frac{d}{dP}f'(P) = \frac{d}{dP}[15(3P+1)^4] = 15 \cdot \frac{d}{dP}(3P+1)^4$$

$$= 15 \cdot 4(3P+1)^3 \cdot \frac{d}{dP}(3P+1) = 60(3P+1)^3(3)$$

$$= 180(3P+1)^3.$$

25. $\frac{d}{dx}(3x^3 - x^2 + 7x - 1) = \frac{d}{dx}(3x^3) + \frac{d}{dx}(-x^2) + \frac{d}{dx}(7x) + \frac{d}{dx}(-1)$

$$= 9x^2 - 2x + 7 + 0$$

$$= 9x^2 - 2x + 7$$

$$\frac{d^2}{dx^2}(3x^2 - x^2 + 7x - 1) = \frac{d}{dx}(9x^2 - 2x + 7)$$

$$= 18x - 2 + 0 = 18x - 2$$

$$\left.\frac{d^2}{dx^2}(3x^2 - x^2 + 7x - 1)\right|_{x=2} = 18(2) - 2 = 36 - 2 = 34.$$

31. $\frac{dR}{dx} = \frac{d}{dx}(1000 + 80x - .02x^2)$

$$= 0 + 80 - (.02)(2x) = 80 - .04x.$$

$$\left.\frac{dR}{dx}\right|_{x=1500} = 80 - .04(1500) = 80 - 60 = 20.$$

37. $C(x)$ is the cost in dollars of manufacturing x bicycles per day. Given $C(50) = 5000$, $x = 50$ so we interpret this as the cost of manufacturing 50 bicycles in one day is \$5000. Given $C'(50) = 45$ this tells us it costs an additional \$45 dollars to manufacture the 51st bicycle in a given day.

43. Enter the given function in Y_1. Then, using the calculator's derivative command, assign the derivative of Y_1 to Y_4 and assign the derivative of Y_4 to Y_5. (You could use Y_2 and Y_3 instead of Y_4 and Y_5, but we recommend keeping Y_2 and Y_3 available for other activities. You will need them later.) On the TI-82 and TI-83, use

$$Y_4 = \text{nDeriv}(Y_1, X, X), \quad Y_5 = \text{nDeriv}(Y_4, X, X)$$

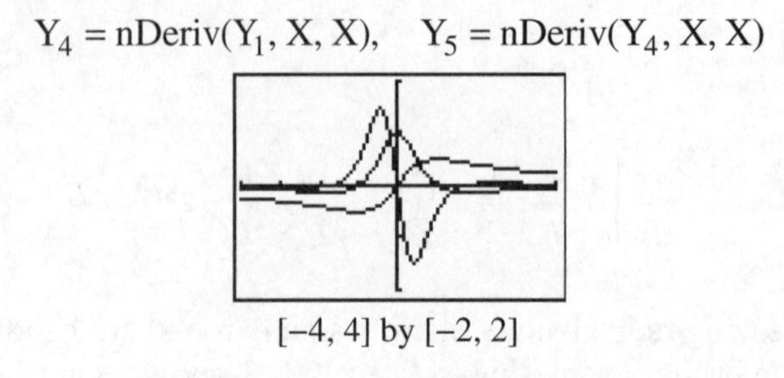

[−4, 4] by [−2, 2]

The TI-85 and TI-86 have the more accurate symbolic derivative operations available, so use

$$y4 = \text{der1}(y1, x, x), \quad y5 = \text{der1}(y2, x, x)$$

Helpful Hint: Keep the definitions of Y_4 and Y_5 permanently in your calculator as the first and second derivatives of Y_1 (as in Exercise 43). Then, whenever you need a function and its first two derivatives, all you have to do is assign the function Y_1. Deselect the derivatives until they are needed. (You may reserve other variables for the derivatives, but in this *Manual* we'll assume they are stored in Y_4 and Y_5.)

On the TI-85 and 86, to define y4 and y5, you need first to (temporarily) assign formulas to y2 and y3. (Setting them equal to 1 will do.) You can clear the formulas after y4 and y5 are defined.

1.8 The Derivative as a Rate of Change

The section is crucial for later work. There are three main categories of problems: (1) the rate of change of some function—either an abstract function or a function in an application where some quantity is changing with respect to time; (2) the rate of change of one economic quantity with respect to another—problems involving the adjective "marginal"; and (3) velocity and acceleration problems.

Problems in one category may seem quite different from those in another category. But this difference is superficial and is due to the terminology involved. Try to discover similarities in the problems. The key to this lies in the Practice Problems. Remember to use the practice problems correctly. Try to answer all six problems yourself before you look at the solutions.

1. **(a)** For $f(x) = 4x^2$, the average rate of change of $f(x)$ over the interval 1 to 2 is

$$\frac{f(2)-f(1)}{2-1} = \frac{4(2)^2 - 4(1)^2}{2-1} = \frac{16-4}{1} = 12.$$

On the interval 1 to 1.5, the average rate of change is

$$\frac{f(1.5)-f(1)}{1.5-1} = \frac{4(1.5)^2 - 4(1)^2}{1.5-1} = \frac{9-4}{.5} = 10.$$

On the interval 1 to 1.1:

$$\frac{f(1.1)-f(1)}{1.1-1} = \frac{4(1.1)^2 - 4(1)^2}{1.1-1} = \frac{4.84-4}{.1} = 8.4.$$

(b) The (instantaneous) rate of change of $4x^2$ at $x = 1$ is

$$\frac{d}{dx}4x^2\bigg|_{x=1} = 8x\bigg|_{x=1} = 8(1) = 8.$$

7. (a) If $s(t)$ represents the position function of an object moving in a straight line, then the velocity $v(t)$ of the object at time t is given by $s'(t)$. Since $s(t) = 2t^2 + 4t$, we know that

$$v(t) = s'(t) = 2(2t) + 4 = 4t + 4.$$
$$v(6) = s'(6) = 4(6) + 4 = 28 \text{ km/hr}$$

(b) In 6 hours, the object has traveled $s(6) = 2(6)^2 + 4(6) = 72 + 24 = 96$ km.

(c) We are asked to determine when $v(t) = s'(t) = 6$.

$$s'(t) = 4t + 4$$
$$6 = 4t + 4 \Rightarrow t = \frac{1}{2}$$

The object is traveling at the rate of 6 km/hr when $t = \frac{1}{2}$ hr.

13. This exercise, along with Exercise 17, is a key problem. It is essential that you try all five parts without looking at the text or the solutions below. If you must have help, look again at the practice problems. After you have tried Exercise 13, compare your answers with those below. Also look again at the practice problems.

(a) Since velocity is the rate of change in position, the initial velocity of the rocket when $t = 0$ is $s'(0)$.

$$s(t) = 160t - 16t^2,$$
$$s'(t) = 160 - 32t,$$
$$s'(0) = 160 - 32(0) = 160.$$

So the velocity at $t = 0$ is 160 ft/sec.

(b) "Velocity after 2 seconds" means velocity when $t = 2$.

$$v(2) = s'(2) = 160 - 32(2) = 96 \text{ ft/sec}.$$

(c) Acceleration involves the rate of change of velocity, that is, the derivative of $v(t) = 160 - 32t$.

$$a(t) = v'(t) = -32.$$

Since the distance is in feet and time is in seconds, the acceleration is -32 feet per second per second or -32 ft/sec^2. If the units were kilometers and hours, for example, the acceleration would be measured in kilometers per hour per hour.

(d) In this part the time is unknown. The rocket hits the ground when the distance above the ground is zero. So set $s(t) = 0$ and solve for t:

$$160t - 16t^2 = 0,$$
$$16t(10 - t) = 0,$$
$$t = 0, \quad \text{and} \quad t = 10.$$

The rocket hits the ground when $t = 10$ seconds.

(e) We know the time from part (d). The velocity at $t = 10$ seconds is

$$v(t) = 160 - 32(10) = -160 \text{ ft/sec.}$$

A negative sign on the velocity indicates that the distance function is decreasing, that is, the rocket is falling down.

Helpful Hint: Exercise 13(e) is a good exam question. If 13(d) is not on the exam, too, here is how to analyze the problem:

1. "At what velocity" means you must find some value of the velocity. To do this you need the velocity function and some specific time t.
2. You can get the velocity function by computing $s'(t)$.
3. Since you aren't given the time, you must determine the time from the fact that the rocket has just smashed into the ground. That is, you must solve a question like #13(d), even though it may not be listed specifically on the exam.

Helpful Hint: Check with your instructor about whether your test answers for velocity and acceleration problems must include the correct units. Almost all instructors require your answer to include the correct units.

19. Suppose $f(100) = 5000$ and $f'(100) = 10$. Then when x is close to 100, the values of $f(x)$ change at the rate of approximately 10 units for each unit change in x. A change in x of h units produces a change of about $f'(100)h = 10h$ units in the values of $f(x)$.

(a) For $f(101)$, x changes by 1 unit, so the values change by about 10 units. Thus $f(101) \approx f(100) + 10 = 5010$.

(b) For $f(100.5)$, x increases by $h = .5$, so the values of $f(x)$ change by about 10(.5) units. Thus $f(100.5) \approx f(100) + 5 = 5005$.

(c) For $f(99)$, x changes by $h = -1$ unit, so the values of $f(x)$ change by about $10(-1)$ units. That is, $f(99) \approx f(100) + (-10) = 4990$.

(d) For $f(98)$, $h = -2$ unit and

$$f(98) \approx f(100) + 10(-2)$$
$$= 5000 - 20 = 4980.$$

(e) For $f(99.75)$, $h = -.25$ unit and

$$f(99.75) \approx f(100) + 10(-.25)$$
$$= 5000 - 2.5 = 4997.5.$$

25. $f(x)$ is the number (in thousands) of computers sold when the price is x hundred dollars per computer. Given $f(12) = 60$, this tells us that when the price of a computer is \$1200 $(x = 12)$, 60,000 computers will be sold. Given $f'(12) = -2$, this tells us that at that price (\$1200), the number of computers sold decreases by 2000 for every \$100 increase in the price of the computer.

$$f(12.5) \approx f(12) + .5f'(12)$$
$$= 60 + .5(-2) = 59$$

About 59,000 computers will be sold if the price increases to \$1250.

31. (a) 1987 corresponds to $t = 7$ years after 1980. The y-coordinate of the point on the graph where $t = 7$ is 500, so about 500 billion dollars were spent in 1987.

(b) The *rate* of expenditures is given by the graph of $f'(t)$. The y-coordinate of the point on this graph where $t = 7$ is about 50. So expenditures were rising at the rate of about 50 billion dollars per year.

(c) A question that asks "when" requires you to find an appropriate value of t. Since the question involves expenditures, look at the graph of $f(t)$. An expenditure of one trillion (which is one thousand billion) dollars corresponds to the point on the graph of $f(t)$ whose y-coordinate is 1000. This appears to be (14 ,1000). That is, the expenditures reached one thousand billion in 1994, about 14 years after 1980.

(d) This question involves the rate of expenditures, so look at the graph of $f'(t)$. Rates of expenditures are the y-coordinates on the graph. The rate of \$100 billion per year occurs at the point on the graph of $f(t)$ whose y-coordinate is 100, which seems to be the point (14,100). Thus, the rate of 100 billion per year occurred in 1994, 14 years after 1980.

Review of Chapter 1

The derivative is presented in three different ways. The derivative is defined *geometrically* in Section 1.3.

The derivative of $f(x)$ is a function $f'(x)$ whose value at $x = a$ gives the slope of the graph of $f(x)$ at $x = a$.

The secant-line approximation describes how derivative formulas are obtained. This is made more precise in Section 1.4 when the derivative is defined *analytically* as the limit of difference quotients.

$$f'(a) = \lim_{h \to 0} \frac{f(a+h) - f(a)}{h}$$

Finally, in Section 1.8 the derivative is described *operationally* as the rate of change of a function. Your review of Chapter 1 should include an attempt to see how these three aspects of the derivative concept are related. At the same time, of course, you should review the basic techniques for *calculating* derivatives (Sections 1.6 and 1.7). The supplementary exercises will help here. Some instructors tend to look at them when they prepare exams. Finally, don't forget to review Section 1.1.

Chapter 1: Supplementary Exercises

1. Use the point-slope equation $y - y_1 = m(x - x_1)$, with $(x_1, y_1) = (0, 3)$ and $m = -2$. That is,

$$y - 3 = -2(x - 0), \quad \text{or} \quad y = -2x + 3.$$

To graph this line, first plot the y-intercept (0, 3). Then move one unit to the right and then two units in the negative y-direction. The new point (1, 1) will also be on the line. Draw the straight line through these two points.

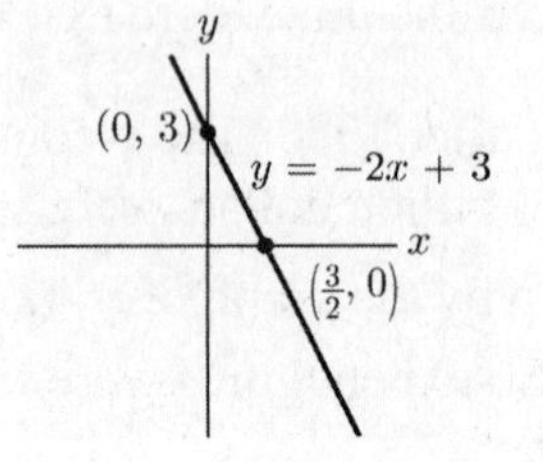

7. Apply Slope Property 2 with $(x_1, y_1) = (-1, 4)$ and $(x_2, y_2) = (3, 7)$. The slope of the line is

$$\frac{y_2 - y_1}{x_2 - x_1} = \frac{7 - 4}{3 - (-1)} = \frac{3}{4}.$$

Set $(x_1, y_1) = (-1, 4)$ and $m = \frac{3}{4}$ in the point-slope equation of a line:

$$y - 4 = \frac{3}{4}(x - (-1)), \quad \text{or} \quad y = \frac{3}{4}x + \frac{19}{4}.$$

To graph this line plot the points $(-1, 4)$ and $(3, 7)$ and draw a straight line through them.

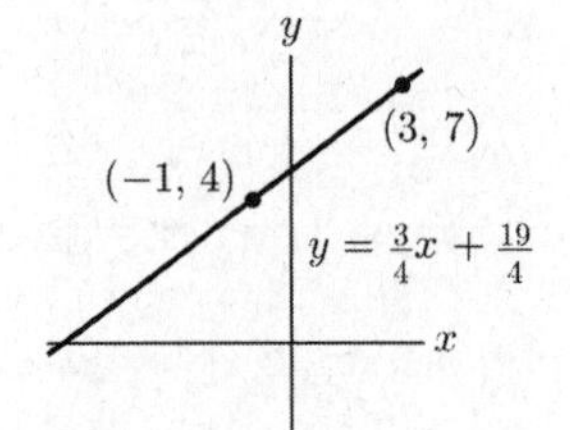

13. The y-axis has the equation $x = 0$.

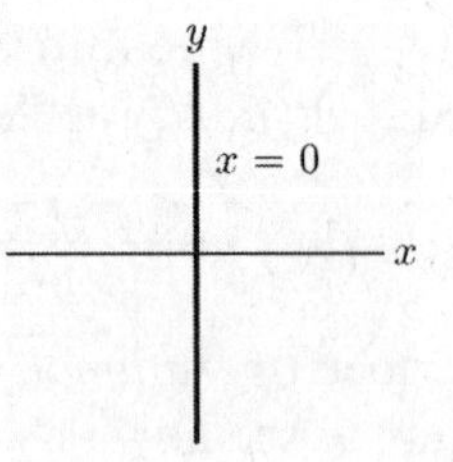

19. $\frac{d}{dx}\left(\frac{3}{x}\right) = \frac{d}{dx}(3 \cdot x^{-1}) = 3 \cdot \frac{d}{dx}(x^{-1})$ (Constant-multiple rule)

$$= 3(-1)x^{-2} = -3x^{-2}, \quad \text{or} \quad -\frac{3}{x^2}.$$

25. $$\frac{d}{dx}\sqrt{x^2+1} = \frac{d}{dx}(x^2+1)^{1/2}$$
$$= \frac{1}{2}(x^2+1)^{-1/2}\cdot\frac{d}{dx}(x^2+1) \qquad \text{(General power rule)}$$
$$= (x^2+1)^{-1/2}(2x) = x(x^2+1)^{-1/2}$$
$$= \frac{x}{\sqrt{x^2+1}}.$$

31. If $f(x) = [x^5 - (x-1)^5]^{10}$, then
$$f'(x) = 10[x^5-(x-1)^5]^9\cdot\frac{d}{dx}[x^5-(x-1)^5] \qquad \text{(General power rule)}$$
$$= 10[x^5-(x-1)^5]^9\left[\frac{d}{dx}(x^5) - \frac{d}{dx}(x-1)^5\right] \qquad \text{(Sum rule)}$$
$$= 10[x^5-(x-1)^5]^9\left[5x^4 - 5(x-1)^4\cdot\frac{d}{dx}(x-1)\right] \qquad \text{(General power rule)}$$
$$= 10[x^5-(x-1)^5]^9[5x^4-5(x-1)^4].$$

37. If $h(x) = \frac{3}{2}x^{3/2} - 6x^{2/3}$, then
$$h'(x) = \frac{3}{2}\frac{d}{dx}(x^{3/2}) - 6\cdot\frac{d}{dx}(x^{2/3})$$
$$= \frac{3}{2}\left(\frac{3}{2}\right)x^{1/2} - 6\left(\frac{2}{3}\right)x^{-1/3}$$
$$= \frac{9}{4}x^{1/2} - 4x^{-1/3}.$$

43. If $f(x) = x^{5/2}$, then
$$f'(x) = \frac{5}{2}x^{3/2} \quad \text{and} \quad f''(x) = \left(\frac{5}{2}\right)\left(\frac{3}{2}\right)x^{1/2} = \frac{15}{4}\sqrt{x}.$$
Hence $$f''(4) = \frac{15}{4}\sqrt{4} = \frac{15}{4}(2) = \frac{15}{2}.$$

49. $$\frac{d}{dP}\sqrt{(1-3P)} = \frac{d}{dP}(1-3P)^{1/2}$$
$$= \frac{1}{2}(1-3P)^{-1/2}\cdot\frac{d}{dP}(1-3P) \qquad \text{(General power rule)}$$
$$= \frac{1}{2}(1-3P)^{-1/2}(-3)$$
$$= -\frac{3}{2}(1-3P)^{-1/2}.$$

55. $$\frac{d}{dt}(t^3+2t^2-t)=\frac{d}{dt}(t^3)+2\cdot\frac{d}{dt}(t^2)-\frac{d}{dt}(t)$$
$$=3t^2+4t-1.$$
$$\frac{d^2}{dt^2}(t^3+2t-t)=\frac{d}{dt}(3t^2+4t-1)=6t+4.$$
$$\left.\frac{d^2}{dt^2}(t^3+2t-t)\right|_{t=-1}=6(-1)+4=-2.$$

61. The slope of the graph of $y=x^2$ at the point (x, y) is $2x$, hence the slope of the tangent line at $\left(\frac{3}{2},\frac{9}{4}\right)$ is $2\left(\frac{3}{2}\right)=3$. Let $(x_1, y_1)=\left(\frac{3}{2},\frac{9}{4}\right)$ and $m=3$, and use the point-slope equation for the tangent line to get

$$y-\frac{9}{4}=3\left(x-\frac{3}{2}\right),\quad \text{or}\quad y=3x-\frac{9}{4}.$$

Now sketch the curve $y=x^2$ by plotting a few points, including the point $\left(\frac{3}{2},\frac{9}{4}\right)$. Move one unit to the right of $\left(\frac{3}{2},\frac{9}{4}\right)$ and then 3 units in the positive y-direction to reach another point on the tangent line. Draw the straight line through these two points.)

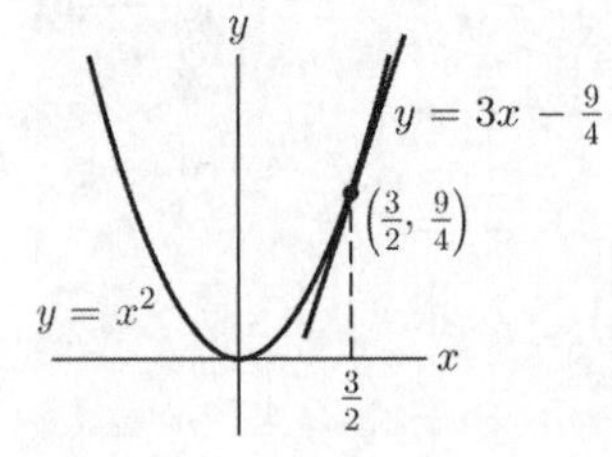

Helpful Hint: Make sure you can work Exercises 61 and 63 without any help. Make a note to work 63 later when you have not just finished reading this *Manual*. If you need help, see the solutions of Exercise 23 in Section 1.2 and Exercise 43 in Section 1.6.

67. $s(t)=-16t^2+32t+128,$ (height of binoculars)
$s'(t)=-32t+32.$ (velocity of binoculars)

To answer the question "How fast..." you must give the value of the velocity, $s'(t)$. But at what time? The time is identified only by the phrase, "when they hit the ground." Can you describe this time using $s(t)$ or $s'(t)$? Yes, since $s(t)=0$ when the binoculars are on the ground. So set $s(t)$ equal to zero and solve to find the time when this happens:

$$-16t^2+32t+128=0,$$
$$-16(t^2-2t-8)=0,$$
$$-16(t-4)(t+2)=0.$$

Thus $t-4=0$ or $t+2=0$, so that $t=4$ or $t=-2$. You may discard the possibility $t=-2$ because the appropriate domain for the function $s(t)$ is $t\geq 0$. The binoculars hit the ground when $t=4$ seconds, and their velocity at that time is $s'(4)=-32(4)+32=-128+32=-96$ feet per second. The negative sign indicates that the distance above the ground is decreasing. So the binoculars are *falling* at the rate of 96 feet per second.

Helpful Hint: Exercise 67 is a typical exam question. The solutions require two steps—finding $t = 4$ and computing $s'(4)$. Students who miss this problem usually try to work the problem in one step and don't know where to begin.

73. (a)

$$C(x) = .1x^3 - 6x^2 + 136x + 200,$$
$$C(21) = .1(21)^3 - 6(21)^2 + 136(21) + 200 = 1336.1,$$
$$C(20) = .1(20)^3 - 6(20)^2 + 136(20) + 200 = 1320.0,$$
$$C(21) - C(20) = 1336.1 - 1320.0 = 16.1 = \$16.10$$

This is the extra cost of raising the production from 20 to 21 units.

(b) The true marginal cost involves the derivative:

$$C'(x) = .3x^2 - 12x + 136,$$
$$C'(20) = .3(20)^2 - 12(20) + 136$$
$$= 16.0 \text{ dollars per unit.}$$

79. $x^2 - 8x + 16 = (x-4)^2$, so

$$\lim_{x \to 4} \frac{x-4}{x^2 - 8x + 16} = \lim_{x \to 4} \frac{x-4}{(x-4)^2}$$
$$= \lim_{x \to 4} \frac{1}{x-4},$$

which does not exist since the denominator is 0 at $x = 4$ and $\frac{1}{0}$ is not defined.

Chapter 2
Applications of the Derivative

This chapter culminates with one of the most powerful applications of the derivative—the solution of optimization problems. Although graphs play a key role in these problems, graphs are important in their own right. Sections 2.1 and 2.2 provide the background for the key topics covered in the remaining sections of the chapter.

2.1 Describing Graphs of Functions

The terms defined in this section are used throughout the text. The words *maximum* and *minimum* are used to refer to both points and values of function. When referring to a point, they are usually preceded with the word "relative." When referring to a value, "maximum value" means the largest value that the function assumes on its domain, and "minimum value" means the smallest value the function assumes on its domain.

1. Functions a, e, and f are increasing for all x since the graphs rise as we move along them from left to right.

7. Since the graph is superimposed on graph paper, we see that (0, 2) is a relative minimum point, (1, 3) is an inflection point, and (2, 4) is a relative maximum point. Therefore, the key features of the graph are as follows:

1. Decreasing for $x < 0$.
2. Relative minimum point at $x = 0$.
3. Increasing for $0 < x < 2$.
4. Relative maximum point at $x = 2$.
5. Decreasing for $x > 2$.
6. Concave up for $x < 1$, concave down for $x > 1$.
7. Inflection point at (1, 3).
8. y-intercept, (0, 2), x-intercept at about (3.6, 0).

13. Take an arbitrary point on the left-hand side of the graph, say (0, 0), and gradually increase its value. Initially the slope of the graph is positive. As x increases from 0 to $x = .5$, this slope becomes less positive, i.e., it decreases until finally at the point (.5, 1) the slope is zero. This point is a relative maximum. Continuing, as x goes from .5 to $x = 1$, the slope is negative and decreases. The graph is concave down for all x in the domain, hence there are no inflection points.

19.

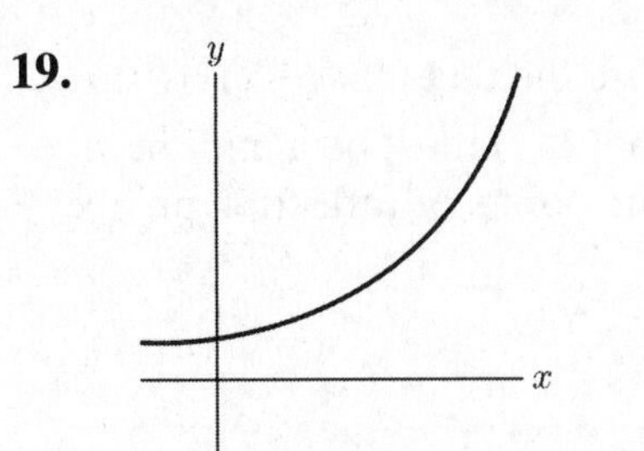

It is clear that both $f(x)$ and its slope increase as x increases, because the curve is rising and it lies above the tangent line at each point.

25.

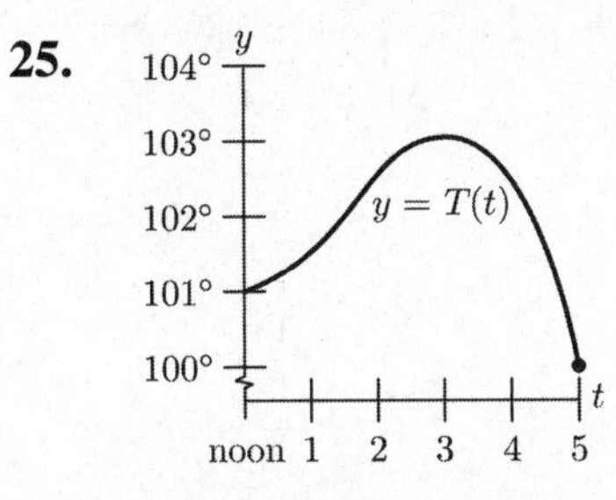

31. The graph is concave down from $t = 1920$ to $t = 1960$ and concave up from $t = 1960$ to $t = 2000$. The term "decreasing most rapidly" graphically, the place of maximum negative slope, and this occurs at the inflection point; i.e., when $t = 1960$. Translating, the number of farms was decreasing most rapidly when $t = 1960$.

37.

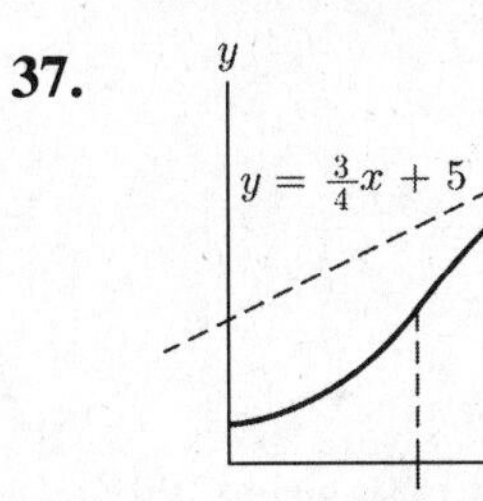

43. Set $Y_1 = 1/(X^\wedge 3 - 2X^\wedge 2 + X - 2)$, specify the window [0, 4) *by* [−15, 15], and graph the function. Use TRACE to determine the value of x at which the vertical asymptote occurs. You should obtain $x = 2$.

2.2 The First and Second Derivative Rules

The exercises in this section provide a warm-up to the problem of sketching the graphs of functions. Here we gain insights into the types of questions to ask in order to determine the main features of the graph of a given function.

Examples 3, 4, and 5 discuss important relations between the graphs of $f(x)$ and $f'(x)$. Study the examples carefully. The discussion in Example 5(e) shows that if the graph of $f'(x)$ is decreasing and crosses the x-axis at, say, $x = 3$, then $f(x)$ has a relative maximum at $x = 3$. Similarly, if the graph of $f'(x)$ is increasing and crosses the x-axis at, say, $x = a$, then $f(x)$ has a relative minimum at $x = a$ (because the value of $f'(x)$ changes from negative to positive at $x = a$ and so the graph of $f(x)$ changes from decreasing to increasing at $x = a$).

1. If a function has a positive first derivative for all x, the First Derivative Rule tells us that the function is increasing for all value of x. This is only true of graph *e*.

7. The only specific point on the graph is (2, 1). Plot this point and then use the fact that $f'(2) = 0$ to sketch the tangent line at $x = 2$. Since the graph is concave up, the point (2, 1) must be a minimum point. Since it is concave up for all x, there are no other relative extreme points or inflection points.

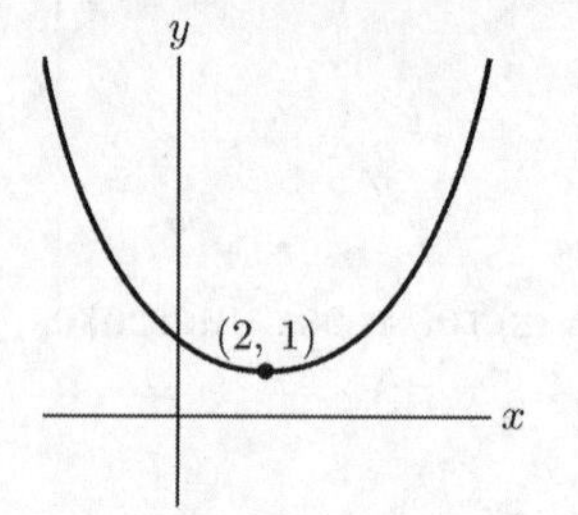

13. $f(3) = 4 \Rightarrow (3, 4)$ on the graph.
$f'(3) = -1/2,$ so the slope of the graph at (3, 4) is $-1/2$.
$f''(3) = 5 > 0 \Rightarrow$ the graph is concave up at $x = 3$.

19.

	f	f'	f''
A	POS	POS	NEG
B	0	NEG	0
C	NEG	0	POS

25. Look at the point (6, 2) on the graph of $f'(x)$. Since the y-coordinate of that point is positive, the *slope* of the graph of $f(x)$ is positive at $x = 6$. So $f(x)$ is increasing at $x = 6$, by the first derivative rule.

31. At $x = 1$, the graph of $f'(x)$ appears to have a relative maximum. That is, the graph of $f'(x)$ changes from increasing to decreasing at $x = 1$. An increasing derivative means that the graph of the original function $f(x)$ is concave up; a decreasing derivative means that the graph of $f(x)$ is concave down. So the graph of $f(x)$ changes at $x = 1$ from concave up to concave down, which means that $f(x)$ has an inflection point at $x = 1$.

Helpful Hint: To find values of x at which $f(x)$ has a relative extreme point, look for places at which the graph of $f'(x)$ *crosses* the x-axis. To find values of x at which $f(x)$ has an inflection point, look for places at which the graph of $f'(x)$ crosses the x-axis. If you are asked to do this on an exam, be sure that you can justify your answers, as in the discussions of Exercises 25 and 31 (above).

37. **(a)** If $h'(100) = \frac{1}{3}$, then at time $t = 100$, the water level is rising at the rate of 1/3 inch per hour. In a half-hour, the water will rise about half that much, namely, 1/6 inch.

(b) The conditions $h'(100) = 2$, and $h''(100) = -5$ say that the water is rising, but the *rate* of rise is declining since the derivative of the rate function is negative. That's good news, because the river may not rise much higher. However, condition (ii), $h'(100) = 2$, says that water level is falling, which is even better news.

Helpful Hint: Exercises 43 and 44 will help you learn how to extract information from graphs of a function and its derivatives. When you look at the graph of $f(t)$, you see how it increases or decreases and how the graph "bends." When you look at a graph of $f'(t)$ or $f''(t)$, focus mainly on the y-coordinates of various points, not on the shapes of the graphs.

43. **(a)** The number of farms is described by the graph of $f(t)$. The year 1990 corresponds to $t = 65$ (years after 1925), and the point (65, 2) appears to be on the graph of $f(t)$. So there were about 2 million farms in 1995.

(b) The *rate* of change in the number of farms is described by $f'(t)$. When $t = 65$, $f'(65)$ is approximately $-.03$. Thus in 1990, the number of farms is decreasing (because $f'(65)$ is negative) at the rate of about .03 million per year, that is, at the rate of about 30,000 farms per year.

(c) 6 million farms corresponds to the point on the graph of $f(t)$ whose y-coordinate is 6. This point is approximately (15, 6), which means that there were about 6 million farms in 1940.

(d) Here a rate is given and the question is "when?" The answer is found on the graph of $f'(t)$, looking for the value of t that corresponds to the given rate of $-60{,}000$ farms per year, that is, $-.06$ million farms per year. The rate is negative because the number of farms is declining. There are two such values of t, namely, 20 and about 52, which correspond to 1945 and 1977.

(e) The number of farms is declining the fastest when the rate is the most negative. This corresponds to the inflection point on the graph of $f(t)$, but that point is hard to find. Looking at the graph of $f'(t)$, you might guess that the minimum rate occurs at about $t = 35$. To be certain, look at $f''(t)$, because the minimum of $f'(t)$ occurs where its derivative crosses the t-axis. This happens at $t = 35$ (that is, 35 years after 1925). Thus, the number of farms was declining the fastest in 1960.

2.3 Curve Sketching (Introduction)

The curves sketched in this section will look like one of the curves shown below or one of these curves turned upside down. The first type of curve is called a *parabola* and is the graph of a quadratic function. The second curve is called a *cubic* and is the graph of a certain type of cubic polynomial. The purpose of this section is to apply the first and second derivative rules to find the graphs of functions. Therefore, you should not automatically assume that the graph will be one of those on the right, but should reason its shape from the two derivative rules. For instance, every time that you claim that a point is an extreme point, you should determine the value of the second derivative at that point and draw your conclusion from the second derivative test. Later in the text, we will graph functions about which we have no prior information.

1. First we find the critical values and critical points of f:

$$f'(x) = 3x^2 - 27 = 3(x^2 - 9)$$
$$= 3(x+3)(x-3).$$

The first derivative $f'(x) = 0$ if $x + 3 = 0$ or $x - 3 = 0$. Thus the critical values are

$$x = -3 \quad \text{and} \quad x = 3.$$

Substituting the critical values into the expression of f:

$$f(-3) = (-3)^3 - 27(-3) = 54$$
$$f(3) = (3)^3 - 27(3) = -54$$

The critical points are $(-3, 54)$ and $(3, -54)$. To determine whether these are relative maximums, minimums, or neither, we will apply the first derivative test. We will use the following chart to study the sign of $f'(x)$.

Critical Points, Intervals	$x < -3$	$-3 < x < 3$	$3 < x$
$x + 3$	$-$	$+$	$+$
$x - 3$	$-$	$-$	$+$
$f'(x)$	$+$	$-$	$+$
$f(x)$	Increasing on $(-\infty, -3)$	Decreasing on $(-3, 3)$	Increasing on $(3, \infty)$

We can see from the chart that the sign of $f'(x)$ changes from positive to negative at $x = -3$. Therefore, according to the first derivative test, f has a local maximum at $x = -3$. Also, the sign of $f'(x)$ changes from negative to positive at $x = 3$. Therefore, according to the first derivative test, f has a local minimum at $x = 3$. We conclude f has a local maximum at $(-3, 54)$ and a local minimum at $(3, -54)$.We can verify this by examining the graph.

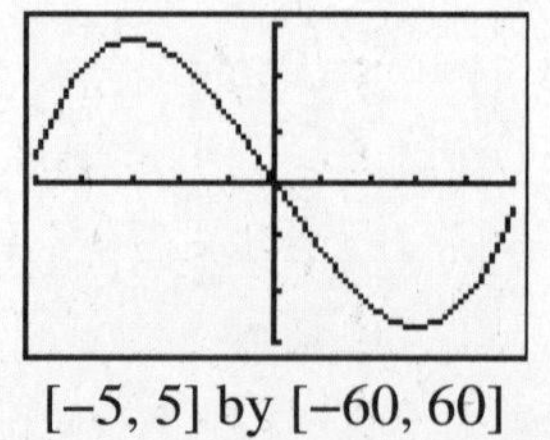

$[-5, 5]$ by $[-60, 60]$

7. First we find the critical values and critical points of f:

$$f'(x) = -3x^2 - 24x = -3x(x+8)$$

The first derivative $f'(x) = 0$ if $-3x = 0$ or $x + 8 = 0$. Thus the critical values are

$$x = 0 \quad \text{and} \quad x = -8.$$

Substituting the critical values into the expression of f:

$$f(-8) = -(-8)^3 - 12(-8)^2 - 2 = -258$$
$$f(0) = -(0)^3 - 12(0)^2 - 2 = -2$$

The critical points are (−8, −258) and (0, −2). To determine whether these are relative maximums, minimums, or neither, we will apply the first derivative test. We will use the following chart to study the sign of $f'(x)$.

Critical Points Intervals	$x < -8$	$-8 < x < 0$	$0 < x$
$x + 8$	−	+	+
$-3x$	+	+	−
$f'(x)$	−	+	−
$f(x)$	Decreasing on $(-\infty, -8)$	Increasing on $(-8, 0)$	Decreasing on $(0, \infty)$

We can see from the chart that the sign of $f'(x)$ changes from negative to positive at $x = -8$. Therefore, according to the first derivative test, f has a local minimum at $x = -8$. Also, the sign of $f'(x)$ changes from positive to negative at $x = 0$. Therefore, according to the first derivative test, f has a local maximum at $x = 0$. We conclude f has a local minimum at (−8, 258) and a local maximum at (0, −2). We can verify this by examining the graph.

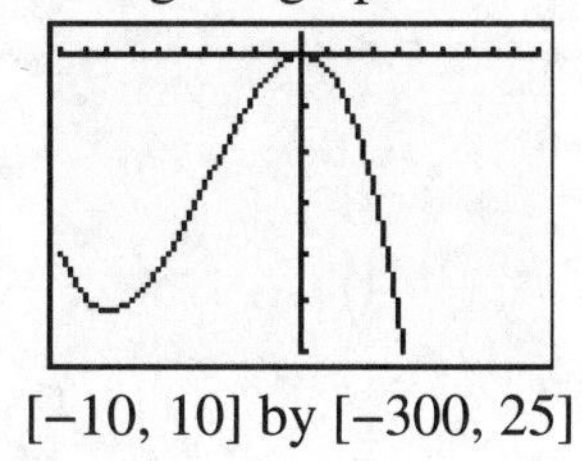

[−10, 10] by [−300, 25]

13. $f(x) = 1 + 6x - x^2$, $f'(x) = 6 - 2x$, $f''(x) = -2$

Set $f'(x) = 0$ and solve for x.

$$6 - 2x = 0 \Rightarrow -2x = -6 \Rightarrow x = 3$$

Substitute this value for x back into $f(x)$ to find the y-coordinate of this possible extreme point.

$$f(3) = 1 + 6(3) - (3)^2$$
$$= 1 + 18 - 9 = 10.$$

The positive extreme point is (3, 10). Since $f''(x) = -2$, which is negative, the graph of $f(x)$ is concave down at $x = 3$, and (3, 10) is a relative maximum point.

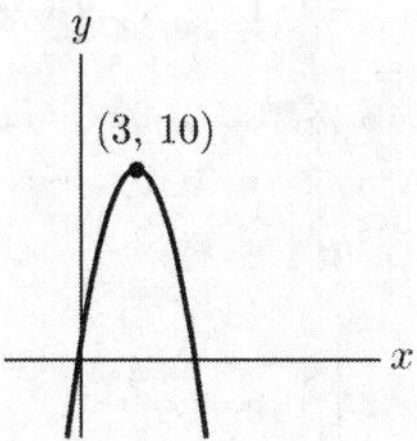

19. $f(x) = x^3 - 12x,\ \ f'(x) = 3x^2 - 12,\ \ f''(x) = 6x$

Set $f'(x) = 0$ and solve for x.

$$3x^2 - 12 = 0$$
$$3(x^2 - 4) = 0$$
$$3(x+2)(x-2) = 0$$
$$x = -2 \quad \text{or} \quad x = 2.$$

Substituting these values of x back into $f(x)$ to find the y-coordinates of these possible relative extreme points:

$$f(-2) = (-2)^3 - 12(-2) = -8 + 24 = 16$$
$$f(2) = (2)^3 - 12(2) = 8 - 24 = -16$$

The possible extreme points are (−2, 16) and (2, −16). Now,

$$f''(-2) = 6(-2) = -12 < 0,$$
$$f''(2) = 6(2) = 12 > 0.$$

The graph of $f(x)$ is concave down at $x = -2$ and concave up at $x = 2$. Therefore, (−2, 16) is a relative maximum point and (2, −16) is a relative minimum point.

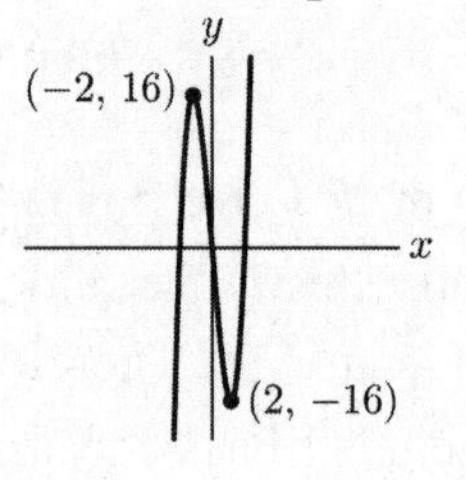

25. $y = x^3 - 3x + 2,\ \ y' = 3x^2 - 3,\ \ y'' = 6x$

Set $y' = 0$ and solve for x.

$$3x^2 - 3 = 0$$
$$3(x^2 - 1) = 0$$
$$3(x+1)(x-1) = 0$$
$$x = -1 \quad \text{or} \quad x = 1.$$

Use these values to find the y-coordinates of the possible relative extreme points:

$$\text{At } x = -1, \quad y = (-1)^3 - 3(-1) + 2 = -1 + 3 + 2 = 4$$
$$\text{At } x = 1, \quad y = (1)^3 - 3(1) + 2 = 1 - 3 + 2 = 0.$$

The possible relative extreme points are (−1, 4) and (1, 0). Now,

At $x = -1$, $y'' = 6(-1) = -6 < 0 \Rightarrow$ graph is concave down at $x = -1$.

At $x = 1$, $y'' = 6(1) = 6 > 0 \Rightarrow$ graph is concave up at $x = 1$.

So the function has a relative maximum at (−1, 4) and a relative minimum at (1, 0). Since the concavity changes somewhere between $x = -1$ and $x = 1$, there must be at least one inflection point. To find this, set $y'' = 0$ and solve for x:

$$6x = 0 \Rightarrow x = 0$$

Substitute $x = 0$ into the equation to find the y-coordinate of this inflection point:

$$y = (0)^3 - 3(0) + 2 = 0 - 0 + 2 = 2.$$

The inflection point is (0, 2).

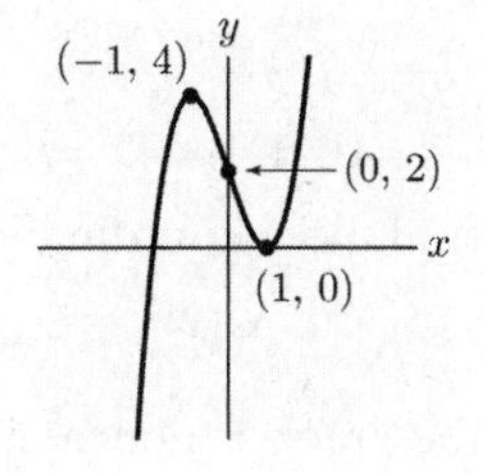

31. $y = 2x^3 - 3x^2 - 36x + 20,$

$y' = 6x^2 - -6x - 36,$

$y'' = 12x - 6,$

Set $y' = 0$ and solve for x.

$$6x^2 - 6x - 3 = 0$$
$$6(x^2 - x - 6) = 0$$
$$3(x + 2)(x - 3) = 0$$
$$x = -2 \quad \text{or} \quad x = 3.$$

Use these values the find the y-coordinates of the possible relative extreme points:

At $x = -2$, $y = 2(-2)^3 - 3(-2)^2 - 36(-2) + 20 = -16 - 12 + 72 + 20 = 64$

At $x = 3$, $y = 2(3)^3 - 3(3)^2 - 36(3) + 20 = 54 - 18 - 108 + 20 = -61.$

The possible relative extreme points are (–2, 64) and (3, –61). Now,

At $x = -2$, $y'' = 12(-2) - 6 = -30 < 0 \Rightarrow$ graph is concave down at $x = -2$.

At $x = 3$, $y'' = 12(3) - 6 = 30 > 0 \Rightarrow$ graph is concave up at $x = 3$.

So the function has a relative maximum at (–2, 64) and a relative minimum at (3, –61). Since the concavity changes somewhere between $x = -2$ and $x = 3$, there must be at least one inflection point. To find this, set $y'' = 0$ and solve for x:

$$12x - 6 = 0 \Rightarrow 12x = 6 \Rightarrow x = \frac{1}{2}$$

Substitute $x = \frac{1}{2}$ into the equation to find the y-coordinate of this inflection point:

$$y = 2\left(\frac{1}{2}\right)^3 - 3\left(\frac{1}{2}\right)^2 - 36\left(\frac{1}{2}\right) + 20 = \frac{2}{8} - \frac{3}{4} - \frac{36}{2} + 20 = \frac{3}{2}.$$

The inflection point is $\left(\frac{1}{2}, \frac{3}{2}\right)$.

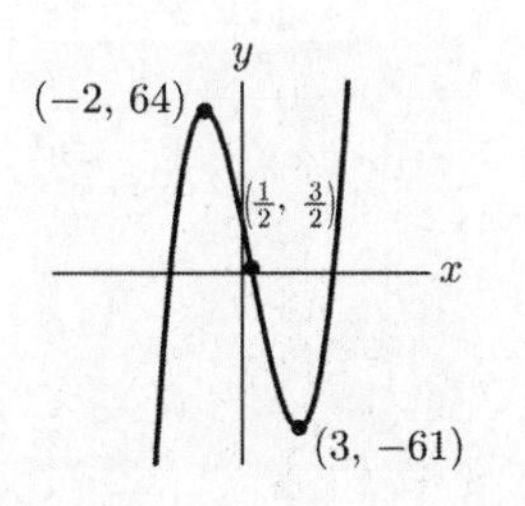

37. $g(x) = 3 + 4x - 2x^2,$
$g'(x) = 4 - 4x,$
$g''(x) = -4,$
Set $g'(x) = 0$ and solve for x:

$$4 - 4x = 0 \Rightarrow 4 = 4x \Rightarrow x = 1$$

Substitute this value back into $g(x)$ to find the y-coordinate of this possible relative extreme point.

$$g(1) = 3 + 4(1) - 2(1)^2 = 3 + 4 - 2 = 5$$

Since $g''(x) = -4$, which is negative, the graph is concave down at $x = 1$, and thus, (1, 5) is a relative maximum point.

43.

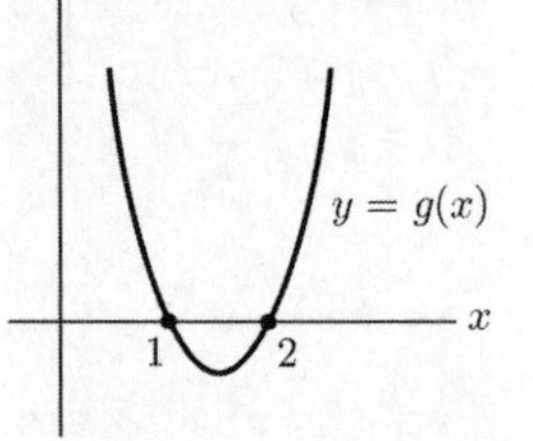

(a) Refer to the figure above. Assume $g(x)$ is the first derivative of $f(x)$. Then, when $x = 2$, $g(2) = 0$, $g(x)$ is negative for $x < 2$ and positive for $x > 2$ (or equivalently; $f'(2) = 0$, $f'(x)$ is negative for $x < 2$ and positive for $x > 2$). This tells us that $f(x)$ is decreasing for $x < 2$ and increasing for $x > 2$ and $f(x)$ has a relative minimum at $x = 2$.

(b) Refer to the figure above. Assume $g(x)$ is the second derivative of $f(x)$. Then, when $x = 2$, $g(2) = 0$, $g(x)$ is negative for $x < 2$ and positive for $x > 2$ (or equivalently; $f''(2) = 0$, $f''(x)$ is negative for $x < 2$ and positive for $x > 2$). This tells us that $f(x)$ is concave down for $x < 2$ and concave up for $x > 2$ and $f(x)$ has an inflection point at $x = 2$.

49.

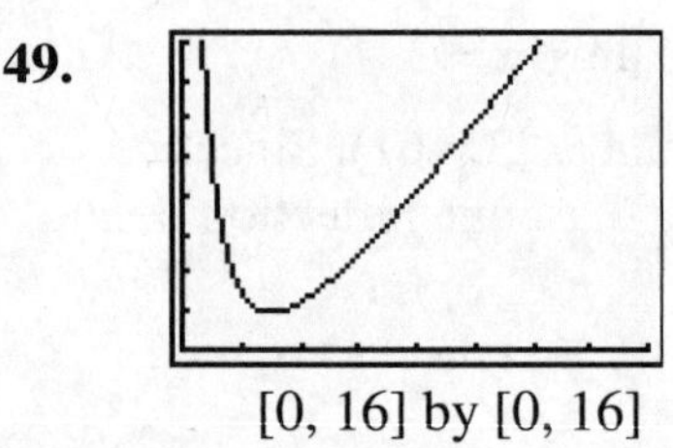

[0, 16] by [0, 16]

This graph is like the graph of a parabola that opens upward because (for $x > 0$) the entire graph is concave up and it has a minimum value. Unlike a parabola, it is not symmetric. Also, this graph has a vertical asymptote ($x = 0$), while a parabola does not have an asymptote.

2.4 Curve Sketching (Conclusion)

The two most important types of curves that are graphed in this section are the cubics without relative extreme points (such as the first curve or the upside down version of this curve) and the curves with asymptotes. These curves are stressed since, along with the curves considered in Section 2.3, they occur frequently in applications. However, there are a few other types of curves that occur in the exercises. As previously, all reasoning should be based on the first and second derivative tests.

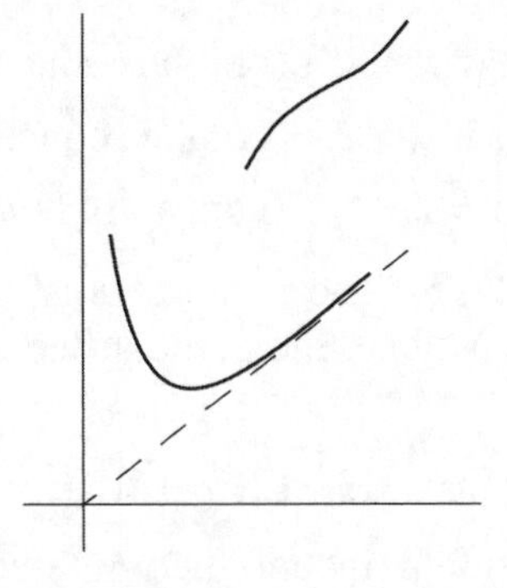

1. To find the x-intercepts of the function $y = x^2 - 3x + 1$, use the quadratic formula to find the values of x for which $f(x) = 0$:

$$y = x^2 - 3x + 1 \quad (a = 1, b = -3, c = 1),$$
$$b^2 - 4ac = (-3)^2 - 4 \cdot 1 \cdot 1 = 5,$$
$$x = \frac{-(-3) \pm \sqrt{5}}{2 \cdot 1} = \frac{3 \pm \sqrt{5}}{2}.$$

Therefore, the x-intercepts are $\left(\frac{3+\sqrt{5}}{2}, 0\right)$ and $\left(\frac{3-\sqrt{5}}{2}, 0\right)$.

7. If $f(x) = \frac{1}{3}x^3 - 2x^2 + 5x$, then

$$f'(x) = x^2 - 4x + 5.$$

If we apply the quadratic formula to $f'(x)$ with $a = 1$, $b = -4$, and $c = 5$, we find that $b^2 - 4ac = (-4)^2 - 4 \cdot 1 \cdot 5 = -4$, a negative number. Since we cannot take the square root of a negative number, there are no values of x for which $f'(x) = 0$. Hence $f(x)$ has no relative extreme points.

13. $f(x) = 5 - 13x + 6x^2 - x^3$,
$f'(x) = -13 + 12x - 3x^2$,
$f''(x) = 12 - 6x$

If we apply the quadratic formula to $f'(x)$, we find that

$$b^2 - 4ac = (12)^2 - 4(-3)(-13) = 144 - 156 = -12,$$

which is negative.

Since we cannot take the square root of a negative number, there are no values of x for which $f'(x)=0$. So $f(x)$ has no relative extreme points. If we evaluate $f'(x)$ at some x, say $x = 0$, we see that the first derivative is negative, and so $f(x)$ is decreasing there. Since the graph of $f(x)$ is a smooth curve with no relative extreme points and no breaks, $f(x)$ must be decreasing for all x. Now, to check the concavity of $f(x)$, we must find where $f''(x)$ is negative, positive, or zero.

$f''(x)=12-6x=6(2-x)$ is:

$$\begin{cases} \text{positive if } x<2, & \text{(graph is concave up)} \\ \text{negative if } x>2, & \text{(graph is concave down)} \\ \text{zero if } x=2. & \text{(concavity reverses)} \end{cases}$$

The inflection point is $(2, f(2)) = (2, -5)$. The y-intercept is $(0, f(0)) = (0, 5)$. To further improve the sketch of the graph, first sketch the tangent line at the inflection point. To do this compute the slope of the graph at $(2, -5)$.

$$f'(2)=-13+12(2)-3(2)^2=-1.$$

To sketch the graph, plot the inflection point and the y-intercept, draw the tangent line at $(2, -5)$, and then draw a curve that has this tangent line and is decreasing for all x, is concave down for $x > 2$, and concave up for $x < 2$.

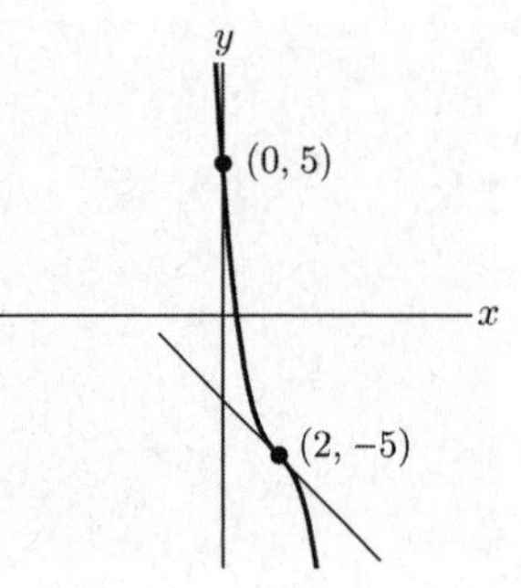

19. $f(x)=x^4-6x^2,$

$f'(x)=4x^3-12x,$

$f''(x)=12x^2-12.$

To find possible relative extreme points, set $f'(x)=0$ and solve for x:

$$4x^3-12x=0$$
$$4x(x^2-3)=0$$
$$4x(x-\sqrt{3})(x+\sqrt{3})=0$$
$$x=0 \quad\text{or}\quad x=\pm\sqrt{3}.$$

Substitute these values into $f(x)$ to find the y-coordinates of these possible relative extreme points:

$$f(0)=(0)^4-6(0)^2=0,$$
$$f(\sqrt{3})=(\sqrt{3})^4-6(\sqrt{3})^2=-9,$$
$$f(-\sqrt{3})=(-\sqrt{3})^4-6(-\sqrt{3})^2=-9$$

The points are $(0, 0)$, $(\sqrt{3},-9)$, and $(-\sqrt{3},-9)$.

Now $f''(0) = 12(0)^2 - 12 = -12$. Therefore, $f(x)$ is concave down at $x = 0$ and (0, 0) is a relative maximum point. For the other two points, compute

$$f''(-\sqrt{3}) = 12(-\sqrt{3})^2 - 12 = 12(3) - 12 = 24,$$
$$f''(\sqrt{3}) = 12(\sqrt{3})^2 - 12 = 12(3) - 12 = 24$$

Hence $f(x)$ is concave up at $x = -\sqrt{3}$ and $x = \sqrt{3}$, and so $(\sqrt{3}, -9)$ and $(-\sqrt{3}, -9)$ are relative minimum points. The concavity of this function reverses twice, so there must be at least two inflection points. To find these, set $f''(x) = 0$ and solve for x:

$$12x^2 - 12 = 0,$$
$$x^2 - 1 = 0,$$
$$(x-1)(x+1) = 0, \quad \text{or} \quad x = \pm 1.$$

The corresponding y-coordinates are given by

$$f(1) = (1)^4 - 6(1)^2 = -5,$$
$$f(-1) = (-1)^4 - 6(-1)^2 = -5.$$

The inflection points are $(1, -5)$.

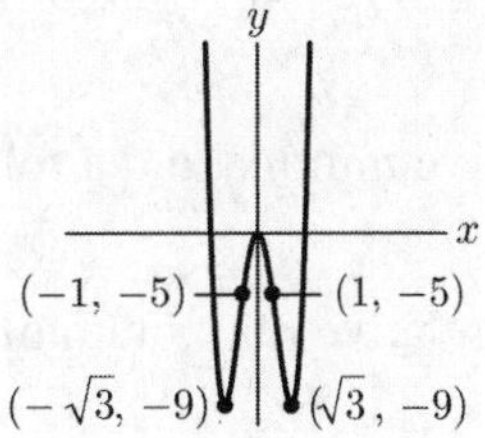

25. $y = \dfrac{9}{x} + x + 1, \quad x > 0$

$$y' = -\frac{9}{x^2} + 1,$$
$$y'' = \frac{18}{x^3}$$

To find possible extrema, set $y' = 0$ and solve for x:

$$-\frac{9}{x^2} + 1 = 0,$$
$$1 = \frac{9}{x^2} \qquad \text{(Multiply both sides by } x^2.\text{)}$$
$$x^2 = 9,$$
$$x = 3. \qquad (\textit{Note}\text{: Only consider } x > 0.)$$

When $x = 3$, $y = \dfrac{9}{3} + 3 + 1 = 7$, and

$$y'' = \frac{18}{3^3} > 0 \qquad \text{(Concave up).}$$

Therefore, (3, 7) is a relative minimum point. Since y'' can never be 0, there are no inflection points. The term $\frac{9}{x}$ in the function y tells us that the y-axis is an asymptote. Also, as x gets large, the graph of y gets arbitrarily close to the straight line $y = x + 1$. Therefore, $y = x + 1$ is an asymptote of the graph.

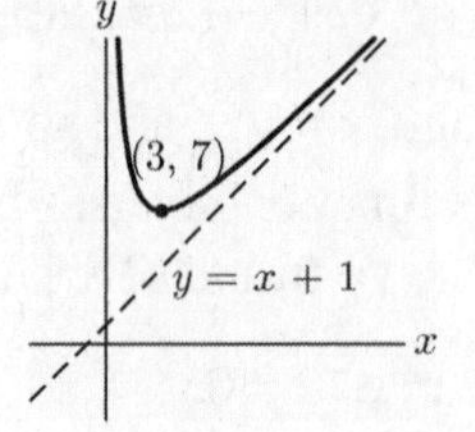

Helpful Hint: In general $f'(a) = 0$ does not necessarily mean that $f(x)$ has an extreme point at $x = a$. Also, $f''(a) = 0$ does not necessarily mean that $f(x)$ has an inflection point at $x = a$. However, if $f(x)$ is quadratic or of the form $hx + \frac{k}{x} + c$, then $f'(a) = 0$ guarantees an extreme point at $x = a$. (Neither of these two types of functions has inflection points.) If $f(x)$ is a cubic polynomial, then there is one inflection point and it occurs where the second derivative is zero. (*Note:* These observations will help you in checking your work. On exams, you must still show that extreme and inflection points have the properties you claim.)

31. $g(x) = f'(x)$, because $f(x)$ has 2 relative minima and a relative maximum and $g(x) = 0$ for these values of x.

$f(x) \neq g'(x)$. To see this, look at the relative maximum of $g(x)$. Since $f(x)$ is not zero for this value of x, $f(x) \neq g'(x)$.

37. **(a)**

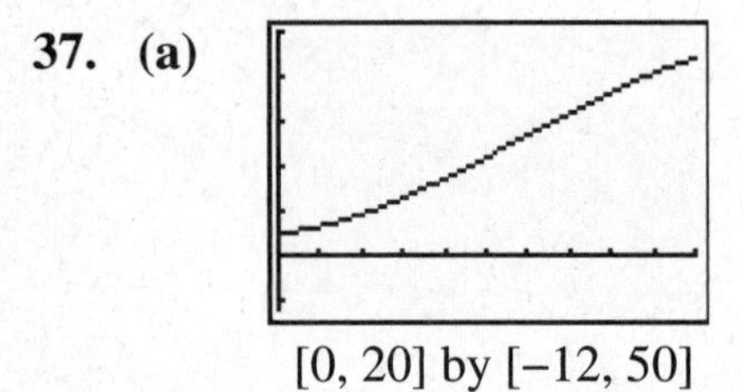

[0, 20] by [−12, 50]

(b) The weight of the rat in grams is given by $f(t) = 4.96 + .48t + .17t^2 - .00048t^3$, after t days. After 7 days, the rat weighed

$$f(7) = 4.96 + .48(7) + .17(7)^2 - .00048(7)^3$$
$$= 15.0036 \text{ grams.}$$

(c) To find when the rat weighed 27 grams, solve the following for t:

$$4.96 + .48t + .17t^2 - .00048t^3 = 27.$$

Use graphing calculator techniques to obtain $t \approx 12.0380$ (one method is to graph both $f(t)$ and $y = 27$ and use the intersect command to find where $f(t) = 27$). Therefore, the rat's weight reached 27 grams after approximately 12 days.

(d) Note that $f'(t) = .48 + .34t - .0144t^2$. Therefore, the rate at which the rat was gaining weight after 4 days is found by

$$f'(4) = .48 + .34(4) - .0144(4)^2 = 1.6096.$$

After 4 days, the rat was gaining weight at a rate of 1.6096 grams per day.

(e) To find when the rat was gaining weight at a rate of 2 grams per day, solve the following for t:

$$.48 + .34t - .0144t^2 = 2$$

Use graphing calculator techniques to obtain $t \approx 5.990$ or $t \approx 17.6207$. (Graph both $f'(t)$ and $y = 2$ and use the intersect command to find where $f'(t) = 2$). The rat was gaining weight at the rate of 2 grams per day after about 6 days and after about 17.6 days.

(f) To find when the rat was gaining wait at the fastest rate, we want to find the maximum value of $f'(t)$. In order to do this, solve $f''(t) = 0$, or

$$f''(t) = .34 - .0288t = 0$$

to obtain $t \approx 11.8056$. Confirm this is a maximum by examining the graph of $f'(t)$, or notice $f'''(t) = -.0288 < 0$. The rat was growing at the fastest rate after about 11.8 days.

2.5 Optimization Problems

The procedure for solving an optimization problem can be thought of as consisting of the following two primary parts:

(a) Find the function to be optimized.

(b) Make a rough sketch of the graph of the function.

Part (a) requires careful reading of the problem. Part (b) relies on techniques presented in Sections 2.3 and 2.4. However, certain shortcuts can be taken when sketching curves in this section. For instance, if you only want to find the x-coordinate of a relative extreme point, you can make a very rough estimate of the y-coordinate when sketching the curve.

One of the common mistakes that students make when working optimization problems on exams is just to set the first derivative equal to 0 and solve for x. They forget to apply the second derivative test at the value of x found.

1. $g(x) = 10 + 40x - x^2,$

$g'(x) = 40 - 2x,$

$g''(x) = -2.$

The graph is obtained by the curve-sketching technique of Section 2.3. Therefore, the maximum value of $g(x)$ occurs at $x = 20$.

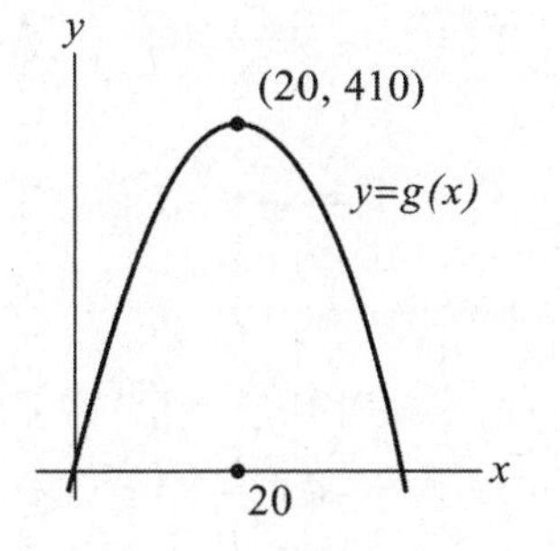

7. The function to be minimized is $Q = x^2 + y^2$. By solving the equation $x + y = 6$ for y, we can write y in terms of x:

$$y = 6 - x,$$

and then substitute in order to express Q as a function of the single variable x:

$$Q = x^2 + (6 - x)^2$$

Simplifying, we have

$$Q = x^2 + (36 - 12x + x^2), \quad \text{or} \quad Q = 2x^2 - 12x + 36.$$

Differentiating,

$$Q' = 4x - 12,$$
$$Q'' = 4$$

Now, solve $Q' = 0$:

$$4x - 12 = 0,$$
$$4x = 12,$$
$$x = 3$$

$Q' = 0$ and Q'' is positive at $x = 3$. Therefore Q has a relative minimum at $x = 3$. Plug in $x = 3$ to the equation for Q in order to find that when $x = 3$, $Q = 18$. Since $x = 3$ is the only critical point, (3, 18) is the global minimum.

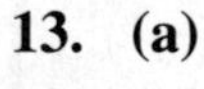

13. (a)

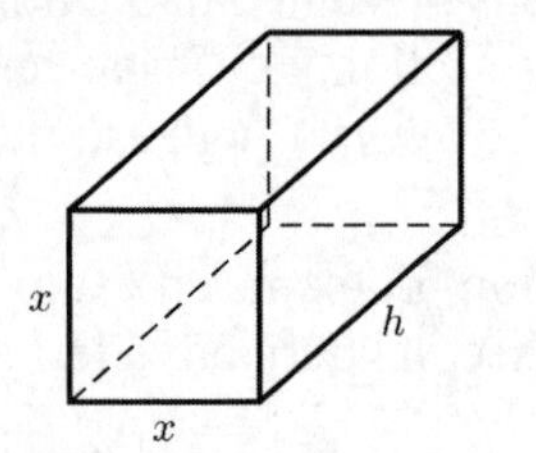

(b) The girth of the box is given by $x \dot{+} x + x + x = 4x$. The formula for the length plus the girth is

$$h + 4x$$

(c) The equation for volume, which must be maximized, is the objective equation,

$$v = x^2 h.$$

The constraint equation is

$$h + 4x = 84,$$

or

$$h = 84 - 4x.$$

(d) Substitute $h = 84 - 4x$ into the objective equation:

$$V = x^2(84 - 4x)$$
$$= -4x^3 + 84x^2.$$

(e) Make a rough sketch of the graph of $V = -4x^3 + 84x^3$ to find the value of x corresponding to the greatest volume.

$$V' = -12x^2 + 168x,$$
$$V'' = -24x + 168.$$

Solve $V' = 0$:

$$-12x^2 + 168x = 0,$$
$$x^2 - 14x = 0,$$
$$x(x-14) = 0,$$
$$x = 0, \quad \text{or} \quad x = 14.$$

V'' is positive at $x = 0$ and negative at $x = 14$. Therefore V has a relative minimum at $x = 0$ and a relative maximum at $x = 14$. When $x = 0$, $V = 0$. When $x = 14$, $V = -4(14)^3 + 84(14)^2 = 5488$. The sketch of the graph shows that the maximum value of V occurs when $x = 14$. From the constraint equation, $h = 84 - 4(14) = 28$.

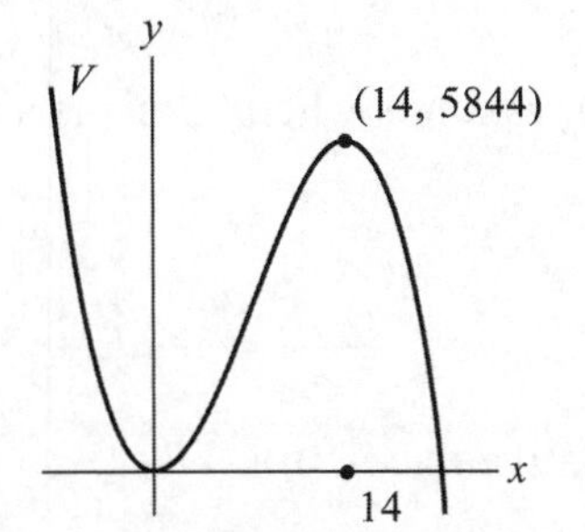

19. The problem asks that total area be maximized. Let A be the area, let x be the length of the fence parallel to the river, and let y be the length of each section perpendicular to the river. The objective equation gives an expression for A in terms of the other variables:

$$A = xy \quad \text{(objective equation).}$$

The cost of the fence parallel to the river is $6x$ dollars (x feet at \$6 per foot) and the cost of the three sections perpendicular to the river is $3 \cdot 5y$ or $15y$ dollars. Since \$1500 is available to build the fence, we must have

$$6x + 15y = 1500 \qquad \text{(constraint equation).}$$

Solving for y, we have

$$15y = -6x + 1500,$$
$$y = -\frac{2}{5}x + 100.$$

Substituting this expression for y into the objective equation, we obtain

$$A = x\left(-\frac{2}{5}x + 100\right) = -\frac{2}{5}x^2 + 100x.$$

The graph of $A = -\frac{2}{5}x^2 + 100x$ is easily obtained by our curve sketching techniques. The maximum value occurs when $x = 125$. Substituting into $y = -\frac{2}{5}x + 100$, this gives us $y = -\frac{2}{5}(125) + 100 = 50$. Therefore, the optimum dimensions are $x = 125$, $y = 50$.

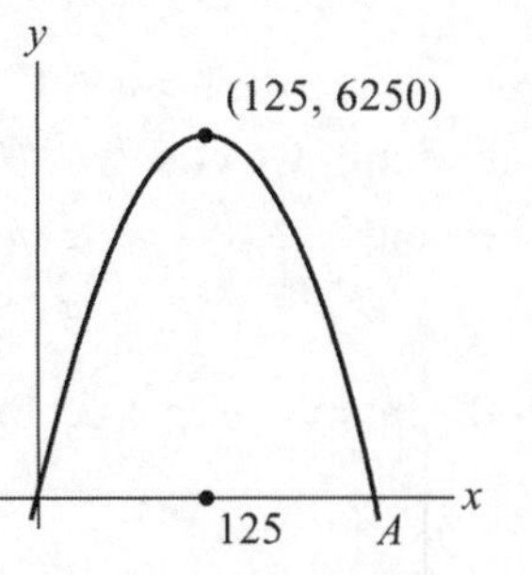

25. $x = 20 - \frac{w}{2}$ (from the constraint equation).

$A = wx$ (the objective function).

Therefore:

$$A = w\left(20 - \frac{w}{2}\right) = -\frac{w^2}{2} + 20w \qquad \text{(has a parabola shape),}$$

$$\frac{dA}{dw} = -\frac{2w}{2} + 20 = -w + 20.$$

A relative minimum or maximum occurs when $\frac{dA}{dw} = 0$. Hence $-w + 20 = 0 \Rightarrow w = 20$. (Since $x = 20 - \frac{w}{2}$, $x = 10$.) The next question is whether this is a relative maximum. Look at the second derivative:

$$\frac{d^2A}{dw^2} = -1 < 0,$$

which shows that $w = 20$ is a relative maximum. To sketch the graph of A, observe that $A = 0$ when $w = 0$ and when $x = 0$: $20 - \frac{w}{2} = 0 \Rightarrow \frac{w}{2} = 20$, or $w = 40$. At $w = 20$, $A = 20\left(20 - \frac{20}{2}\right) = 200$.

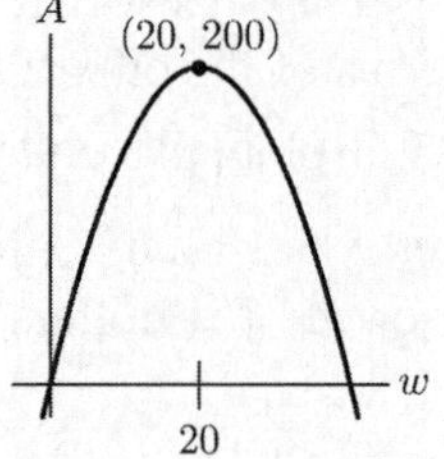

31. The distance from any point (x, y) on the line $y = -2x + 5$ to the origin is given by

$$\text{Distance} = \sqrt{x^2 + y^2} = \sqrt{x^2 + (-2x + 5)^2} = \sqrt{5x^2 - 20x + 25}.$$

The distance has its smallest value when $5x^2 - 20x + 25$ does, so we need to minimize $D(x) = 5x^2 - 20x + 25$:

$$D(x) = 5x^2 - 20x + 25$$
$$D'(x) = 10x - 20$$
$$D'(x) = 0 \Rightarrow 10x - 20 = 0$$
$$x = 2$$

To confirm $x = 2$ is a minimum, we check $D''(x)$: $D''(x) = 10 > 0$, so $x = 2$ is in fact a minimum of $D(x)$. Substituting $x = 2$ into $y = -2x + 5$, we find

$$y = -2(2) + 5 = 1$$

Therefore, the point on the line $y = -2x + 5$ closest to the origin is (2, 1).

2.6 Further Optimization Problems

The exercises in this section are of four types.

1. Variations of the types of exercises considered in Section 2.5.
2. Inventory problems. These problems all have the same constraint equation, $r \cdot x =$ [number of items used or manufactured during the year], and similar objective equations, [costs] = ...
3. Orchard problems (that is, problems similar to the first practice problem or to Example 1). These problems have the same type of objective equation, [profit or revenue] = [amount of money] · [quantity]. The constraint equations are found by reading the problem to see if [amount of money] depends on [quantity] or vice versa, and expressing the dependent variable in terms of the other.
4. Original problems. Situations that use the same type of machinery as the examples worked in the text but require a fresh approach.

1. **(a)** In each order-reorder period, the amount of cherries in inventory decreases linearly from 180 pounds at the beginning of the period to 0 pounds at the end of the period. Thus the average amount of cherries in inventory in one order-reorder period is the average of 0 and 180, i.e., 90 pounds.

(b) The maximum amount of cherries in inventory during a given order-reorder period is 180 pounds. This occurs at the beginning of the period.

(c) 6 orders were placed during the year; each order-recorder period is represented by one of the 6 line segments on the graph.

(d) Since 180 pounds of cherries were sold during each order-reorder period, and there were 6 order-reorder periods during the year, a total of $180 \cdot 6 = 1080$ pounds of cherries were sold during the year.

7. Let r represent the number of production runs and x the number of microscopes manufactured per run. Hence the total number of microscopes manufactured is

$$\begin{bmatrix}\text{number of microscopes per} \\ \text{production run}\end{bmatrix} \cdot [\text{number of runs}] = xr.$$

The constraint equation is

$$xr = 1600, \quad \text{or} \quad x = \frac{1600}{r}.$$

We wish to minimize inventory expenses. There are three expenses which make up the total cost. The storage costs (SC), based on the maximum number of microscopes in the warehouse, will be

$$\text{SC} = 15x,$$

since the warehouse is the most full just after a run, which will produce x microscopes. Since the insurance costs are based on the average number of microscopes in the warehouse, and the average number of microscopes is given by $\frac{x}{2}$, the insurance costs (IC) are

$$\text{IC} = 20\left(\frac{x}{2}\right).$$

Each production run costs \$2500, so the total production costs (PC) are given by

$$\text{PC} = 2500r.$$

Hence the objective equation is

$$\begin{aligned} C &= \text{PC} + \text{SC} + \text{IC} \\ &= 2500r + 15x + 20\left(\frac{x}{2}\right) \\ &= 2500r + 25x. \end{aligned}$$

Substituting $x = 1600/r$ into the objective equation we have,

$$\begin{aligned} C &= 2500r + 25\left(\frac{1600}{r}\right) \\ &= 2500r + \frac{40,000}{r}. \end{aligned}$$

Use our curve sketching techniques to make a rough sketch of this function. Only positive values of r are relevant; that is, $r > 0$.

$$C' = 2500 - \frac{40,000}{r^2},$$

$$C'' = \frac{80,000}{r^3}.$$

Set $C' = 0$ and solve for r:

$$2500 - \frac{40,000}{r^2} = 0,$$

$$2500 = \frac{40,000}{r^2},$$

$$r^2 = \frac{40,000}{2,500} = 16, \quad \text{or} \quad r = 4.$$

When $r = 4$, $C'' > 0$. There is a relative minimum at $r = 4$. Therefore, there should be 4 production runs. *Note:* The value of C when $r = 4$ need not be calculated. After establishing the existence of relative minimum point when $r = 4$, use your familiarity with graphs of functions of the form $y = \frac{a}{x} + bx$ to make a rough sketch.

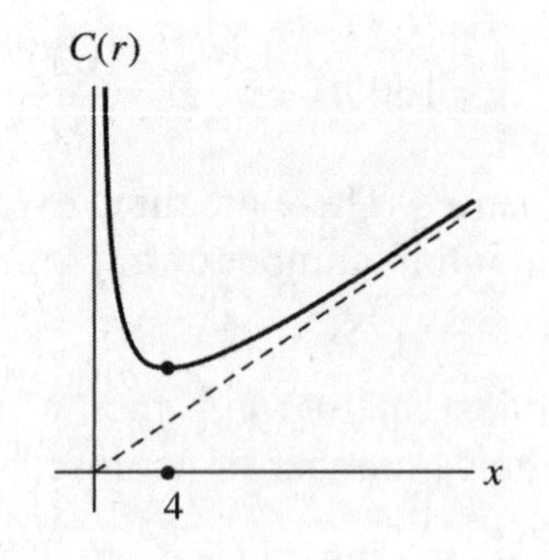

13. 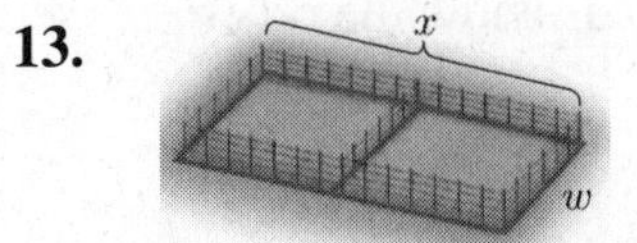

Let w and x be the dimensions of the corral. The corral is to have an area of 54 square meters, hence the constraint equation is

$$xw = 54 \quad \text{or} \quad w = \frac{54}{x} \qquad \text{(Constraint)}.$$

The total amount of fencing needed is two pieces of length x and three pieces of length w. Hence the objective equation is

$$F = 2x + 3w \qquad \text{(Objective)}.$$

Substituting $w = \frac{54}{x}$ into the objective equation we have,

$$F = 2x + 3\left(\frac{54}{x}\right) = 2x + \frac{162}{x}.$$

Next, use curve sketching techniques to make a rough sketch of the graph.

$$F = 2x + \frac{162}{x}, \quad x > 0,$$

$$F' = 2 - \frac{162}{x^2},$$

$$F'' = \frac{324}{x^3}.$$

Set $F' = 0$ and solve for x:

$$2 - \frac{162}{x^2} = 0 \Rightarrow 2 = \frac{162}{x^2} \Rightarrow x^2 = \frac{162}{2} = 81, \quad \text{or} \quad x = 9.$$

Notice that $F'' > 0$ for all positive x, so the graph of F is concave up, and F has a relative minimum at $x = 9$. In fact the formula for F has the form $y = \frac{a}{x} + bx$, and so the graph has the basic shape shown below.

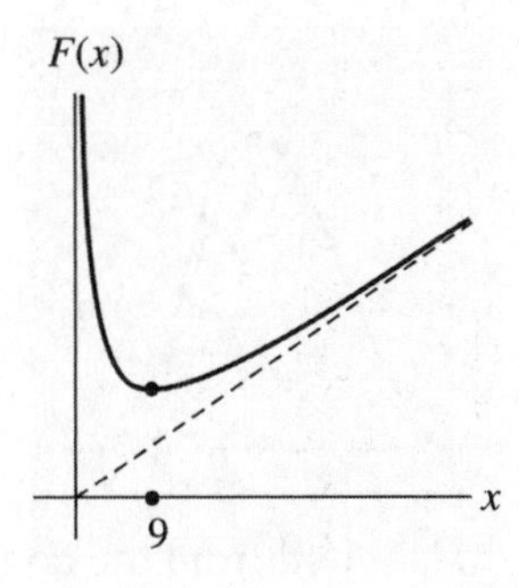

From the constraint equation,

$$w = \frac{54}{x} = \frac{54}{9} = 6.$$

Therefore, $w = 6$, $x = 9$ are the optimum dimensions.

19. First consider the diagram. Let x be the length of each edge of the square ends and h the other dimension.

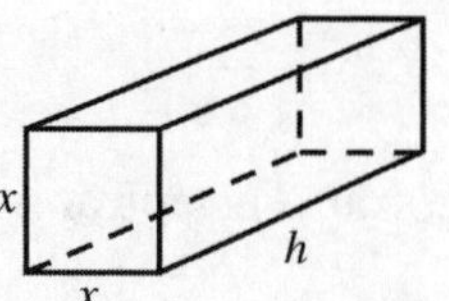

Since the sum of the three dimensions can be at most 120 centimeters, the constraint equation is

$$2x + h = 120, \quad h = 120 - 2x.$$

Since we wish to maximize the volume of the package, the objective equation is

$$V = x^2h.$$

Substitute $h = 120 - 2x$ into the objective equation and compute

$$V = x^2(120 - 2x) = 120x^2 - 2x^3,$$
$$V' = 240x - 6x^2,$$
$$V'' = 240 - 12x.$$

Set $V' = 0$ and solve for x:

$$240x - 6x^2 = 0,$$
$$6x(40 - x) = 0,$$
$$x = 0, \quad \text{and} \quad x = 40.$$

Check concavity: At $x = 0$,

$$V'' = 420 - 12(0) = 240 > 0 \qquad \text{(graph concave up).}$$

At $x = 40$,

$$V'' = 240 - 12(40) = -240 < 0 \qquad \text{(graph concave down).}$$

Therefore, the graph has a relative minimum at $x = 0$ and a relative maximum at $x = 40$. When $x = 0$, $V = 0$; at $x = 40$, $V = 120(40)^2 - 2(40)^3 = 64{,}000$. A rough sketch of the graph

$$V = 120x^2 - 2x^3, \quad x \geq 0,$$

reveals a maximum value at $x = 40$.

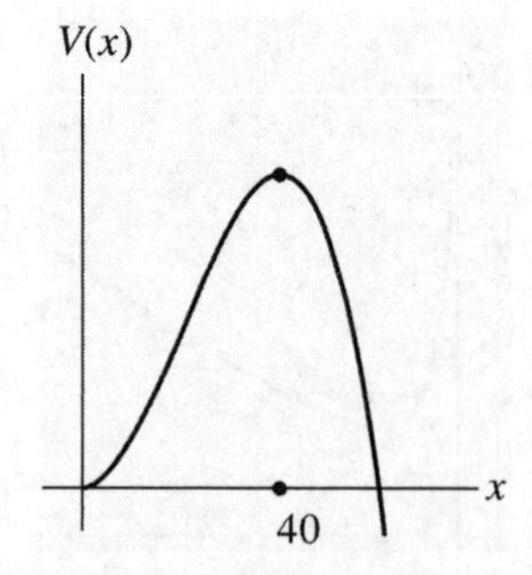

The value of h corresponding to $x = 40$ is found from the constraint equation

$$h = 120 - 2x = 120 - 2(40) = 40.$$

Thus to achieve maximum volume, the package should be 40 cm × 40 cm × 40 cm.

25. First, consider the diagram below. The base of the window has length $2x$. The area of the window is given by

$$\begin{aligned} A &= 2xy = 2x(9 - x^2) \\ &= 18x - 2x^3. \end{aligned}$$

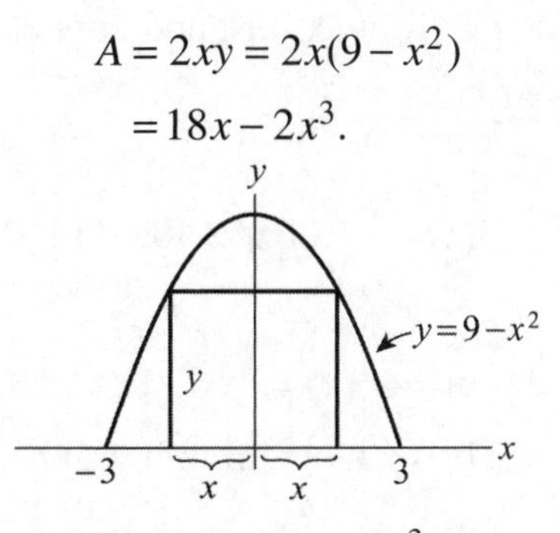

Next, make a rough sketch of the graph of $A = 18x - 2x^3$, for $x > 0$:

$$\begin{aligned} A' &= 18 - 6x^2, \\ A'' &= -12x. \end{aligned}$$

Set $A' = 0$:

$$\begin{aligned} 18 - 6x^2 &= 0, \\ 18 &= 6x^2, \\ 3 &= x^2, \quad \text{or} \quad x = \sqrt{3}. \end{aligned}$$

When $x = \sqrt{3}$, $A'' = -12(\sqrt{3})$ is negative. Therefore the graph of $A = 18x - 2x^3$ has a relative maximum value at $x = \sqrt{3}$. When $x = \sqrt{3}$,

$$\begin{aligned} A &= 18\sqrt{3} - 2(\sqrt{3})^3 \\ &= 18\sqrt{3} - 2 \cdot 3 \cdot \sqrt{3} \qquad \left[\text{since } (\sqrt{3})^3 = (\sqrt{3} \cdot \sqrt{3}) \cdot \sqrt{3} = 3\sqrt{3}\right] \\ &= 12\sqrt{3}. \end{aligned}$$

A rough sketch of the graph reveals a maximum value at $x = \sqrt{3}$.

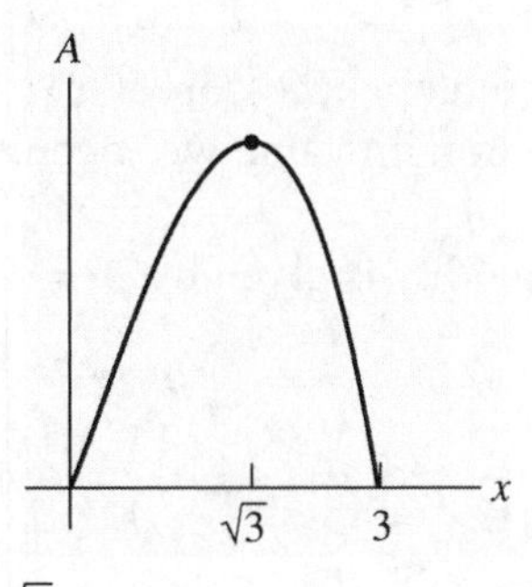

The height y, corresponding to $x = \sqrt{3}$ is found from the equation of the parabola

$$\begin{aligned} y &= 9 = x^2 \\ &= 9(\sqrt{3})^2 = 9 - 3 = 6. \end{aligned}$$

The window of maximum area should be 6 units high and $2\sqrt{3}$ units wide. (The value of x gives only half the width of the base of the window.)

2.7 Applications of Derivatives to Business and Economics

This section explores cost, revenue, and profit functions, and applies calculus to the optimization of profits of a firm.

1. If the cost function is

$$C(x) = x^3 - 6x^2 + 13x + 15,$$

the marginal cost function is

$$M(x) = C'(x) = 3x^2 - 12x + 13.$$

We find the minimum value of $M(x)$ by making a rough graph, a simple matter since $M(x)$ is quadratic function.

$$M'(x) = 6x - 12,$$
$$M''(x) = 6.$$

Set $M'(x) = 0$ to locate the x where the marginal cost is minimized.

$$6x - 12 = 0$$
$$6x = 12, \quad \text{or} \quad x = 2.$$

Now, $M(2) = 3(2)^2 - 12(2) + 13 = 1$ and $M''(2)$ is positive. Therefore, $M(x)$ has a relative minimum at $x = 2$. A sketch of the graph reveals a minimum value of 1. That is, the minimum marginal cost is \$1.

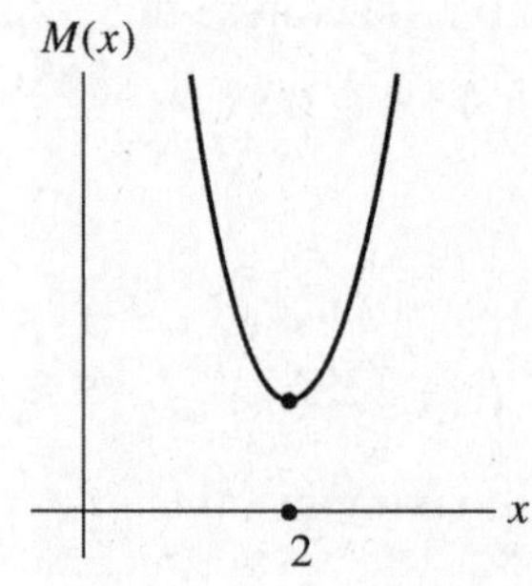

Note: Sketching the graph is not absolutely necessary. Once you realize that it is a parabola opening upward, you should know that its minimum value will occur where the first derivative is zero.

7. If the demand equation for the commodity is given by $p = \frac{1}{12}x^2 - 10x + 300,\ 0 \le x \le 60,$ then the revenue function is

$$R(x) = x \cdot p = x\left(\frac{1}{12}x^2 - 10x + 300\right) = \frac{1}{12}x^3 - 10x^2 + 300x.$$

To find the maximum value of $R(x)$, first make a rough sketch of its graph.

$$R(x) = \frac{1}{12}x^3 - 10x^2 + 300x,$$

$$R'(x) = \frac{1}{4}x^2 - 20x + 300,$$

$$R''(x) = \frac{1}{2}x - 20.$$

Set $R'(x) = 0$ and solve for x:

$$\frac{1}{4}x^2 - 20x + 300 = 0,$$

$$x^2 - 80x + 1200 = 0, \qquad \text{(multiply by 4)}$$

$$(x-20)(x-60) = 0.$$

$$x = 20 \quad \text{or} \quad x = 60.$$

[*Note:* To see how to factor $x^2 - 80x + 1200$, first try to factor $x^2 - 8x + 12$.] Now,

$$R(20) = \frac{1}{12}(20)^3 - 10(20)^2 + 300(20) = 2666\frac{2}{3},$$

$$R(60) = \frac{1}{12}(60)^3 - 10(60)^2 + 300(60) = 0,$$

$$R''(20) = \frac{1}{4}(20) - 20 = -10 < 0,$$

$$R''(60) = \frac{1}{2}(60) - 20 = 10 > 0.$$

A rough sketch of the graph reveals that revenue is maximized when $x = 20$. From the demand equation,

$$p = \frac{1}{12}(20)^2 - 10(20) + 300 = 300\frac{1}{3}.$$

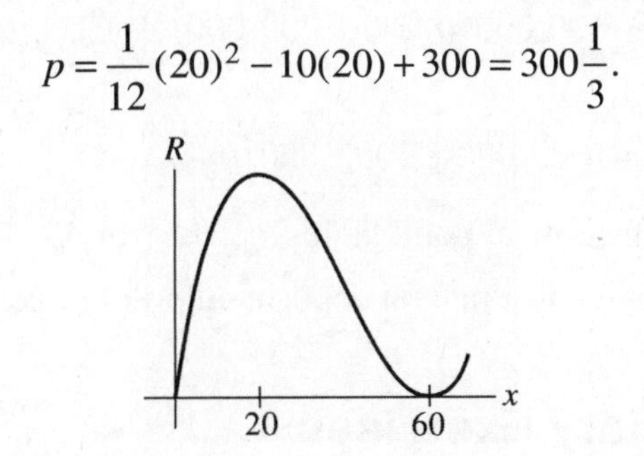

13. Let x represent the number of prints that the artist makes and let $R(x)$ represent the artist's total revenue. Since she can charge \$400 per print if she offers 50 prints for sale, and this price will decrease by \$5 for each print offered in excess of 50, the price per print if she offers x prints for sale is:

$$p(x) = 400 - 5(x - 50).$$

Thus

$$R(x) = x \cdot p(x) = x \cdot (400 - 5(x - 50)),$$

i.e.,

$$R(x) = 650x - 5x^2.$$

Differentiating,

$$R'(x) = 650 - 10x,$$

$$R''(x) = -10$$

Setting $R'(x) = 0$ and solving for x yields $x = 65$.

Since R'' is negative, $x = 65$ is a maximum. When $x = 65$, the artist's revenue is

$$R(x) = 650(65) - 5(65)^2 = 21{,}125.$$

The point (65, 21125) is the maximum of $R(x)$. In order to attain the maximum revenue of \$21,125, the artist should make 65 prints.

19. The savings and loan association spends money by paying interest to customers with savings accounts, and earns money by charging interest to customers taking loans. The association's profit is determined by the formula $P = R - C$, where R is the revenue from interest charged on loans and C is the cost of interest paid to savings accounts. Let

$$i_d = \text{interest rate on deposits, and}$$
$$i_l = \text{interest rate on loans.}$$

Then

$$R = (\text{amount loaned}) \cdot (i_l), \text{ and}$$
$$C = (\text{amount deposited in savings}) \cdot (i_d).$$

From the information given in the problem, we have:

$$i_l = .1, \text{ and}$$
$$(\text{amount loaned}) = (\text{amount deposited})$$
$$= (1{,}000{,}000) \cdot (i_d).$$

Thus

$$P = (1{,}000{,}000)(i_d) \cdot (.1) - (1{,}000{,}000) \cdot (i_d) \cdot (i_d),$$

i.e.,

$$P = (100{,}000)(i_d) - (1{,}000{,}000)(i_d)^2.$$

Differentiating,

$$P' = 100{,}000 - (2{,}000{,}000)(i_d).$$

Setting P' equal to 0 to find the critical point yields $i_d = \frac{1}{20}$, or 5%. Thus the savings and loan association should offer a 5% interest rate on deposits in order to generate the most profit.

Chapter 2: Supplementary Exercises

The most difficult problems in this chapter are the optimization problems. These problems cannot be solved all at once, but require that you break them down into small manageable pieces. Although some of them fit into neat categories, others require patient analysis.

1. **(a)** The graph of $f(x)$ is increasing for values of x at which $f'(x)$ is positive. The graph of $f'(x)$ shows that this happens for $-3 < x < 1$ and $x > 5$. The graph of $f(x)$ is decreasing for $x < -3$ and $1 < x < 5$, because the values of $f'(x)$ are negative.

(b) The graph of $f(x)$ is concave up for values of x at which the slopes of the graph are increasing, that is, for values of x at which the graph of $f'(x)$ is *increasing*. Be careful to distinguish between "increasing" and "positive". The graph of $f'(x)$ is increasing for $x < -1$ and for $x > 3$, even though the values of $f'(x)$ are not always positive there.

The graph of $f(x)$ is concave down for values of x at which the slopes of the graph are decreasing, that is, for values of x at which the graph of $f'(x)$ is *decreasing*. This happens for $-1 < x < 3$.

7. $f'(x)$ is positive at d and e.

13. Since $f(1) = 2$, the graph goes through the point (1, 2). Since $f'(1) > 0$, the graph is increasing $x = 1$.

19. Since $g(5) = -1$, the graph goes through the point $(5, -1)$. Since $g'(5) = -2 < 0$, the graph is decreasing at $x = 5$. The fact that $g''(5) = 0$ is inconclusive since it is possible for $g''(x)$ to be 0 without x being an inflection point. For example, if $g(x)$ is a straight line, $g''(x) = 0$ for all x but $g(x)$ has no inflection points.

25. If $y = x^2 + 3x - 10$, then

$$y' = 2x + 3,$$
$$y'' = 2.$$

Set $y' = 0$ and solve for x to find the relative extreme point:

$$2x + 3 = 0,$$
$$x = -\frac{3}{2}.$$

If $x = -\frac{3}{2}$, then $y = \left(-\frac{3}{2}\right)^2 + 3\left(-\frac{3}{2}\right) - 10 = \frac{9}{4} - \frac{9}{2} - 10 = -\frac{49}{4}$. Since $y'' = 2 > 0$, the parabola is concave up and the point $\left(-\frac{3}{2}, -\frac{49}{4}\right)$ is a minimum. When $x = 0$, $y = 0^2 + 3(0) - 10 = -10$. Hence the y-intercept is $(0, -10)$. To find the x-intercepts, set $y = 0$ and solve for x:

$$x^2 + 3x - 10 = 0,$$
$$(x + 5)(x - 2) = 0,$$
$$x = -5 \quad \text{and} \quad x = 2.$$

So the x-intercepts are $(-5, 0)$ and $(2, 0)$.

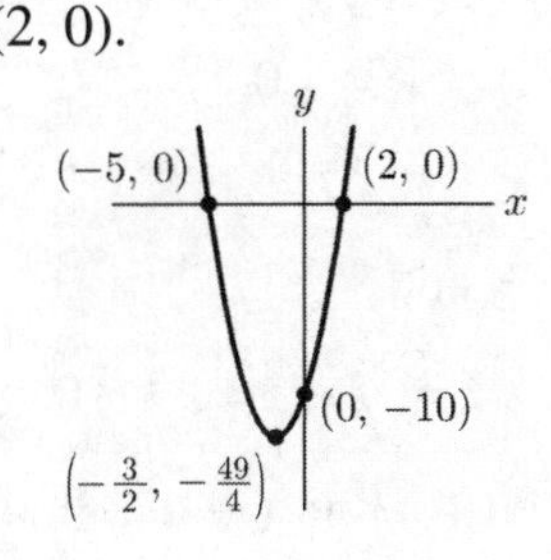

31. If $y = -x^2 + 20x - 90$, then

$$y' = -2x + 20,$$
$$y'' = -2.$$

Set $y' = 0$ and solve for x to find the relative extreme point:

$$-2x + 20 = 0,$$
$$x = 10.$$

If $x = 10$, $y = -(10)^2 + 20(10) - 90 = -100 + 200 - 90 = 10$. Since $y'' = -2 < 0$, the parabola is concave down and $(10, 10)$ is the relative maximum point. When $x = 0$, $y = -0^2 + 20(0) - 90 = -90$. Hence the y-intercept is $(0, -90)$. To find the x-intercept, use the quadratic formula:

$$b^2 - 4ac = (20)^2 - 4(-1)(-90) = 40,$$
$$\sqrt{b^2 - 4ac} = \sqrt{40} = \sqrt{4 \cdot 10} = \sqrt{4} \cdot \sqrt{10} = 2\sqrt{10}.$$
$$x = \frac{-20 \pm 2\sqrt{10}}{-2} = 10 \pm \sqrt{10}.$$

So the x-intercepts are $(10 + \sqrt{10},\ 0)$ and $(10 - \sqrt{10},\ 0)$.

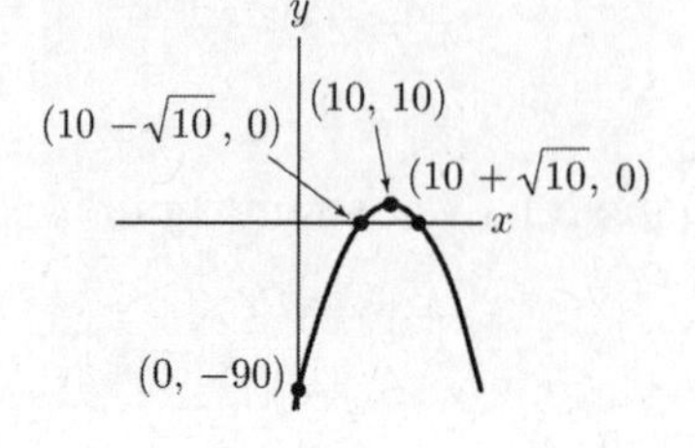

37. $y = \frac{11}{3} + 3x - x^2 - \frac{1}{3}x^3,$

$y' = 3 - 2x - x^2,$

$y'' = -2 - 2x.$

Set $y' = 0$ and solve for x to find the possible relative extreme points:

$$3 - 2x - x^2 = 0.$$

Multiple by -1 and rearrange terms.

$$x^2 + 2x - 3 = 0,$$
$$(x-1)(x+3) = 0,$$
$$x = 1, \quad \text{or} \quad x = -3.$$

If $x = 1$, $y = \frac{11}{3} + 3(1) - (1)^2 - \frac{1}{3}(1)^3 = \frac{16}{3}$, and

$$y'' = -2 - 2 = -4 < 0.$$

Hence the graph is concave down and has a relative maximum at $\left(1,\ \frac{16}{3}\right)$.

At $x = -3$, $y = \frac{11}{3} + 3(-3) - (-3)^2 - \frac{1}{3}(-3)^3 = -\frac{16}{3}$ and $y'' = -2 - 2(-3) = 4 > 0$. Hence the graph is concave up and has a relative minimum at $\left(-3,\ -\frac{16}{3}\right)$. To find the inflection point, set $y'' = 0$ and solve for x:

$$-2 - 2x = 0,$$
$$-2x = 2,$$
$$x = -1.$$

If $x = -1$, $y = \frac{11}{3} + 3(-1) - (-1)^2 - \frac{1}{3}(-1)^3 = 0$. Hence the inflection point is $(-1,\ 0)$.

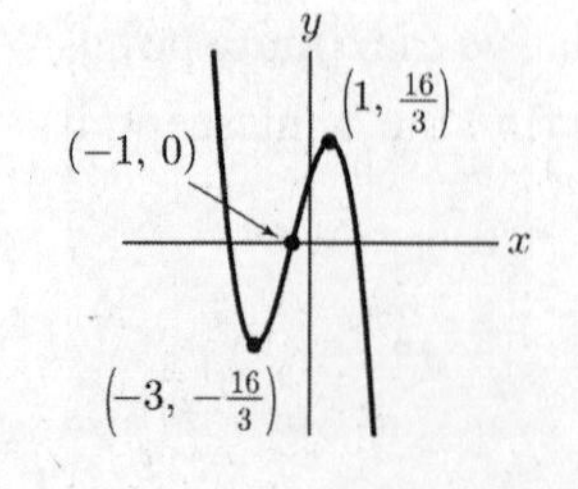

43. This function has the form $y = g(x) + mx + b$, where

$$g(x) = \frac{20}{x} \quad \text{and} \quad mx + b = \frac{x}{5} + 3 \quad \left(= \frac{1}{5}x + 3, \text{ a straight line of slope } \frac{1}{5} \right).$$

$$y = \frac{x}{5} + \frac{20}{x} + 3, \quad y' = \frac{1}{5} - \frac{20}{x^2}, \quad y'' = \frac{40}{x^3}.$$

Set $y' = 0$ and solve for x:

$$\frac{1}{5} - \frac{20}{x^2} = 0 \Rightarrow \frac{1}{5} = \frac{20}{x^2} \Rightarrow x^2 = 100$$

(The last equation was obtained by cross multiplication). Thus $x = 10$, since only $x > 0$ is being considered. When $x = 10$, $y = \frac{10}{5} + \frac{20}{10} + 3 = 2 + 2 + 3 = 7$, and y'' is obviously positive. Therefore the graph is concave up at $x = 10$, and so (10.7) is a relative minimum point. Since y'' can never be 0, there are no inflection points. The graph has the *y*-axis as a vertical asymptote and approaches the straight line $y = \frac{x}{5} + 3$ as *x* gets large.

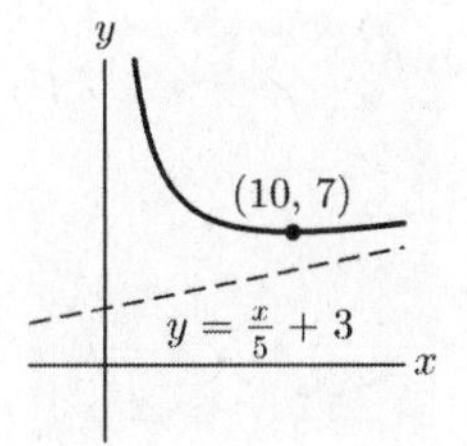

49. *A* and *c*, because a constant function corresponds to no change in the position of the car.
B and *e*, because a positive derivative corresponds to an increase of the distance from the reference point, and the fact that $s'(t)$ is constant means that the velocity is a "steady rate".
C and *f*, because information about $s'(a)$ gives information about $f(t)$ for t close to $t = a$, and a positive derivative corresponds to "moving forward."
D and *b*, for the same reason that *C* matches *f*, except that a negative derivative goes with "backing up".
E and *a*. A positive **first** derivative corresponds to "moving forward" and a positive **second** derivative goes with increasing velocity ("speeding up").
F and *d*, for the same reason that *E* matches *a*, except that a negative **second** derivative corresponds to decreasing velocity.

55. First consider the figure. Let *x* be the width and *h* be the height of the box.

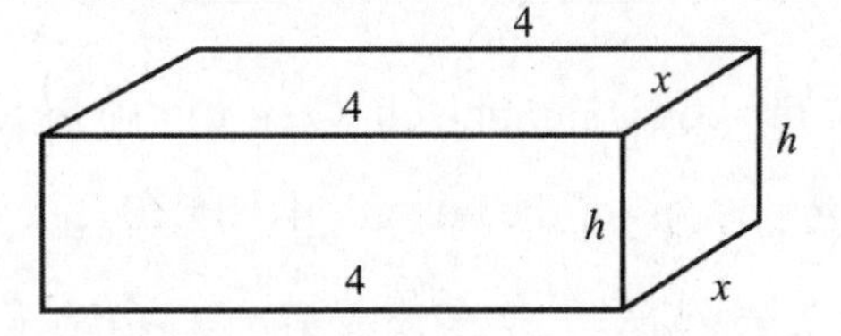

The box is to have a volume of 200 cubic feet, hence the constraint equation is

$$4xh = 200 \quad \text{or} \quad h = \frac{50}{x}.$$

We minimize the amount of material by minimizing the amount of surface area. The objective equation is

$$A = \underbrace{4x}_{\substack{\text{area of}\\ \text{base}}} + \underbrace{2xh}_{\substack{\text{area of 2}\\ \text{ends}}} + \underbrace{8h}_{\substack{\text{area of 2}\\ \text{sides}}}$$

Substitute the value $h = \dfrac{50}{x}$ into the objective equation:

$$A = 4x + 2x\left(\frac{50}{x}\right) + 8\left(\frac{50}{x}\right)$$

$$= 4x + 100 + \frac{400}{x} \quad (x > 0).$$

Next, make a rough sketch of the graph of this function:

$$A = 4x + 100 + \frac{400}{x}, \quad x > 0,$$

$$A' = 4 - \frac{400}{x^2},$$

$$A'' = \frac{800}{x^3}.$$

Set $A' = 0$ and solve for x:

$$4 - \frac{400}{x^2} = 0,$$

$$4 = \frac{400}{x^2},$$

$$x^2 = \frac{400}{4} = 100, \quad \text{or} \quad x = 10.$$

When $x = 10$, $A = 4(10) + 100 + \frac{400}{10} = 180$. Since the second derivative $\dfrac{800}{x^3}$ is positive for all $x > 0$, the graph is concave up for $x > 0$; also, the point (10, 180) is a minimum point.

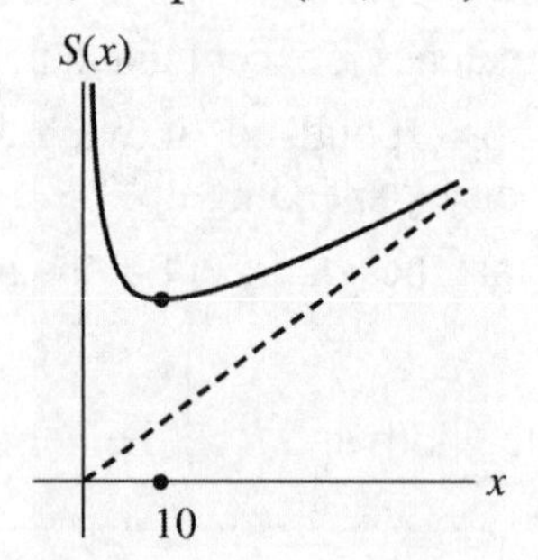

So the amount of material of the box is minimized when x is 10 feet. The other dimension is $h = \dfrac{50}{x} = \dfrac{50}{10} = 5$ feet. The dimensions of the box are $4 \times 5 \times 10$ feet.

61. Since the profit function is

$$P(x) = R(x) - C(x) \qquad \text{(revenue minus cost)},$$

and since

$$R(x) = x \cdot p = x(150 - .02x) = 150x - .02x^2$$

the profit function is

$$\begin{aligned} P(x) &= (150x - .02x^2) - (10x + 300) \\ &= 140x - .02x^2 - 300 \\ &= -.02x^2 + 140x - 300. \end{aligned}$$

The graph of $P(x)$ is a parabola opening downward (since the coefficient of x^2 is negative) and therefore assumes its maximum value where the first derivative is zero.

$$P'(x) = -.04x + 140.$$

Setting $P'(x) = 0$ and solving for x

$$\begin{aligned} -.04x + 140 &= 0 \\ -.04x &= -140 \\ x &= \frac{-140}{-.04} = 3500. \end{aligned}$$

[*Note:* One way to evaluate $\frac{140}{.04}$ is to first multiply the numerator and denominator by 100 and then divide. That is, $\frac{140}{.04} = \frac{14,000}{4} = 3500$.] Thus the profit is maximized when the sales level x is 3500 units.

Chapter 3
Techniques of Differentiation

3.1 The Product and Quotient Rules

The first twenty-eight exercises are routine drill. You will be able to tell from the answers if you are mastering the use of the product and quotient rules. In each case, try to obtain the simplified version of the answer shown in the answer section. Most students find this difficult, but now is a good time to practice a skill that may be important for the next exam!

1. To differentiate the product $(x+1)(x^3+5x+2)$, let $f(x)=(x+1)$ and $g(x)=(x^3+5x+2)$ and apply the product rule. Then

$$\begin{aligned}\frac{d}{dx}[(x+1)(x^3+5x+2)] &= (x+1)\cdot\frac{d}{dx}(x^3+5x+2)\\ &\quad+(x^3+5x+2)\cdot\frac{d}{dx}(x+1)\\ &=(x+1)(3x^2+5)+(x^3+5x+2)(1).\end{aligned}$$

Carry out the multiplication in the first product above and combine like terms:

$$\begin{aligned}\frac{d}{dx}[(x+1)(x^3+5x+2)] &= (3x^3+3x^2+5x+5)+(x^3+5x+2)\\ &=4x^3+3x^2+10x+7.\end{aligned}$$

7. To differentiate $(x^2+3)(x^2-3)^{10}$, let $f(x)=(x^2+3)$ and $g(x)=(x^2-3)^{10}$ and apply the product rule. To compute $g'(x)$, use the general power rule.

$$\begin{aligned}\frac{d}{dx}[(x^2+3)(x^2-3)^{10}] &= (x^2+3)\cdot\frac{d}{dx}(x^2-3)^{10}+(x^2-3)^{10}\cdot\frac{d}{dx}(x^2+3)\\ &=(x^2+3)10(x^2-3)^9 2x+(x^2-3)^{10}2x.\end{aligned}$$

To simplify the answer, factor $2x(x^2-3)^9$ out of each term of the derivative:

$$\begin{aligned}\frac{d}{dx}[(x^2+3)(x^2-3)^{10}] &= 2x(x^2-3)^9[(x^2+3)10+(x^2-3)]\\ &=2x(x^2-3)^9(11x^2+27).\end{aligned}$$

Helpful Hint: Here is a good way to remember the order of the terms in the quotient rule:

1. Draw a long fraction bar and write $[g(x)]^2$ in the denominator:

$$\frac{\qquad\qquad\qquad}{[g(x)]^2}$$

2. While you are still thinking about $g(x)$, write it in the numerator:

$$\frac{g(x)}{[g(x)]^2}$$

3. The rest is easy because each term in the numerator of the derivative involves one function and one derivative. Since $g(x)$ is already written, it must go with $f'(x)$. Next comes the minus sign and then the function and derivative that have not yet been used.

$$\frac{g(x)f'(x)-f(x)g'(x)}{[g(x)]^2}$$

Think of $g(x)$ and $g'(x)$ as being "around the outside" of this formula.

13. To differentiate $\frac{x^2-1}{x^2+1}$, let $f(x)=x^2-1$ and $g(x)=x^2+1$ and apply the quotient rule:

$$\frac{d}{dx}\left(\frac{x^2-1}{x^2+1}\right)=\frac{(x^2+1)\cdot\frac{d}{dx}(x^2-1)-(x^2-1)\cdot\frac{d}{dx}(x^2+1)}{(x^2+1)^2}$$

$$=\frac{(x^2+1)2x-(x^2-1)2x}{(x^2+1)^2}.$$

If the products in the numerator are expanded, this derivative simplifies:

$$\frac{d}{dx}\left(\frac{x^2-1}{x^2+1}\right)=\frac{2x^3+2x-2x^3+2x}{(x^2+1)^2}$$

$$=\frac{4x}{(x^2+1)^2}.$$

19. To differentiate $\frac{3x^2+5x+1}{3-x^2}$, let $f(x)=3x^2+5x+1$ and $g(x)=3-x^2$ and apply the quotient rule.

$$\frac{d}{dx}\left(\frac{3x^2+5x+1}{3-x^2}\right)=\frac{(3-x^2)\frac{d}{dx}(3x^2+5x+1)-(3x^2+5x+1)\frac{d}{dx}(3-x^2)}{(3-x^2)^2}$$

$$=\frac{(3-x^2)(6x+5)-(3x^2+5x+1)(-2x)}{(3-x^2)^2}.$$

If the products in the numerator are expanded, we have

$$\frac{d}{dx}\left(\frac{3x^2+5x+1}{3-x^2}\right)=\frac{(18x+15-6x^3-5x^2)+(6x^3+10x^2+2x)}{(3-x^2)^2}$$

$$=\frac{5x^2+20x+15}{(3-x^2)^2}$$

$$=\frac{5(x+1)(x+3)}{(3-x^2)^2}.$$

Warning: Some students prefer to work #19 using the product rule, but this may cause them more work than they realize. It is possible to write

$$y = \frac{3x^2+5x+1}{3-x^2} = (3x^2+5x+1)(3-x^2)^{-1},$$

and

$$\begin{aligned}\frac{dy}{dx} &= (3x^2+5x+1)\cdot\frac{d}{dx}(3-x^2)^{-1} + (3-x^2)^{-1}\cdot\frac{d}{dx}(3x^2+5x+1)\\ &= (3x^2+5x+1)\cdot(-1)(3-x^2)^{-2}\cdot(-2x) + (3-x^2)^{-1}\cdot(6x+5).\end{aligned}$$

This method is not really shorter, but it does avoid the quotient rule. The main difficulty arises when you try to simplify the derivative, for you must factor out terms that involve negative exponents. However, if you really prefer this method, the next paragraph explains how to handle negative exponents.

How to simplify a sum involving negative exponents: Suppose that various powers of some quantity Q occur in each term of the sum. Find the exponent on Q that lies farthest to the left on the number line and factor out this power of Q from each term in the sum.

In the solution above, notice that $\frac{dy}{dx}$ involves $(3-x^2)^{-2}$ and $(3-x^2)^{-1}$. Since -2 is to the left of -1 on the number line, factor out $(3-x^2)^{-2}$. In the second term below we need $(3-x^2)^1$, because $(3-x^2)^{-1} = (3-x^2)^{-2}(3-x^2)^{+1}$.

$$\begin{aligned}\frac{dy}{dx} &= (3-x^2)^{-2}[(3x^2+5x+1)(-1)(-2x) + (3-x^2)^1(6x+5)\\ &= (3-x^2)^{-2}[(6x^3+10x^2+2x) + (18x+15-6x^3-5x^2)]\\ &= (3-x^2)^{-2}[5x^2+20x+15]\\ &= 5(3-x^2)^{-2}(x+1)(x+3).\end{aligned}$$

Moral of this story: If you have to simplify the derivative of a quotient, use the quotient rule for the differentiation step.

Helpful Hint: A problem such as #22 can be worked with the quotient rule, but since the numerator is a constant, the differentiation is easier if you rewrite the quotient with the notation $(\cdots)^{-1}$. See Practice Problem #2.

25. Simplify first:

$$\frac{x^4-4x^2+3}{x} = x^3-4x+\frac{3}{x} = x^3-4x+3x^{-1}.$$

Then,

$$\begin{aligned}\frac{d}{dx}(x^3-4x+3x^{-1}) &= 3x^2-4-3x^{-2} = 3x^2-4-\frac{3}{x^2}\\ &= \frac{3x^4-4x^2-3}{x^2}.\end{aligned}$$

31. The tangent line is horizontal when $\frac{dy}{dx} = 0$. Using the quotient rule, we find

$$\begin{aligned}\frac{dy}{dx} &= \frac{(x-4)^3(5)(x-2)^4-(x-2)^5(3)(x-4)^2}{[(x-4)^3]^2}\\ &= \frac{(x-4)^2(x-2)^4(5x-20-3x+6)}{(x-4)^6}\\ &= \frac{(x-2)^4(2x-14)}{(x-4)^4}.\end{aligned}$$

$\frac{dy}{dx} = 0$ when $(x-2)^4(2x-14) = 0$, or when $x = 2$ or $x = 7$. Thus, the tangent line is horizontal for $x = 2$ or $x = 7$.

Helpful Hint: Do you try the Practice Problems before starting your homework? You should. If you worked Practice Problem #1 in this section, you may have already learned to think about simplifying a function before starting to differentiate. Often there is no need to simplify first, but in Exercise 35 this really helps. Notice that

$$y = \frac{x^2+3x-1}{x} = x+3-\frac{1}{x},$$

so that

$$\frac{dy}{dx} = 1+\frac{1}{x^2}.$$

37. Here we are asked to determine $\frac{d^2y}{dx^2}$, the second derivative. So, we differentiate the function, and then differentiate the result.

$$\begin{aligned}\frac{dy}{dx} &= 4\left(x^2+1\right)^3(2x) &&\text{Use the general power rule. (See section 1.6.)}\\ &= 8x\left(x^2+1\right)^3 &&\text{Simplify.}\end{aligned}$$

Now use the product rule to differentiate $8x\left(x^2+1\right)^3$.

$$\begin{aligned}\frac{d^2y}{dx^2} &= 8x(3)\left(x^2+1\right)^2(2x)+\left(x^2+1\right)^3(8)\\ &= 8\left(x^2+1\right)^2\left(3x(2x)+\left(x^2+1\right)\right)\\ &= 8\left(x^2+1\right)^2\left(7x^2+1\right)\end{aligned}$$

43. Applying the quotient rule, we have

$$h'(x) = \frac{d}{dx}\left[\frac{f(x)}{x^2+1}\right] = \frac{\left(x^2+1\right)f'(x)-f(x)(2x)}{\left(x^2+1\right)^2}$$

49. Recall that the marginal revenue, MR, is defined by

$$MR = R'(x).$$

The average revenue is maximized when $\frac{d}{dx}(AR) = 0.$

$$\frac{d}{dx}(AR) = \frac{d}{dx}\left(\frac{R(x)}{x}\right) = \frac{xR'(x) - R(x)(1)}{x^2}$$

If $\frac{xR'(x) - R(x)}{x^2} = 0$, then $xR'(x) - R(x) = 0 \Rightarrow xR'(x) = R(x) \Rightarrow R'(x) = \frac{R(x)}{x} \Rightarrow MR = AR.$

Helpful Hint: Remember that when you have to determine when the derivative of a quotient is zero, the algebra is usually easier if you use the quotient rule.

55. We use the first derivative rule to determine the x-coordinate of the maximum point.

$$y = \frac{10x}{1 + .25x^2}$$

$$\frac{dy}{dx} = \frac{(1 + .25x^2)10 - 10x(.5x)}{(1 + .25x^2)^2}$$

$$= \frac{10 - 2.5x^2}{(1 + .25x^2)^2}$$

Now solve $\frac{dy}{dx} = 0.$

$$\frac{dy}{dx} = 0$$

$$\frac{10 - 2.5x^2}{(1 + .25x^2)^2} = 0$$

$$10 = 2.5x^2$$

$$x = \pm 2$$

The function is defined only for $x \geq 0$, so the x-coordinate of the maximum point is $x = 2$. Now determine $f(2)$ to find the y-coordinate of the maximum point.

$$y = \frac{10(2)}{1 + .25(2)^2} = \frac{20}{2} = 10$$

Thus, the coordinates of the maximum point are (2, 10).

61. In exercise 60, we are given $f(1) = 2,\ f'(1) = 3,\ g(1) = 4,$ and $g'(1) = 5.$

$$\frac{d}{dx}\left[\frac{f(x)}{g(x)}\right]_{x=1} = \frac{g(1)f'(1) - f(1)g'(1)}{g(1)^2}$$

$$= \frac{(4)(3) - (2)(5)}{4^2} = \frac{1}{8}$$

67. $b(t) = \dfrac{w(t)}{[h(t)]^2}$

Using the quotient rule, we have

$$\begin{aligned} b'(t) &= \frac{d}{dt}\left[\frac{w(t)}{[h(t)]^2}\right] \\ &= \frac{[h(t)]^2 w'(t) - w(t)(2)h(t)h'(t)}{[h(t)]^4} \\ &= \frac{h(t)w'(t) - 2h'(t)w(t)}{[h(t)]^3} \end{aligned}$$

3.2 The Chain Rule and the General Power Rule

The chain rule is used in subsequent chapters to derive formulas for the derivatives of composite functions where the outer function is either a logarithm function, an exponential function, or one of the trigonometric functions. In each case there will be a formula that has the same feel as the general power rule. These formulas will become so automatic that you will sometimes forget that they are special cases of the chain rule.

Many of the exercises in this section may be solved with the general power rule, but you should approach the exercises with the abstract chain rule in mind. The practice you gain here will make it easier to use and remember the later versions of the chain rule.

1. If $f(x) = \dfrac{x}{x+1}$, $g(x) = x^3$, then $f(g(x)) = \dfrac{x^3}{x^3+1}$.

7. If $f(g(x)) = \sqrt{4-x^2}$, then $f(x) = \sqrt{x}$ and $g(x) = 4 - x^2$.

13. If $y = 6x^2(x-1)^3$,

$$\begin{aligned} \frac{dy}{dx} &= \left[6x^2 \cdot \frac{d}{dx}(x-1)^3\right] + (x-1)^3 \cdot \frac{d}{dx}(6x^2) && \text{[Product Rule]} \\ &= 6x^2 3(x-1)^2(1) + (x-1)^3 12x && \text{[General Power Rule]} \end{aligned}$$

Factor $6x(x-1)^2$ out of each term to get

$$\begin{aligned} \frac{dy}{dx} &= 6x(x-1)^2[3x + 2(x-1)] \\ &= 6x(x-1)^2(5x-2). \end{aligned}$$

19. To differentiate $\left(\frac{4x-1}{3x+1}\right)^3$, let $f(x)=x^3$, $g(x)=\frac{4x-1}{3x+1}$, and use the chain rule. Now $f'(x)=3x^2$ and

$$g'(x)=\frac{(3x+1)4-(4x-1)3}{(3x+1)^2} \qquad \text{[Quotient Rule]}$$

Hence

$$\begin{aligned}\frac{d}{dx}f(g(x)) &= f'(g(x))g'(x)\\ &= 3\left(\frac{4x-1}{3x+1}\right)^2 \cdot \frac{(3x+1)4-(4x-1)3}{(3x+1)^2}.\end{aligned}$$

This simplifies to

$$\begin{aligned}\frac{d}{dx}f(g(x)) &= \frac{3(4x-1)^2(12x+4-12x+3)}{(3x+1)^4}\\ &= \frac{21(4x-1)^2}{(3x+1)^4}.\end{aligned}$$

25. Using the quotient rule:

$$\begin{aligned}h'(x) &= \left[\frac{f(x^2)}{x}\right]\\ &= \frac{x\cdot\frac{d}{dx}[f(x^2)]-f(x^2)\cdot\frac{d}{dx}(x)}{x^2}\\ &= \frac{x\cdot\frac{d}{dx}[f(x^2)]-f(x^2)\cdot(1)}{x^2}\end{aligned}$$

Now apply the chain rule to in order to find $\frac{d}{dx}[f(x^2)]$:

$$\frac{d}{dx}[f(x^2)] = f'(x^2)\cdot\frac{d}{dx}(x^2) = f'(x^2)\cdot(2x).$$

Putting these results together, we have

$$h'(x)=\frac{x\cdot f'(x^2)\cdot(2x)-f(x^2)\cdot(1)}{x^2}=\frac{2x^2\cdot f'(x^2)-f(x^2)}{x^2}.$$

31. If $f(x)=\frac{1}{x}=x^{-1}$ and $g(x)=1-x^2$, then $f'(x)=-x^{-2}=\frac{-1}{x^2}$ and $g'(x)=-2x$. Hence

$$\begin{aligned}\frac{d}{dx}f(g(x)) &= f'(g(x))\cdot g'(x)\\ &= \frac{-1}{(1-x^2)^2}\cdot(-2x)=\frac{2x}{(1-x^2)^2}.\end{aligned}$$

37. If $y = u^{3/2}$ and $u = 4x+1$, then

$$\frac{dy}{du} = \frac{3}{2}u^{1/2}, \quad \text{and} \quad \frac{du}{dx} = 4.$$

Hence
$$\frac{dy}{dx} = \frac{dy}{du}\cdot\frac{du}{dx} = \frac{3}{2}u^{1/2}\cdot(4) = 6u^{1/2}.$$

To express $\frac{dy}{dx}$ as a function of x alone, substitute $4x+1$ for u to obtain

$$\frac{dy}{dx} = 6\cdot(4x+1)^{1/2}.$$

43. If $y = \frac{x+1}{x-1}$ and $x = \frac{t^2}{4} = \frac{1}{4}t^2$, then

$$\frac{dy}{dx} = \frac{(x-1)(1)-(x+1)(1)}{(x-1)^2} \qquad \text{[Quotient Rule]}$$

$$= \frac{-2}{(x-1)^2}$$

and
$$\frac{dx}{dt} = \frac{1}{4}(2t) = \frac{t}{2}.$$

Therefore,
$$\frac{dy}{dx} = \frac{dy}{dx}\cdot\frac{dx}{dt} = \frac{-2}{(x-1)^2}\cdot\frac{t}{2}.$$

To express $\frac{dy}{dt}$ as a function of t alone, substitute $\frac{t^2}{4}$ for x to obtain

$$\frac{dx}{dt} = \frac{-2}{\left(\frac{t^2}{4}-1\right)^2}\cdot\frac{t}{2}.$$

Now,
$$\left.\frac{dy}{dt}\right|_{t_0=3} = \left.\frac{-2}{\left(\frac{t^2}{4}-1\right)^2}\cdot\frac{t}{2}\right|_{t_0=3} = \frac{-2}{\left(\frac{9}{4}-1\right)^2}\cdot\frac{3}{2}$$

$$= \frac{-32}{25}\cdot\frac{3}{2} = -\frac{48}{25}.$$

49. **(a)** The volume of the cube is a function of the length, x, of the edge of the cube, and the length x is a function of time. Thus by the chain rule we have:

$$\frac{dV}{dt} = \frac{dV}{dx}\cdot\frac{dx}{dt}$$

While the formula for the function describing x as a function of time is not given, the function that describes V as a function of x is $V = x^3$. Thus $\frac{dV}{dx} = 3x^2$. Substituting, we have

$$\frac{dV}{dt} = 3x^2\cdot\frac{dx}{dt}.$$

(b) When $\frac{dV}{dx} = 12$, we have

$$\frac{dV}{dt} = 12 \cdot \frac{dx}{dt},$$

which says that the volume of the cube is increasing twelve times as quickly as the length of the edge of the cube. To find the value of x which makes this the case, we must solve the equation $3x^2 = 12$ for x. Choosing the positive solution since x represents a length, we arrive at $x = 2$.

55. (a) Since $L = 10 + .4x + .0001x^2$, differentiating gives

$$\frac{dL}{dx} = .4 + .0002x.$$

(b) Similarly, since $x = 752 + 23t + .5t^2$, differentiating gives

$$\frac{dx}{dt} = 23 + t.$$

Evaluating at time $t = 2$ years,

$$\left.\frac{dx}{dt}\right|_{t=2} = 23 + 2 = 25.$$

The population is increasing at the rate of 25 thousand people per year.

(c) Use the chain rule to find $\frac{dL}{dt}$:

$$\frac{dL}{dt} = \frac{dL}{dx} \cdot \frac{dx}{dt} = (.4 + .0002x) \cdot (23 + t).$$

At time $t = 2$ years, $x = 752 + 23(2) + .5 \cdot (2)^2 = 800$. Thus

$$\left.\frac{dL}{dt}\right|_{t=2} = (.4 + .0002(800)) \cdot (25) = (.56)(25) = 14.$$

The carbon monoxide level is increasing at the rate of 14 parts per million per year.

61. (a) Referring to Figure 1(a) in the text, we see that the value of a share of the company's stock at time $t = 1.5$ months is \$40. Next, referring to Figure 1(b), we see that the total value of the company when the value of a single share is \$40 is approximately 78 million dollars. Similarly, at $t = 3.5$ months, the value of one share of the stock is \$30. At this share price, the company is valued at approximately 76 million dollars.

(b) The graph in Figure 1(a) in the text shows a straight line segment between the points (0, 10) and (2, 50). Thus, for each value of t between 0 and 2, the value of $\frac{dx}{dt}$ is the slope of this line segment. In particular, the value of $\frac{dx}{dt}$ at time $t = 1.5$ months is

$$\left.\frac{dx}{dt}\right|_{t=1.5} = \frac{50 - 10}{2 - 0} = \frac{40}{2} = 20.$$

This means that 1.5 months after the company went public, the value of a share of the company's stock was increasing at a rate of \$20 per month.

At time $t = 3.5$ months,

$$\left.\frac{dx}{dt}\right|_{t=3.5} = \frac{30-30}{5-3} = \frac{0}{2} = 0.$$

(Also note that the graph is horizontal; therefore, the slope is 0.) This means that 3.5 months after the company went public, the value of a share of the company's stock was holding constant (neither increasing nor decreasing) at $30.

3.3 Implicit Differentiation and Related Rates

The material in this section is not used in the rest of the text. However, the two techniques introduced here are fundamental concepts of calculus and are commonly found in applications. Both techniques involve using the chain rule in situations where we don't know the specific formula for the inner part of a composite function.

1. Differentiate $x^2 - y^2 = 1$ term by term. The first term x^2 has derivative $2x$. Think of the second term y^2 as having the form $[g(x)]^2$. Use the chain rule to differentiate.

$$\frac{d}{dx}[g(x)]^2 = 2[g(x)]g'(x).$$

Hence,

$$\frac{d}{dx}y^2 = 2y \cdot \frac{dy}{dx}.$$

On the right side of the original equation, the derivative of the constant function 1 is zero. Thus implicit differentiation of $x^2 - y^2 = 1$ yields

$$2x - 2y \cdot \frac{dy}{dx} = 0.$$

Solving for $\frac{dy}{dx}$,

$$-2y \cdot \frac{dy}{dx} = -2x.$$

If $y \neq 0$,

$$\frac{dy}{dx} = \frac{x}{y}.$$

7. Differentiate $2x^3 + y = 2y^3 + x$ term by term. The derivative of $2x^3$ is $6x^2$. Write $\frac{dy}{dx}$ for the derivative of y. On the right side of the equation, $2y^3$ has derivative $6y^2 \cdot \frac{dy}{dx}$ and x has derivative 1. Thus implicit differentiation of $2x^3 + y = 2y^3 + x$ yields

$$6x^2 + \frac{dy}{dx} = 6y^2 \cdot \frac{dy}{dx} + 1.$$

Solving for $\frac{dy}{dx}$,

$$(1-6y^2)\frac{dy}{dx} = 1 - 6x^2,$$

$$\frac{dy}{dx} = \frac{1-6x^2}{1-6y^2}.$$

13. To differentiate x^3y^2, use the product rule, treating y as a function of x:

$$\frac{d}{dx}x^3y^2 = x^3\cdot\frac{d}{dx}(y^2)+y^2\cdot\frac{d}{dx}(x^3)$$
$$= x^3 2y\cdot\frac{dy}{dx}+y^2(3x^2)$$
$$= 2x^3y\frac{dy}{dx}+3x^2y^2.$$

Hence implicit differentiation of $x^3y^2-4x^2=1$ yields

$$2x^3y\frac{dy}{dx}+3x^2y^2-8x=0.$$

Solving for $\frac{dy}{dx}$,

$$2x^3y\frac{dy}{dx}=8x-3x^2y^2,$$
$$\frac{dy}{dx}=\frac{8x-3x^2y^2}{2x^3y}=\frac{8-3xy^2}{2x^2y}.$$

19. Implicit differentiation of $4y^3-x^2=-5$ yields

$$12y^2\cdot\frac{dy}{dx}-2x=0.$$

Solving for $\frac{dy}{dx}$,

$$12y^2\frac{dy}{dx}=2x,$$
$$\frac{dy}{dx}=\frac{2x}{12y^2}.$$

Hence, $$\left.\frac{dy}{dx}\right|_{\substack{x=3\\y=1}}=\frac{2(3)}{12(1)^2}=\frac{1}{2}.$$

25. A two-step procedure is required.

(a) Find the slope at each point.

(b) Use the point-slope formula to obtain the equation of the tangent line.

For (a), note that

$$\frac{d}{dx}x^2y^4=x^2\cdot\frac{d}{dx}y^4+y^4\cdot\frac{d}{dx}x^2=x^2\cdot 4y^3\frac{dy}{dx}+y^4\cdot 2x.$$

Thus implicit differentiation of $x^2y^4 = 1$ yields

$$4y^3x^2\frac{dy}{dx} + 2y^4x = 0,$$

$$4y^3x^2\frac{dy}{dx} = -2y^4x,$$

$$\frac{dy}{dx} = \frac{-2y^4x}{4y^3x^2} = \frac{-y}{2x}.$$

At the point $\left(4, \frac{1}{2}\right)$, the slope of the curve is given by

$$\left.\frac{dy}{dx}\right|_{\substack{x=4\\y=1/2}} = \frac{-\frac{1}{2}}{2(4)} = -\frac{1}{16}.$$

Hence the equation of the tangent line at $\left(4, \frac{1}{2}\right)$ is

$$y - \frac{1}{2} = -\frac{1}{16}(x-4).$$

At the point $\left(4, -\frac{1}{2}\right)$, the slope of the curve is given by

$$\left.\frac{dy}{dx}\right|_{\substack{x=4\\y=-1/2}} = \frac{-\left(-\frac{1}{2}\right)}{2(4)} = \frac{1}{16}.$$

Hence the equation of the tangent line at $\left(4, -\frac{1}{2}\right)$ is

$$y + \frac{1}{2} = \frac{1}{16}(x-4).$$

31. Differentiate $x^4 + y^4 = 1$ term by term. Since x is a function of t, the general power rule gives

$$\frac{d}{dx}x^4 = 4x^3 \cdot \frac{dx}{dt}.$$

Similarly,
$$\frac{d}{dt}y^4 = 4y^3 \cdot \frac{dy}{dt}.$$

And
$$\frac{d}{dx}(1) = 0.$$

Hence
$$4x^3\frac{dx}{dt} + 4y^3\frac{dy}{dt} = 0.$$

Solving for $\frac{dy}{dt}$,

$$4y^3\frac{dy}{dt} = -4x^3\frac{dx}{dt},$$

$$\frac{dy}{dt} = -\frac{x^3}{y^3}\frac{dx}{dt}.$$

37. First compute $\frac{dy}{dt}$. Differentiating $x^2 - 4y^2 = 9$ term by term,

$$2x\frac{dx}{dt} - 8y\frac{dy}{dt} = 0.$$

Solving for $\frac{dy}{dt}$,

$$-8y\frac{dy}{dt} = -2x\frac{dx}{dt},$$
$$\frac{dy}{dt} = \frac{x}{4y}\frac{dx}{dt}.$$

The problem says that at the point (5, −2), the x-coordinate is increasing at the rate of 3 units per second, that is $\frac{dx}{dt} = 3$. Hence,

$$\frac{dy}{dt} = \frac{5}{4(-2)}(3) = -\frac{15}{8}$$

The y-coordinate is decreasing at $\frac{15}{8}$ units per second.

43. $\frac{d}{dt}P^5V^7 = P^5 \cdot \frac{d}{dt}V^7 + V^7 \cdot \frac{d}{dt}P^5$

$$= P^5 7V^6\frac{dV}{dt} + V^7 5P^4\frac{dP}{dt}.$$

$\frac{d}{dx}(k) = 0$ (since k is a constant).

Hence $7P^5V^6\frac{dV}{dt} + 5P^4V^7\frac{dP}{dt} = 0$. Solving for $\frac{dV}{dt}$,

$$7P^5V^6\frac{dV}{dt} = -5P^4V^7\frac{dP}{dt},$$
$$\frac{dV}{dt} = \frac{-5P^4V^7}{7P^5V^6}\frac{dP}{dt},$$
$$= \frac{-5V}{7P}\frac{dP}{dt}.$$

The problem says that when $V = 4$ liters, $P = 200$ units and $\frac{dP}{dt} = 5$ units per second. Hence

$$\frac{dV}{dt} = \frac{-5(4)}{7(200)}(5) = -\frac{1}{14}.$$

Therefore the volume is decreasing at the rate of $\frac{1}{14}$ liters per second.

Chapter 3: Supplementary Exercises

The techniques of differentiation presented in the first two sections of this chapter must be mastered. They will be used throughout the text.

You may need to review curve sketching and optimization problems because these problems can be more difficult when they involve the product rule or quotient rule. On an exam that covers both Chapters 2 and 3 (or on the final exam) you may see a problem similar to Exercises 45 and 46 in Section 3.1 or Exercises 27 and 28 in Section 3.2.

1. $\frac{d}{dx}[(4x-1)(3x+1)^4] = (4x-1)\cdot\frac{d}{dx}(3x+1)^4 + (3x+1)^4\cdot\frac{d}{dx}(4x-1)$ [Product Rule]

$$= (4x-1)4(3x+1)^3\cdot(3) + (3x+1)^4\cdot(4).$$

Factor $4(3x+1)^3$ from each term to obtain

$$\frac{d}{dx}[(4x-1)(3x+1)^4] = 4(3x+1)^3[3(4x-1)+(3x+1)]$$
$$= 4(3x+1)^3[12x-3+3x+1]$$
$$= 4(3x+1)^3(15x-2).$$

7. If $y = 3(x^2-1)^3(x^2+1)^5$, then

$$\frac{dy}{dx} = 3(x^2-1)^3\cdot\frac{d}{dx}(x^2+1)^5 + (x^2+1)^5\cdot\frac{d}{dx}3(x^2-1)^3 \quad \text{[Product Rule]}$$
$$= 3(x^2-1)^3\cdot 5(x^2+1)^4(2x) + (x^2+1)^5\cdot 9(x^2-1)^2(2x).$$

Factor $6x(x^2-1)^2(x^2+1)^4$ from each term to obtain

$$\frac{dy}{dx} = 6x(x^2-1)^2(x^2+1)^4[5(x^2-1)+3(x^2+1)]$$
$$= 6x(x^2-1)^2(x^2+1)^4(8x^2-2)$$
$$= 12x(x^2-1)^2(x^2+1)^4(4x^2-1).$$

13. If $f(x) = (3x+1)^4(3-x)^5$, then

$$f'(x) = (3x+1)^4\cdot\frac{d}{dx}(3-x)^5 + (3-x)^5\cdot\frac{d}{dx}(3x+1)^4 \quad \text{[Product Rule]}$$
$$= (3x+1)^4\cdot 5(3-x)^4(-1) + (3-x)^5\cdot 4(3x+1)^3(3)$$
$$= (3x+1)^3(3-x)^4[-5(3x+1)+12(3-x)]$$
$$= (3x+1)^3(3-x)^4(-27x+31).$$

Set $f'(x) = 0$ and solve for x:

$$(3x+1)^3(3-x)^4(-27x-31) = 0.$$

$$x = -\frac{1}{3}, \quad x = 3, \quad \text{or} \quad x = \frac{31}{27}.$$

19. We want to evaluate $\frac{dC}{dt}$. The problem tells us that daily sales are rising at the rate of three lamps per day. This means $\frac{dx}{dt} = 3$. Now if $C = 40x + 30$, then $\frac{dC}{dx} = 40$. Hence by the chain rule,

$$\frac{dC}{dt} = \frac{dC}{dx} \cdot \frac{dx}{dt} = 40(3) = 120.$$

Costs are rising by $120 per day.

25. $h(x) = f(g(x))$, $h(1) = f(g)(1))$. From the graph of $g(x)$, you can see that $g(1) = 2$, so $h(1) = f(2)$. From the graph of $f(x)$, you can see that $f(2) = 1$, so $h(1) = f(2) = 1$. Next, by the Chain Rule,

$$\begin{aligned} h'(x) &= f'(g(x))g'(x) \\ h'(1) &= f'(g(1))g'(1) \\ &= f'(2)g'(1). \end{aligned}$$

The slope of the tangent line at $x = 2$ on the graph of $f(x)$ is -1. This is clear from an examination of the grid background in Figure 2. Therefore, $f'(2) = -1$. Similarly, Figure 2 shows that the slope of the tangent line at $x = 1$ on the graph of $g(x)$ is $\frac{3}{2}$, so

$$\begin{aligned} g'(1) &= \frac{3}{2}, \\ h'(1) &= (-1)\frac{3}{2} = -\frac{3}{2}. \end{aligned}$$

31. If $g(x) = \sqrt{x} = x^{1/2}$, then $g'(x) = \frac{1}{2}x^{-1/2}$. Since $f'(x) = x\sqrt{1-x^2}$,

$$\begin{aligned} \frac{d}{dx} f(g(x)) = f'(g(x))g'(x) &= (x^{1/2})\sqrt{1-x}\left(\frac{1}{2}x^{-1/2}\right) \\ &= \frac{1}{2}\sqrt{1-x}. \end{aligned}$$

37. If $u = \sqrt{x} = x^{1/2}$, then

$$\frac{du}{dx} = \frac{1}{2}x^{-1/2}.$$

Since

$$\begin{aligned} \frac{dy}{du} &= \frac{u}{\sqrt{1+u^4}}, \\ \frac{du}{dx} = \frac{dy}{du} \cdot \frac{du}{dx} &= \frac{u}{\sqrt{1+u^4}}\left(\frac{1}{2}x^{-1/2}\right). \end{aligned}$$

To express $\frac{dy}{dx}$ as a function of x alone, substitute $x^{1/2}$ for u to obtain

$$\frac{dy}{dx} = \frac{x^{1/2}}{(1+x^2)^{1/2}}\left(\frac{1}{2}x^{-1/2}\right) = \frac{1}{2\sqrt{1+x^2}}.$$

43. By the product rule,

$$\frac{d}{dx}x^2y^2 = x^2 \cdot \frac{d}{dx}y^2 + y^2 \cdot \frac{d}{dx}x^2$$
$$= x^2 2y\frac{dy}{dx} + y^2 2x.$$

Hence, implicit differentiation of $x^2y^2 = 9$ yields

$$2x^2y\frac{dy}{dx} + 2xy^2 = 0.$$

Solving for $\frac{dy}{dx}$,

$$2x^2y\frac{dy}{dx} = -2xy^2$$
$$\frac{dy}{dx} = \frac{-2xy^2}{2x^2y} = -\frac{y}{x}.$$

If $x = 1$ and $y = 3$, then $\frac{dy}{dx} = -\frac{3}{1} = -3.$

49. First, to find $\frac{dx}{dt}$, differentiate $6p + 5x + xp = 50$ term by term, with respect to *t*, using the product rule on *xp*.

$$6\frac{dp}{dt} + 5\frac{dx}{dt} + x\frac{dp}{dt} + p\frac{dx}{dt} = 0.$$

Solving for $\frac{dx}{dt}$,

$$5\frac{dx}{dt} + p\frac{dx}{dt} = -6\frac{dp}{dt} - x\frac{dp}{dt},$$
$$(5+p)\frac{dx}{dt} = -(6+x)\frac{dp}{dt},$$
$$\frac{dx}{dt} = -\left(\frac{6+x}{5+p}\right)\frac{dp}{dt}.$$

If $x = 4$, $p = 3$, and $\frac{dp}{dt} = -2$, then

$$\frac{dx}{dt} = \left(\frac{6+4}{5+3}\right)(-2) = -\frac{10}{4} = -\frac{5}{2} = -2.5.$$

Therefore, the quantity *x* is decreasing at the rate of 2.5 units per unit time.

Chapter 4
The Exponential and Natural Logarithm Functions

4.1 Exponential Functions

The exercises in Section 4.1 focus on the specific skills you will use later in Chapter 4. If you will see more practice, go back to Section 0.5.

It will be helpful to make a table that lists the laws of exponents shown on page 230 and to add to it the "reverse" laws listed in the *Manual* notes for Section 0.5. (Use x and y in the exponents now instead of r and s.) You definitely need to memorize this enlarged list of laws. Working scores of problems will help, of course, because you will have to refer to your list. Here is an additional method. Write only the left side of each law in the table on page 230. Maybe mix up the order of the laws. On a separate sheet of paper, write the left side of the laws listed in the *Manual* for Section 0.5 (with x, y in place of r, s). Put these partial lists aside for a day or two while you work exercises from Section 4.1. Later, try to fill in the lists of laws without any help. You may need to do this more than once.

1. $4^x = (2^2)^x = 2^{2x}$,

$(\sqrt{3})^x = (3^{1/2})^x = 3^{(1/2)x}$,

$\left(\frac{1}{9}\right)^x = (3^{-2})^x = 3^{-2x}$.

7. $6^x \cdot 3^{-x} = (2 \cdot 3)^x \cdot 3^{-x} = 2^x \cdot 3^x \cdot 3^{-x} = 2^x \cdot 3^0 = 2^x \cdot 1 = 2^x$,

$\frac{15^x}{5^x} = \left(\frac{15}{5}\right)^x = 3^x$, or $\frac{15^x}{5^x} = (3 \cdot 5)^x \cdot 5^{-x} = 3^x \cdot 5^x \cdot 5^{-x} = 3^x$,

$\frac{12^x}{2^{2x}} = 12x \cdot 2^{-2x} = (3.2^2)^x \cdot 2^{-2x} = 3^x \cdot 2^{2x} \cdot 2^{-2x} = 3^x \cdot 2^0 = 3^x$, or $\frac{12^x}{2^{2x}} = \frac{12^x}{4^x} = \left(\frac{12}{4}\right)^x = 3^x$.

13. $(2^{-3x} \cdot 2^{-2x})^{2/5} = (2^{-5x})^{2/5} = 2^{(-5x)(2/5)} = 2^{-2x}$,

$(9^{1/2} \cdot 9^4)^{x/9} = 9^{(9/2)(x/9)} = 9^{x/2} = (3^2)^{x/2} = 3^x$,

or $(9^{1/2} \cdot 9^4)^{x/9} = (3 \cdot (3^2)^4)^{x/9} = (3^9)^{x/9} = 3^x$.

19. If $(2.5)^{2x+1} = (2.5)^5$, then equating exponents,

$$2x+1=5 \Rightarrow 2x=4 \Rightarrow x=2.$$

25. If $(2^{x+1} \cdot 2^{-3})^2 = 2$, then

$$(2^{(x+1)-3})^2 = 2, \quad (2^{x-2})^2 = 2, \quad 2^{2x-4} = 2^1.$$

Equating exponents, we have $2x - 4 = 1$ and hence $x = \dfrac{5}{2}$.

31. If $2^x - \dfrac{8}{2^{2x}} = 0$, then

$$\begin{aligned} 2^x &= \frac{8}{2^{2x}} \\ 2^x \cdot 2^{2x} &= 8 \\ 2^{x+2x} &= 8 \\ 2^{3x} &= 8. \end{aligned}$$

But $8 = 2^3$, thus $2^{3x} = 2^3$. Equating exponents,

$$3x = 3, \quad x = 1.$$

37. Since $2^{3+h} = 2^3(2^h)$, the missing factor is 2^h.

43. Set Y_1=2^X. Review the **INCORPORATING TECHNOLOGY** material in section 1.2 (page 69) to approximate the slope of a graph at a point.

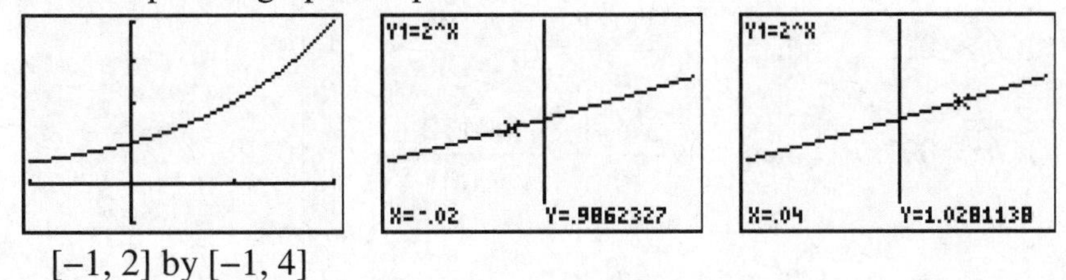

[−1, 2] by [−1, 4]

Alternatively, if Y_4 is the derivative of Y_1, then evaluate $Y_4(0)$, as described in **INCORPORATING TECHNOLOGY in section 0.2** (and apply the steps to Y_4 in place of Y_1). You should obtain .69314724.

```
Plot1 Plot2 Plot3
\Y1=2^X
\Y2=
\Y3=
\Y4=nDeriv(Y1,X,
X)
\Y5=
\Y6=
```

```
Y4(0)
        .6931472361
```

45. Apply the methods for Exercise 43 to b^X in place of 2^X. Try different numbers for b, such as 2.5 and 3.0. By trial and error you should obtain $b = 2.7$.

4.2 The Exponential Function e^x

The purpose of this brief section and its exercise is to introduce the function e^x and to help you get used to working with it. Here are the main facts about e^x. The number e is between 2 and 3; the graph of the function e^x is similar to the graphs of 2^x and 3^x and lies between them. The number e is chosen so that the slope of $y = e^x$ is 1 at $x = 0$. The slope of $y = e^x$ at an arbitrary point (x, e^x) on the graph has the same numerical value as the y-coordinate of the point. That is

$$[\text{slope at } (x, e^x)] = [y\text{-coordinate of } (x, e^x)]$$

$$\frac{d}{dx}e^x = e^x.$$

See Figure 3 on page 236.

1. If $y = 3^x$, then the slope of the secant line through (0, 1) and $(h, 3^h)$ is

$$\frac{3^h - 1}{h}.$$

As h approaches zero, the slope of the secant line approaches the slope of $y = 3^x$ at $x = 0$; that is,

$$\left.\frac{d}{dx}3^x\right|_{x=0}.$$

If $h = .1$, then $\dfrac{3^h - 1}{h} = \dfrac{1.11612 - 1}{.1} = 1.1612.$

If $h = .01$, then $\dfrac{3^h - 1}{h} = \dfrac{1.01105 - 1}{.01} = 1.105.$

If $h = .001$, then $\dfrac{3^h - 1}{h} = \dfrac{1.00110 - 1}{.001} = 1.10.$

Therefore, $\left.\dfrac{d}{dx}3^x\right|_{x=0} = \lim\limits_{x \to 0} \dfrac{3^h - 1}{h} \approx 1.1.$

Helpful Hint: The calculations in Exercise 1 can be automated in various ways on a graphing calculator. Even if such calculators are not required in your course, you should consider learning how to perform simple computations. For instance, set $Y_1 = (3\wedge X - 1)/X$ and then evaluate $Y_1(.1)$, $Y_1(.01)$, and $Y_1(.001)$. Or, in the home screen, compute $(3\wedge.1 - 1)/.1$. Then use the key that recalls the last computation and edit the command, replacing .1 with .01. Repeat, with .001 in place of .01

7. At $h = .01$, the slope of the secant line through the points (0, 1) and $(.01, e^{.01})$ is:

$$\frac{e^h - 1}{h} = \frac{e^{.01} - 1}{.01} \approx \frac{1.01005 - 1}{.01} = 1.005.$$

At $h = .001$, the slope of the secant line through the points (0, 1) and $(.001, e^{.001})$ is:

$$\frac{e^h - 1}{h} = \frac{e^{.001} - 1}{.001} \approx \frac{1.001001 - 1}{.001} = 1.001.$$

At $h = .0001$, the slope of the secant line through the points (0, 1) and $(.0001, e^{.0001})$ is:

$$\frac{e^h - 1}{h} = \frac{e^{.0001} - 1}{.0001} \approx \frac{1.00010 - 1}{.0001} = 1.000.$$

Using this, we estimate the slope of e^x at $x = 0$ to be 1.

13. $(e^2)^x = e^{2x} \Rightarrow k = 2.$ $\left(\frac{1}{e}\right)^x = (e^{-1}) = e^{-x} \Rightarrow k = -1.$

19. To solve $e^{5x} = e^{20}$, equate exponents:

$$5x = 20, \quad x = 4.$$

25. $\frac{d}{dx}\left(3e^x - 7x\right) = \frac{d}{dx}\left(3e^x\right) - \frac{d}{dx}(7x) = 3e^x - 7$

31. By the quotient rule,

$$\frac{d}{dx}\frac{e^x}{x+1} = \frac{(x+1)e^x - e^x \cdot 1}{(x+1)^2} = \frac{xe^x}{(x+1)^2}.$$

37. The tangent line is horizontal when $\frac{dy}{dx} = 0$. Using the product rule we find

$$\frac{dy}{dx} = (1+x^2)e^x + e^x(2x)$$
$$= e^x(x^2 + 2x + 1).$$

$\frac{dy}{dx} = 0$ when $x^2 + 2x + 1 = (x+1)^2 = 0$, or when $x = -1$. To find the y-coordinate, substitute $x = -1$ into the equation:

$$y = (1 + (-1)^2)e^{-1} = 2e^{-1}$$

Thus, the tangent line is horizontal at the point $(-1, 2e^{-1})$.

43. $f(x) = e^x(1+x)^2$. Use the product rule along with the chain rule to find $f'(x)$:

$$f'(x) = e^x 2(1+x)(1) + (1+x)^2 e^x$$
$$= e^x(2 + 2x + 1 + 2x + x^2)$$
$$= e^x(x^2 + 4x + 3).$$

Use the product rule to find $f''(x)$:

$$f''(x) = e^x(2x+4) + (x^2 + 4x + 3)e^x$$
$$= e^x(2x + 4 + x^2 + 4x + 3)$$
$$= e^x(x^2 + 6x + 7).$$

49.

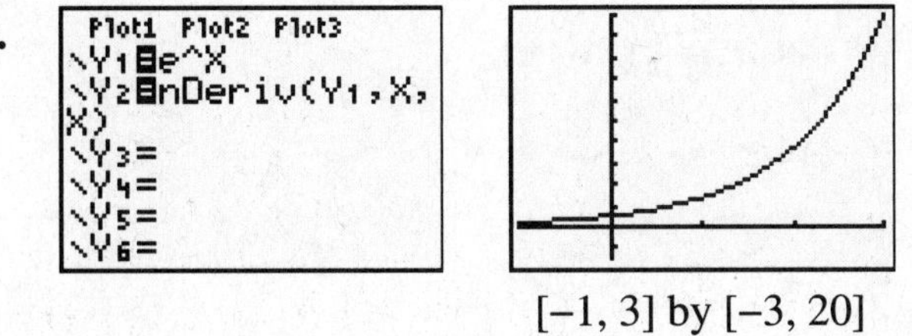

[−1, 3] by [−3, 20]

4.3 Differentiation of Exponential Functions

Everything in this section is important. It is essential that you not fall behind the class at this point. Make every effort to completely finish the work in Sections 4.1–4.3 before the class begins Section 4.4. You may have difficulty with the material on logarithms in Sections 4.4–4.6 if you try to learn it while you are still uncertain about exponential functions. Be sure to read the boxes on pages 240–241. The importance of the differential equation $y' = ky$ will not be apparent until you reach Chapter 5, but it is desirable to think a little bit now about questions such as Exercises 41–44.

1. $\frac{d}{dx}4e^{2x} = 4\cdot\frac{d}{dx}e^{2x} = 4(2e^{2x}) = 8e^{2x}.$

7. Begin with the general power rule.

$$\frac{d}{dx}(e^x + e^{-x})^3 = 3(e^x + e^{-x})^2 \cdot \frac{d}{dx}(e^x + e^{-x}) = 3(e^x + e^{-x})(e^x - e^{-x})^2.$$

13. Use the chain rule for $e^{g(x)}$:

$$\begin{aligned}\frac{d}{dx}(e^{x^2-5x+4}) &= e^{x^2-5x+4} \cdot \frac{d}{dx}(x^2 - 5x + 4)\\ &= e^{x^2-5x+4} \cdot (2x - 5).\end{aligned}$$

19. Use the product rule, then the chain rule for e^{2t} and $(t+1)^2$:

$$\begin{aligned}\frac{d}{dt}(t+1)^2 e^{2t} &= (t+1)^2 \cdot \frac{d}{dx}e^{2t} + e^{2t} \cdot \frac{d}{dx}(t+1)^2\\ &= (t+1)^2 \cdot 2e^{2t} + e^{2t} \cdot 2 \cdot (t+1) \cdot 1\\ &= 2e^{2t}(t+1)^2(t+2)\end{aligned}$$

25. Write $\frac{1}{x}$ as x^{-1} and use the product rule:

$$\begin{aligned}\frac{d}{dx}(x^{-1} + 3)e^{2x} &= (x^{-1} + 3) \cdot \frac{d}{dx}e^{2x} + e^{2x} \cdot \frac{d}{dx}(x^{-1} + 3)\\ &= (x^{-1} + 3) \cdot 2e^{2x} + e^{2x} + e^{2x} \cdot (-x^{-2})\\ &= (6 + 2x^{-1} - x^{-2})e^{2x} \text{ or } \left(-\frac{1}{x^2} + \frac{2}{x} + 6\right)e^{2x}\end{aligned}$$

31. $f(x) = (5x - 2)e^{1-2x}$. Use the product rule:

$$\begin{aligned}f'(x) &= (5x - 2) \cdot \frac{d}{dx}e^{1-2x} + e^{1-2x} \cdot \frac{d}{dx}(5x - 2)\\ &= (5x - 2) \cdot e^{1-2x} \cdot (-2) + e^{1-2x} \cdot 5.\end{aligned}$$

Simplify $f'(x)$ first, and then find $f''(x)$:

$$\begin{aligned}f'(x) &= (-10x + 4)e^{1-2x} + 5e^{1-2x}\\ &= (-10x + 4 + 5)e^{1-2x}\\ &= (9 - 10x)e^{1-2x}\end{aligned}$$

$$f''(x) = (9-10x)\cdot e^{1-2x}(-2) + e^{1-2x}\cdot(-10)$$

To find values of x at which $f(x)$ has a possible relative max or min, set the derivative equal to zero, and solve for x:

$$f'(t)\ (9-10x)e^{1-2x} = 0 \Rightarrow 9-10x = 0 \quad \text{and} \quad x = \frac{9}{10}$$

Here we used the fact that e^{1-2x} is never zero. To check the concavity of the graph of $f(x)$ at $x = .9$, compute $f''(.9)$. Looking at $f''(x)$, notice that the first term is zero when $9 - 10x = 0$, so

$$f''(.9) = 0 + e^{1-2(.9)}\cdot(-10) < 0,$$

because every value of e^x is positive. The graph is concave down at $x = .9$, and so $f(x)$ has relative maximum point at $x = .9$ or 9/10.

Helpful Hint: Simplify $f'(x)$ before computing $f''(x)$, but don't simplify $f''(x)$. When you find an x that makes $f'(x) = 0$, this value of x may satisfy some equation that will make part of the formula for $f''(x)$ equal 0. In Exercise 31 above, the equation was $9 - 10x = 0$. This fact simplifies calculations when you check $f''(x)$ for concavity.

37. The phrase "after 7 weeks" means that 7 weeks have passed since t was zero; that is, $t = 7$. The rate of growth involves the derivative, so compute:

$$f(t) = (.05 + e^{-.4t})^{-1}$$

$$f'(t) = (-1)(.05 + e^{-.4t})^{-2}\cdot\frac{d}{dt}(.05 + e^{-.4t})$$

$$= -(.05 + e^{-.4t})^{-2}\cdot e^{-.4t}(-.4).$$

There is no need to simplify $f'(t)$, because further differentiation is not required.

$$f'(7) = -(.05 + e^{-2.8})^{-2}\cdot e^{-2.8}(-.4) \approx 1.98.$$

The plant was growing at the rate of nearly 2 inches per week after 7 weeks.

43. This question relies on the basic fact discussed just before Example 5. If $y' = -.5y$, then y must be a function of the form $f(t) = Ce^{-.5t}$, for some constant C. If, in addition, $f(0) = 1$, then $Ce^{-.5(0)} = 1$, which shows that $C = 1$ (because $e^0 = 1$). Thus, $f(x) = e^{-.5x}$.

49. Set $Y_1 = 1.825 \wedge 3 * (1 - 1.6 * e \wedge (-.4196X)) \wedge 3$, set Y_4 to be the derivative (possibly a numerical derivative) of Y_1, and set Y_5 to be the derivative of Y_4.

(a) A suitable window for graphing is [1, 15] *by* [0, 8]. The graph seems to have a horizontal asymptote close to $y = 6$ milliliters. As X grows, the negative exponential goes to zero, and values of Y_1 approach 1.825^3, which equals 6.0784, to four decimal places. This means that the tumor's volume appears to stabilize around 6 ml.

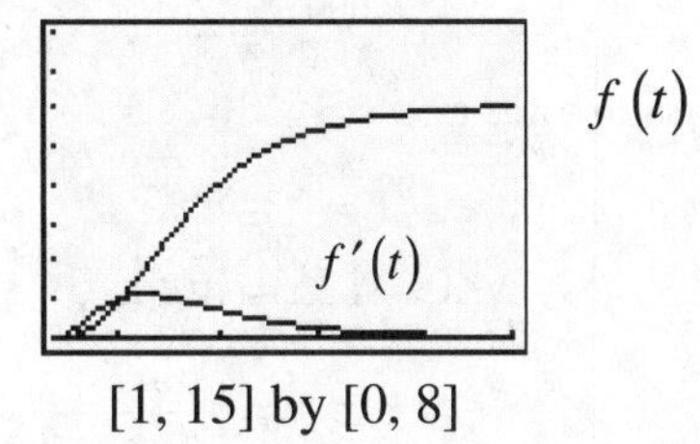

[1, 15] by [0, 8]

(b) Use **TRACE** or function evaluation to find $Y_1(5) \approx 3.16$ milliliters at $x = 5$ weeks.

(c) Set $Y_2 = 5$ (a constant function), and use **intersect** to find the X at which the graphs of Y_1 and Y_2 cross. You should obtain $X \approx 7.71$ weeks.

(d) Graph Y_4 and use **TRACE**, or evaluate $Y_4(5)$, to find the rate of growth when $X = 5$. You should the find the $Y_4(5)$ is about .97 milliliters per week.

(e) The growth is fastest when the derivative has a maximum, which can happen only if the second derivative is zero. So use a **Solver** command to find when $Y_5 = 0$. You should obtain $X \approx 3.74$ weeks.

(f) Evaluate Y_4 at the value of X found in Part (e). You should obtain $Y_4(3.74) \approx 1.13$ ml/wk.

4.4 The Natural Logarithm Function

The natural logarithm function is defined by its graph, and this is what you should think of when you are trying to understand what the ln x function really is. Study Figure 2 in Section 4.4. The natural logarithm is a function whose graph is defined only for $x > 0$; the graph is increasing, passes through (1, 0), and is concave downward. The graph of ln x is obtained by interchanging the x- and y-coordinates of points on the graph of $y = e^x$. This fact leads immediately to the fundamental relations (numbered below as in the text):

$$e^{\ln x} = x \qquad \text{(all positive } x) \qquad (2)$$

$$\ln e^x = x \qquad \text{(all } x) \qquad (3)$$

The main purpose of Section 4.4 is to teach you how to use these relations to solve equations involving ln x and e^x.

1. $\ln(\sqrt{e}) = \ln e^{1/2} = \frac{1}{2}$ [Relation (3)]

7. $\ln e^{-3} = -3$ [Relation (3)]

13. $e^{2\ln x} = (e^{\ln x})^2 = (x)^2 = x^2$ [Relation (2)]

19. If $e^{2x} = 5$, take the (natural) logarithm of each side to obtain:

$$\ln e^{2x} = \ln 5,$$

$$2x = \ln 5, \qquad \text{[R elation (3)]}$$

$$x = \frac{1}{2}\ln 5.$$

25. The first step is to isolate the term $e^{-.00012x}$ in order to use to the logarithm function to undo the effect of the exponentiation. Given $6e^{-.00012x} = 3,$ divide by 6 to obtain:

$$e^{-.00012x} = .5.$$

Take the natural logarithm of each side and solve:

$$\ln e^{-.00012x} = \ln .5,$$

$$-.00012x = \ln .5,$$

$$x = \frac{\ln .5}{-.00012}.$$

31. If $2e^{x/3} - 9 = 0$ then

$$2e^{x/3} = 9,$$

$$e^{x/3} = \frac{9}{2},$$

$$\ln(e^{x/3}) = \ln\frac{9}{2},$$

$$\frac{x}{3} = \ln\frac{9}{2}, \quad \text{or} \quad x = 3\ln\frac{9}{2}.$$

37. Given $4e^{x} \cdot e^{-2x} = 6$, observe that x occurs in two places. Use a property of exponents to write

$$4e^{-x} = 6.$$

Now proceed as in Exercise 25.

$$e^{-x} = \frac{6}{4} = \frac{3}{2},$$

$$\ln e^{-x} = \ln\frac{3}{2},$$

$$-x = \ln\frac{3}{2}, \quad \text{or} \quad x = -\ln\frac{3}{2}.$$

Warning: A slight change in Exercise 37 can make it seem harder. Consider the equation

$$e^{x} = \frac{3}{2}e^{2x}.$$

Taking logarithms a this point is of no help because the equation $x = \ln\left(\frac{3}{2}e^{2x}\right)$ expresses x in terms of something involving x. The correct step is to get x on only one side of the equation. There are two ways to do this:

$$\text{(i)} \quad e^{x} = \frac{3}{2}e^{2x}, \quad \text{or} \quad \text{(ii)} \quad e^{x} = \frac{3}{2}e^{2x},$$

$$e^{x} - \frac{3}{2}e^{2x} = 0. \qquad e^{x} \cdot e^{-2x} = \frac{3}{2}e^{2x}e^{-2x}.$$

Equation (i) leads nowhere because you can't take the logarithm of each side, since ln 0 is not defined. Also, $\ln\left(e^{x} - \frac{3}{2}e^{2x}\right)$ cannot be simplified. Equation (ii) is the correct approach, since a property of exponents leads to $e^{-x} = \frac{3}{2}e^{0}$, that is, $e^{-x} = \frac{3}{2}$. This can be solved for x as in Exercise 37, yielding $x = -\ln\frac{3}{2}$.

43. $f(x) = e^{-x} + 3x;\quad f'(x) = -e^{-x} + 3;\quad f''(x) = e^{-x}$

Set $f'(x) = 0$ and solve for x.

$$\begin{aligned} -e^{-x} + 3 &= 0 \\ -e^{-x} &= -3 \\ e^{-x} &= 3 \\ \ln e^{-x} &= \ln 3 \\ -x &= \ln 3 \\ x &= -\ln 3. \end{aligned}$$

Substitute this value for x back into $f(x)$ to find the y-coordinate of this possible extreme point.

$$\begin{aligned} f(3) &= e^{-(-\ln 3)} + 3(-\ln 3) \\ &= e^{\ln 3} - 3\ln 3 \\ &= 3 - 3\ln 3. \end{aligned}$$

The possible extreme point is $(-\ln 3, 3 - 3\ln 3)$. Since $f''(-\ln 3) = e^{-(-\ln 3)} = e^{-(-\ln 3)} = e^{\ln 3} = 3$, which is positive, the graph of $f(x)$ is concave up at $x = -\ln 3$ and $(-\ln 3, 3 - 3\ln 3)$ is a relative minimum point.

49. Graph of $y = \ln(e^x)$: Graph of $y = e^{\ln x}$:

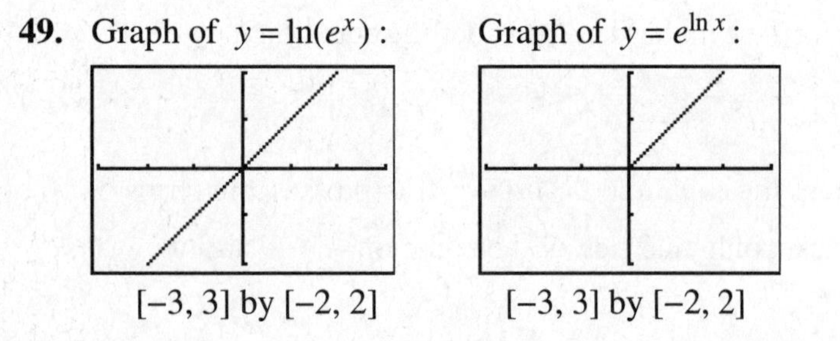

[−3, 3] by [−2, 2] [−3, 3] by [−2, 2]

The graph of $y = e^{\ln x}$ is the same as the graph of $y = x$ for $x > 0$.

4.5 The Derivative of ln *x*

The exercises in this section combine the derivative formula for $\ln x$ with a review of the product rule, the quotient rule, and the chain rule. You should memorize the chain rule for the logarithm function:

$$\frac{d}{dx}[\ln g(x)] = \frac{1}{g(x)} \cdot g'(x) = \frac{g'(x)}{g(x)}$$

This rule is used in Exercises 1, 7, 13, and 19, discussed below.

1. $\dfrac{d}{dx}\ln(2x) = \dfrac{1}{2x} \cdot \dfrac{d}{dx}(2x) = \dfrac{1}{2x}(2) = \dfrac{1}{x}$

7. $\dfrac{d}{dx}e^{\ln x + x} = e^{\ln x + x} \cdot \dfrac{d}{dx}(\ln x + x) = \left(\dfrac{1}{x} + 1\right)e^{\ln x + x}$

13. $\dfrac{d}{dx}\ln(kx) = \dfrac{1}{kx} \cdot \dfrac{d}{dx}(kx) = \dfrac{1}{kx}(k) = \dfrac{1}{x}$

19. $\frac{d}{dx}\ln(e^{5x}+1) = \frac{1}{e^{5x}+1}\cdot\frac{d}{dx}(e^{5x}+1) = \frac{1}{e^{5x}+1}\cdot e^{5x}(5) = \frac{5e^{5x}}{e^{5x}+1}$

25. For the equation of the tangent line, you need a point on the line and the slope of the line. Use the original equation $y = \ln(x^2+e)$ to find a point. If $x = 0$, then $y = \ln(0^2 + e) = \ln e = 1$, and hence (0, 1) is on the line. For the slope of the tangent line, first find the general slope formula:

$$\begin{aligned}\frac{dy}{dx} &= \frac{1}{x^2+e}\cdot\frac{d}{dx}(x^2+e)\\ &= \frac{1}{x^2+e}(2x) \qquad [e \text{ is a constant}]\\ &= \frac{2x}{x^2+e}.\end{aligned}$$

The slope of the tangent line when $x = 0$ is $\frac{2(0)}{0^2+e} = 0$. Therefore the tangent line is the horizontal line passing through the point (0, 1). The equation of this line is $y-1 = 0(x-0)$; that is, $y = 1$.

Warning: Exercises 25–30 make good exam questions because they review important concepts and skills from Chapters 1–3 as well as testing your ability to differentiate ln $g(x)$.

31. The marginal cost at $x = 10$ is $C'(10)$. Using the quotient, first find

$$\begin{aligned}C'(x) &= \frac{d}{dx}\left(\frac{100\ln x}{40-3x}\right)\\ &= \frac{(40-3x)\cdot\frac{d}{dx}(100\ln x) - (100\ln x)\cdot\frac{d}{dx}(40-3x)}{(40-3x)^2}\\ &= \frac{(40-3x)\left(\frac{100}{x}\right) - (100\ln x)(-3)}{(40-3x)^2}.\end{aligned}$$

Now

$$C'(10) = \frac{(40-3(10))\frac{100}{10} + 300\ln 10}{(40-3(10))^2} = \frac{100+300\ln 10}{(10)^2} = 1+3\ln 10.$$

37. Set Y_1 = ln (abs(X)) and set Y_4 to be the derivative of Y_1. The "abs" command is located in the MATH NUM menu. Alternatively, use the **CATALOG** command to find "abs" in the list of all commands. Graph both functions in the **ZDecimal** window (found in the **ZOOM** menu). Use **TRACE** on the graph of Y_4 to check the values of Y_4 at −4, −2, −1, 1, 2, and 4. (You should obtain −.25, −.5, −1, 1, .5, and .25.) Using the graph verifies the derivative formula. What happens at $x = 0$?

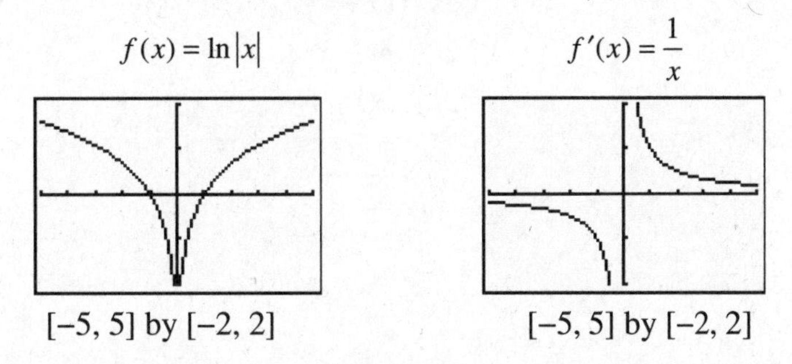

[−5, 5] by [−2, 2]   [−5, 5] by [−2, 2]

4.6 Properties of the Natural Logarithm Function

Just as with laws of exponents, it is essential to know the properties of logarithms "backwards and forwards." You will find it helpful to add the following properties to the list on page 252.

LI′	$\ln x + \ln y = \ln xy$
LII′	$-\ln x = \ln\left(\frac{1}{x}\right)$
LIII′	$\ln x - \ln y = \ln\left(\frac{x}{y}\right)$
LIV′	$b \ln x = \ln(x^b)$

Unfortunately, many students make up additional "laws" that are not true. A study of the following facts will help you avoid the most common incorrect "laws".

(a) ln $(x + y)$ is not equal to ln + ln y.

(b) ln $(e + e^y)$ is not equal to + y

(c) (ln x) (ln y) is not equal to ln x + ln y.

(d) $\frac{\ln x}{\ln y}$ is not equal to ln x – ln y.

1. $\ln 5 + \ln x = \ln 5x$. (LI′)

7. $e^{2\ln x} = e^{\ln x^2} = x^2$. (LIV′)

13. Property (LIV′) shows that $2 \ln 5 = \ln 5^2 = \ln 25$ and $3 \ln 3 = \ln 3^3 = \ln 27$. Since $25 < 27$, and since the graph of the natural logarithm function is increasing, $\ln 25 < \ln 27$. Therefore, $3 \ln 3$ is larger.

19. Using property (LIV′), we see $4 \ln 2x = \ln(2x)^4 = \ln(2^4x^4) = \ln(16x^4)$. The answer is (d).

25. $\ln x^4 - 2\ln x = 1 \Rightarrow 4\ln x - 2\ln x = 1 \Rightarrow 2\ln x = 1 \Rightarrow \ln x = \frac{1}{2} \Rightarrow x = e^{1/2} = \sqrt{e}$.

31. $\ln(x+1) - \ln(x-2) = 1 \Rightarrow \ln\left(\frac{x+1}{x-2}\right) = 1$. Exponentiate each side to obtain

$$\begin{aligned} \frac{x+1}{x-2} &= e^1 = e \\ x+1 &= e(x-2) \\ x+1 &= ex - 2e \\ 1+2e &= ex - x \\ 1+2e &= x(e-1) \\ x &= \frac{1+2e}{e-1}. \end{aligned}$$

Helpful Hint: Whenever you have a function to differentiate, you should pause and check if the form of the function can be simplified *before* you begin to differentiate. Exercises 33–40 illustrate how much this will simplify your work.

37. $y = \ln\left[\sqrt{xe^{x^2+1}}\right] = \ln\left(xe^{x^2+1}\right)^{1/2} = \frac{1}{2}\ln\left(xe^{x^2+1}\right) = \frac{1}{2}\ln x + \frac{1}{2}\ln e^{x^2+1} = \frac{1}{2}\ln x + \frac{1}{2}\left(x^2+1\right)$

Therefore,

$$\frac{dy}{dx} = \frac{d}{dx}\left(\frac{1}{2}\ln x\right) + \frac{d}{dx}\frac{1}{2}\left(x^2+1\right)$$
$$= \frac{1}{2}\cdot\frac{1}{x} + \frac{1}{2}\cdot 2x = \frac{1}{2x} + x$$

Warning: Remember that the natural logarithm converts products and quotients into sums and differences. But a *product* of logarithms *cannot* be simplified by a standard logarithm rule. Be on guard when you work Exercises 41 and 42.

43. Let $f(x) = (x+1)^4(4x-1)^2$. Then,

$$\ln f(x) = \ln\left[(x+1)^4(4x-1)^2\right]$$
$$= \ln(x+1)^4 + \ln(4x-1)^2$$
$$= 4\ln(x+1) + 2\ln(4x-1).$$

Differentiate both sides with respect to x:

$$\frac{f'(x)}{f(x)} = 4\cdot\frac{1}{x+1} + 2\cdot\frac{1}{4x-1}(4)$$
$$= \frac{4}{x+1} + \frac{8}{4x-1}.$$

So,

$$f'(x) = f(x)\left(\frac{4}{x+1} + \frac{8}{4x-1}\right) = (x+1)^4(4x-1)^2\left(\frac{4}{x+1} + \frac{8}{4x-1}\right).$$

49. Let $f(x) = x^x$. Then,

$$\ln f(x) = \ln x^x = x\ln x.$$

Differentiate both sides with respect to x:

$$\frac{f'(x)}{f(x)} = x\cdot\frac{1}{x} + \ln x\cdot(1) = 1 + \ln x.$$

So,

$$f'(x) = f(x)(1+\ln x) = x^x(1+\ln x).$$

Review of Chapter 4

There are many facts in this chapter to remember and keep straight. The learning process requires time and lots of practice. Don't wait until the last minute to review. Your efforts to master this chapter will be rewarded later, since the exponential and natural logarithm functions appear in nearly every section in the rest of the text.

By now you should have constructed the expanded lists of properties of exponents and logarithms. To these lists add notes about common mistakes—yours and the ones mentioned in the *Manual* notes for this chapter. Here is another common error: if you take logarithms of each side of an equation of the form

$$A = B + C$$

you *cannot* write $\ln A = \ln B + \ln C$. The correct form is

$$\ln A = \ln(B + C).$$

Similarly, if you exponentiate each side of $A = B + C$, you *cannot* write $e^A = e^B + e^C$. The correct form is

$$e^A = e^{(B+C)}.$$

Here are four problems that contain "traps" for the unwary student. Try them now. You will find answers later in the manual as you work through the supplementary exercises. *Please* don't look for the answers until you have done your best to work the problems.

(A) Solve for y in terms of x: $\ln y - \ln x^2 = \ln 5$.

(B) Solve for y in terms for x: $e^y - e^{-3} = e^{2x}$.

(C) Solve for x: $\dfrac{\ln 10x^3}{\ln 2x^2} = 1$.

(D) Simplify, if possible: $\ln(x^3 - x^2)$.

Chapter 4: Supplementary Exercises

In Exercises 9–14 and 39–44, "*simplify*" means to use the laws of exponents and logarithms to write the expression in another form that either is less complicated or at least is more useful for some purposes.

1. $27^{4/3} = (27^{1/3})^4 = 3^4 = 81.$

7. $\dfrac{9^{5/2}}{9^{3/2}} = 9^{5/2-3/2} = 9^{2/2} = 9.$

13. $(e^{8x} + 7e^{-2x})e^{3x} = e^{8x} \cdot e^{3x} + 7e^{-2x} \cdot e^{3x} = e^{11x} + 7e^x.$

19. $\dfrac{d}{dx}10e^{7x} = 10\dfrac{d}{dx}e^{7x} = 10e^{7x}(7) = 70e^{7x}.$

25. $\dfrac{d}{dx}\left(\dfrac{x^2 - x + 5}{e^{3x} + 3}\right) = \dfrac{(e^{3x} + 3)(2x - 1) - (x^2 - x + 5) \cdot 3e^{3x}}{(e^{3x} + 3)^2}$ [Quotient Rule]

Helpful Hint: In Exercises 27–30, the solution is a *function,* not a number. An equation in which both the unknown function and its (unknown) derivative appear is called a *differential equation*. The differential equations that appear here are solved with the boxed result (3) of Section 4.3 (page 241).

31. $y = e^{-x} + x, \quad y' = -e^{-x} + 1, \quad y'' = e^{-x}.$

Now set $y' = 0$ and solve for x:

$$\begin{aligned} -e^{-x} + 1 &= 0, \\ e^{-x} &= 1, \\ \ln e^{-x} &= \ln(1), \\ -x &= 0, \quad \text{or} \quad x = 0 \end{aligned}$$

If $x = 0$, then $y = e^{-0} + 0 = 1$ and $y'' = e^{-0} = 1 > 0$, hence the graph is concave up at (0, 1). Thus the graph has a relative minimum at (0, 1). Since $y'' = e^{-x} > 0$ for all x, the graph is concave up for all x, and there are no inflection points. As x becomes large, e^{-x} approaches 0, and for this case, $e^{-x} + x$ is only slightly larger than x. Hence for large positive values of x, the graph of $y = e^{-x} + x$ has $y = x$ as an asymptote.

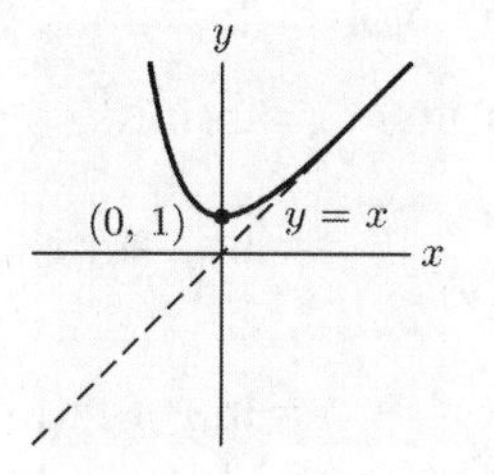

37. The slope of the tangent line to the graph of $y = \frac{e^x}{1+e^x}$ at (0, .5) is the value of the derivative $\frac{dy}{dx}$ when $x = 0$. By the quotient rule,

$$\begin{aligned} \frac{d}{dx}\left(\frac{e^x}{1+e^x}\right)\bigg|_{x=0} &= \frac{(1+e^x)\cdot e^x - (e^x)(e^x)}{(1+e^x)^2}\bigg|_{x=0} \\ &= \frac{(1+e^0)\cdot e^0 - (e^0)(e^0)}{(1+e^0)^2} \\ &= \frac{(1+1)\cdot 1 - (1)(1)}{(1+1)^2} \\ &= \frac{2-1}{2^2} \\ &= \frac{1}{4}. \end{aligned}$$

The tangent line is thus a line with slope $\frac{1}{4}$ passing through the point (0, .5). Since the line passes through (0, .5), the y-intercept is $\frac{1}{2}$ and the equation of the line is:

$$y = \frac{1}{4}x + \frac{1}{2}.$$

Warning: Be careful with Exercise 41. Don't use an incorrect property of logarithms, and don't look at the answer until you have tried the problem!

43. $e^{-5\ln 1} = e^{-5(0)} = e^0 = 1.$

49. The equation $2\ln t = 5$ leads to $\ln t = \frac{5}{2}$ and $t = e^{5/2}$.

55. $\frac{d}{dx}(\ln x)^2 = 2(\ln x)\cdot\frac{d}{dx}(\ln x) = 2(\ln x)\left(\frac{1}{x}\right) = \frac{2\ln x}{x}.$

61.
$$\frac{d}{dx}\ln(\ln\sqrt{x}) = \frac{1}{\ln\sqrt{x}}\cdot\frac{d}{dx}\ln\sqrt{x} = \frac{1}{\ln x^{1/2}}\cdot\frac{d}{dx}\ln x^{1/2}$$
$$= \frac{1}{(1/2)\ln x}\cdot\frac{d}{dx}\left[\frac{1}{2}\ln x\right]$$
$$= \frac{2}{\ln x}\cdot\frac{1}{2}\cdot\frac{d}{dx}\ln x = \frac{1}{\ln x}\cdot\frac{1}{x} = \frac{1}{x\ln x}.$$

Helpful Hint: Always be alert to the possibility of simplifying a function before you differentiate it. This is particularly important when the function involves $\ln x$ or e^x. Observe that

$$\ln(\ln\sqrt{x}) = \ln(\ln x^{1/2})$$
$$= \ln\left(\frac{1}{2}\ln x\right)$$
$$= \ln\frac{1}{2} + \ln(\ln x).$$

Thus

$$\frac{d}{dx}\ln(\ln\sqrt{x}) = \frac{d}{dx}\ln\frac{1}{2} + \frac{d}{dx}\ln(\ln x)$$
$$= 0 + \frac{1}{\ln x}\cdot\frac{d}{dx}\ln x = \frac{1}{\ln x}\cdot\frac{1}{x} = \frac{1}{x\ln x}.$$

Hints for Extra Problems

(A) Get $\ln y$ by itself on the left, and use a property of logarithms to simplify $\ln 5 + \ln x^2$.

(B) Write $e^y = e^{2x} + e^{-3}$ and then take the logarithm of each side. Be careful!

(C) You made a mistake if you obtained any of the following equations from problem (C):

$$\frac{10x^3}{2x^2} = 1, \quad \ln\left(\frac{10x^3}{2x^2}\right) = 1, \quad \ln(10x^3 - 2x^2) = 1.$$

None of these equations is equivalent to the original equation. The correct first step is to multiply both sides of the equation in (C) by $\ln 2x^2$ and obtain $\ln 10x^3 = \ln 2x^2$. *Now* can you solve for x?

(D) Is it true that $\ln(x^3 - x^2)$ is the same as $\frac{\ln x^3}{\ln x^2}$?

The correct solutions to (A)–(D) are at the end of this supplementary exercise set.

67. First, simplify the original function using the properties of the logarithm from Chapter 4.

$$\ln\left(\frac{e^{x^2}}{x}\right) = \ln(e^{x^2}) - \ln(x) \qquad \text{(LIII)}$$

$$= x^2 - \ln(x) \qquad \text{(Equation 3, Section 4.4)}$$

Now take the derivative of the simplified function.

$$\frac{d}{dx}(x^2 - \ln(x)) = 2x - \frac{1}{x}$$

Note that this problem could also be solved using the quotient rule and multiple applications of the chain rule. However, using the properties of the natural logarithm to simplify before differentiating simplifies the solution substantially.

73. First, simplify the original function using the properties of the logarithm from this chapter.

$$\ln\left(\frac{1}{e^{\sqrt{x}}}\right) = -\ln(e^{\sqrt{x}}) \qquad \text{(LII)}$$

$$= -\sqrt{x}. \qquad \text{(Equation 3, Section 4.4)}$$

Now take the derivative of the simplified function.

$$\frac{d}{dx}(-\sqrt{x}) = \frac{d}{dx}(-x^{1/2}) = \frac{-1}{2}x^{-1/2} = \frac{-1}{2\sqrt{x}}$$

Again, this problem could also be solved using the quotient rule and multiple applications of the chain rule, but it is much easier to use the properties of the natural logarithm to simplify before differentiating.

79. First take the natural logarithm of each side.

$$\ln f(x) = \ln[(x^2+5)^6(x^3+7)^8(x^4+9)^{10}]$$

$$= \ln(x^2+5)^6 + \ln(x^3+7)^8 + \ln(x^4+9)^{10} \qquad \text{(LI)}$$

$$= 6\cdot\ln(x^2+5) + 8\cdot\ln(x^3+7) + 10\cdot\ln(x^4+9). \qquad \text{(LIV)}$$

Now, take the derivative of each side and solve for $f'(x)$.

$$\frac{f'(x)}{f(x)} = 6\left[\frac{1}{x^2+5}\cdot 2x\right] + 8\left[\frac{1}{x^3+7}\cdot 3x^2\right] + 10\left[\frac{1}{x^4+9}\cdot 4x^3\right]$$

$$= \frac{12x}{x^2+5} + \frac{24x^2}{x^3+7} + \frac{40x^3}{x^4+9}.$$

$$f'(x) = f(x)\left[\frac{12x}{x^2+5} + \frac{24x^2}{x^3+7} + \frac{40x^3}{x^4+9}\right]$$

$$= (x^2+5)^6(x^3+7)^8(x^4+9)^{10}\left[\frac{12x}{x^2+5} + \frac{24x^2}{x^3+7} + \frac{40x^3}{x^4+9}\right].$$

85. Given

$$f(x) = e^{x+1}(x^2+1)(x)$$
$$= (e^{x+1})(x^3+x),$$

first take the natural logarithm of each side:

$$\ln(f(x)) = \ln\left((e^{x+1})(x^3+x)\right)$$
$$= \ln(e^{x+1}) + \ln(x^3+x) \qquad \text{(LI)}$$
$$= x+1+\ln(x^3+x). \qquad \text{(Equation 3, Section 4.4)}$$

Now, take the derivative of each side and solve for $f'(x)$.

$$\frac{f'(x)}{f(x)} = 1 + \frac{3x^2+1}{x^3+x}.$$
$$f'(x) = f(x)\left(1+\frac{3x^2+1}{x^3+x}\right)$$
$$= (e^{x+1})(x^3+x)\left(1+\frac{3x^2+1}{x^3+x}\right)$$
$$= e^{x+1}(x^3+x) + \frac{e^{x+1}(x^3+x)(3x^2+1)}{x^3+x}$$
$$= e^{x+1}(x^3+x) + e^{x+1}(3x^2+1)$$
$$= e^{x+1}(x^3+3x^2+x+1).$$

91. $y = (\ln x)^2,$

$$y' = 2(\ln x)\cdot\frac{d}{dx}\ln x = \frac{2\ln x}{x}.$$
$$y'' = \frac{x\frac{d}{dx}(2\ln x) - (2\ln x)\frac{d}{dx}x}{x^2}$$
$$= \frac{x\cdot 2(\frac{1}{x}) - 2\ln x}{x^2} = \frac{2-2\ln x}{x^2}.$$

Set $y' = 0$ and solve for x:

$$\frac{2\ln x}{x} = 0$$

A fraction is zero only when its numerator is zero. Thus $2\ln x = 0$, and $\ln x = 0$. Then $e^{\ln x} = e^0 = 1$, so $x = 1$. If $x = 1$, then $y = (\ln 1)^2 = 0$, and

$$y'' = \frac{2-2\ln 1}{1^2} = 2 > 0.$$

Hence, the curve is concave up at (1, 0). Now, set $y'' = 0$ and solve for x to find the possible inflection points.

$$\frac{2-2\ln x}{x^2}=0,$$
$$2-2\ln x=0,$$
$$-2\ln x=-2, \quad \text{or} \quad \ln x=1.$$

So,

$$e^{\ln x}=e^1, \quad \text{or} \quad x=e.$$

If $x = e$, then $y = (\ln e)^2 = 1$, hence $(e, 1)$ is the only possible inflection point. We have seen that the second derivative is positive at $x = 0$. When x is large, $2 - 2 \ln x$ is negative, so the second derivative is negative. Thus the concavity must change somewhere, which shows that $(e, 1)$ is the inflection point.

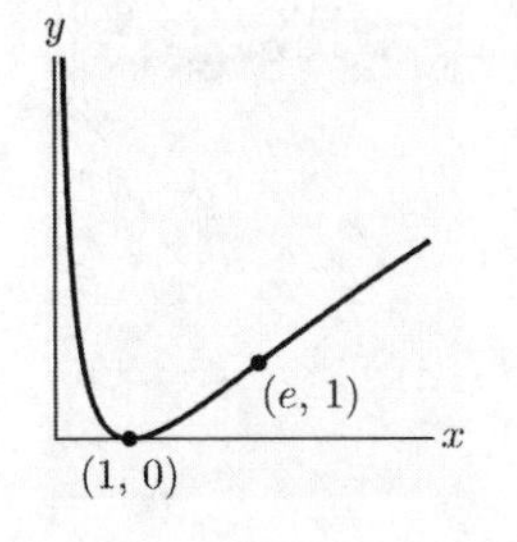

97. First, take the natural logarithm of both sides in order to solve for y.

$$\ln(e^{xy})=\ln(x)$$
$$xy=\ln(x) \qquad \text{(Equation (3), Section 4.4)}$$
$$y=\frac{\ln(x)}{x}.$$

Next, use the quotient rule to find $\frac{dy}{dx}$.

$$\frac{dy}{dx}=\frac{x\left(\frac{1}{x}\right)-(1)\ln(x)}{x^2}=\frac{1-\ln(x)}{x^2}.$$

Finally, evaluate $\frac{dy}{dx}$ at $x = 1$.

$$\left.\frac{dy}{dx}\right|_{x=1}=\frac{1-\ln(1)}{1^2}=\frac{1-0}{1}=1.$$

Solutions to Extra Problems

(A) $\ln y-\ln x^2=\ln 5,$

$\ln y=\ln 5+\ln x^2,$

$\ln y=\ln 5x^2,$

Exponentiate each side to obtain $y = 5x^2$.

(B) $e^y=e^{2x}+e^{-3},$

$\ln e^y=\ln(e^{2x}+e^{-3}),$ [*Note*: The right side cannot be simplified.]

$y=\ln(e^{2x}+e^{-3}).$

(C) $\ln 10x^3 = \ln 2x^2$. Exponentiate to obtain $10x^3 = 2x^2$.

Notice that x cannot be zero, since $\ln 10x^3$ is not defined when $x = 0$. So you may divide by $2x^2$ and obtain $5x = 1$. Hence $x = 1/5$.

(D) If you answered "yes" to the "Hint", then you have made one of the most common mistakes involving logarithms. The logarithm of a sum of difference *cannot be simplified,* unless that sum of difference can be written somehow as a product or quotient. For instance, you *may* write

$$\begin{aligned}\ln(x^3 - x^2) &= \ln[x^2(x-1)] \quad \text{[Now the expression inside the logarithm is a product.]} \\ &= \ln x^2 + \ln(x-1) \\ &= 2\ln x + \ln(x-1).\end{aligned}$$

Whether or not this answer is "simplified" depends on the use to be made of the expression. See our remark at the beginning of the Supplementary Exercises.

Chapter 5
Applications of the Exponential and Natural Logarithm Functions

5.1 Exponential Growth and Decay

The basic differential equation $y' = ky$ was introduced in Section 4.3, just before Example 5. You should review that now. All the exercises in this section refer either to that differential equation or to the general form of functions that satisfy the equation, namely, $f(x) = Ce^{kx}$. In this chapter, we usually use t in place of x, because most of the applications involve functions of time.

The first question to ask when considering an exponential growth and decay problem is, "Does the quantity increase or decrease with time?" For instance, populations increase and radioactive substances decrease. The formula for the quantity present after t units of time will have the form $P(t) = P_0 e^{kt}$ if increasing and $P(t) = P_0 e^{-\lambda t}$ if decreasing. The problems are primarily of the following types:

1. Given P_0 and k (or λ), find the value of $P(t)$ for a specific time t. These problems are solved by just substituting the value of t into the formula.
2. Given P_0 and k (or λ), find the time when $P(t)$ assumes a specific value, call it A. That is, solve $P(t) = A$ for t. Problems of this type are solved by replacing $P(t)$ by its formula, dividing by P_0 and using logarithms. Sometimes A is given as a multiple of P_0 in which case the P_0's cancel each other.
3. Given P_0 and the value of $P(t)$ at some specific time t, solve for k (or λ). This problem reduces to one of the form "solve $P_0 e^{kt} = A$" where A is the value of $P(t)$. This equation is solved by the method in 2 above.
4. The problem gives enough information to solve for k (or λ), but only asks for the value of $P(t)$ at a future time or asks for the time at which $P(t)$ attains a given size. The important point here is that the problem really consists of two parts. First, find k (or λ) and second, answer the question asked. Exercises 9 and 19 are of this type. Students usually have difficulty with such problems on exams since they try to answer the question posed immediately without doing the intermediate step.
5. Determine the rate of change of $P(t)$ at a time when $P(t)$ is a certain size. The actual time is not specified, so you cannot compute $P'(t)$ and evaluate at t. Instead, you must use the differential equation $P'(t) = k \cdot P(t)$. Assuming that k is known, you can determine $P'(t)$ from $P(t)$ (for the same t), and vice versa. You'll find this type of problem in Exercises 1, 13, and elsewhere.

1. **(a)** The differential equation $P'(t) = .02P(t)$ has the form $y' = ky$, with $k = .02$, so the general solution $P(t)$ has the form $P(t) = P_0e^{.02t}$. Also,

$$P(0) = P_0e^{.02(0)} = P_0,$$

so the condition $P(0) = 3$ shows that $P(t) = 3e^{.02t}$.

(b) *Initial population* is the population when $t = 0$, which in this problem refers to 1990: 3 million persons.

(c) The growth constant is the constant .02 that appears in the differential equation $P'(t) = .02P(t)$.

(d) 1998 corresponds to $t = 8$. The problem then was $P(8) = 3e^{.02(8)} = 3.52053261298$ million. Of course, writing the answer in this form is not realistic. Even the answer 3520533 persons is inappropriate. A problem such as this exercise purports to be a model of population growth, but it is only an approximation at best. Since the initial population is given with only the single digit "3", a rounded answer such as 3.5 million, or perhaps 3.52 million, would be more realistic.

(e) Assume that t_0 is the time when $P(t_0) = 4$ (million). At that value of t, the differential equation says that

$$P'(t_0) = .02P(t_0) = .02 \cdot 4 = .08.$$

The population is growing at the rate of .08 million (or 80,000) people per year.

(f) The unit of measurement of the population is millions of people, so convert 70,000 to .07 million. If t represents the time when $P'(t) = .07$ million people per year, then the differential equation

$$P'(t) = .02P(t)$$

shows that $.07 = .02P(t)$ and hence $P(t) = .07/.02 = 3.5$. The population is approximately 3.5 million.

Helpful Hint: In Exercises 5, 6, 9, and 10, the phrase "growing at a rate proportional to its size" describes the differential equation $y' = ky$. You may assume that the solution of this equation has the form $y = P_0e^{kt}$, or $P(t) = P_0e^{kt}$. This fact was discussed at the beginning of the chapter.

7. See the notes about problem type #2 on page 109 in this *Manual*. The phrase "growing exponentially" is another way of saying that the model for the growth is $P(t) = P_0e^{kt}$. Hence k is .05, so $P(t) = P_0e^{.05t}$. The population will be triple its initial size when $P(3) = 3P_0$. To find the time when this happens, set the formula for $P(t)$ equal to $3P_0$ and solve for t:

$$P_0e^{.05t} = 3P_0$$

$$e^{.05t} = 3 \qquad \text{(dividing both sides by } P_0\text{)}.$$

Apply the natural logarithm to each side:

$$\ln e^{.05t} = \ln 3$$

$$.05t = \ln 3$$

$$t = \frac{\ln 3}{.05} \approx 22 \text{ years} \quad \text{(rounding off 21.97 to 22)}.$$

13. **(a)** From the differential equation $P'(t) = -.021P(t)$, you know that $P(t)$ has the form $P(t) = P_0e^{-.021t}$. The initial amount is 8 grams, so $P(t) = 8e^{-.021t}$.

(b) $P(0) = P_0 = 8$ grams.

(c) The decay constant is .021 (not –.021)

(d) Compute $P(10) - 8e^{-.021(10)} \approx 6.5$ grams.

(e) $P'(t) = -.021P(t)$. If $P(t) = 1$, then at this same time, $P'(t) = -.021(1) = -.021$ grams/year. The sample is disintegrating at the rate of .021 grams/year.

(f) The time is not specified directly, but it described by the property that P(t) is "disintegrating at the rate of .105 grams per year," that is, $P'(t_0) = -.105$ for some particular time t_0. From the differential equation, you can conclude that

$$P'(t_0) = -.021P(t_0)$$

that is,

$$-.105 = -.021 \cdot P(t_0).$$

Hence, $P(t_0) = \frac{-.105}{-.021} = 5$. Thus 5 grams of material remain at the time when the disintegration rate is .105 grams per year.

(g) A half-life of 33 years means that half of any given amount will remain in 33 years. Of the original 8 grams, 4 will remain in 33 years; of that amount, 2 grams will remain after another 33 years, and 1 gram will remain after yet another 33 years (a total of 99 years).

19. This problem is the problem type #4 on page 109. The first step is to find the decay constant, using the fact that 5 grams decay to 2 grams in 100 days. You may assume that $P(0) = 5$, and $P(t) = 5e^{-\lambda t}$. Set P(100) = 2 and solve for λ:

$$\begin{aligned} 5e^{-\lambda(100)} &= 2 \\ e^{-100\lambda} &= \frac{2}{5} = .4 \\ -100\lambda &= \ln(.4) \\ \lambda &= \frac{\ln(.4)}{-100} \\ &\approx .00916291 \quad \text{or} \quad .00916. \end{aligned}$$

Once λ is determined, write the explicit formula for $P(t)$, that is, $P(t) = 5e^{-.00916t}$. The second step is to find the value of t at which $P(t)$ is 1 gram. Set $P(t) = 1$ and solve for t:

$$\begin{aligned} 5e^{-.00916t} &= 1 \\ e^{-.00916t} &= .2 \\ -.00916t &= \ln(.2) \\ t &= \frac{\ln(.2)}{-.00916} \\ &\approx 175.70. \end{aligned}$$

The material will decay to 1 gram in about 176 days.

Helpful Hint: It is wise to keep at least two or three significant figures in a decay constant or growth constant. Otherwise, the answers to other parts of the problem can vary somewhat. In the solution to Exercise 19, a value of .009 for λ leads to the answer that 1 gram will remain in 179 days.

25. Let P_0 be the original level of ^{14}C in the charcoal. Since the decay constant for ^{14}C is $\lambda = .00012$, the amount of ^{14}C remaining after t years is $P(t) = P_0 e^{-.00012t}$. The discovery in 1947 found that the amount of ^{14}C found in the charcoal is $.20P_0$, assuming that the original ^{14}C level in the charcoal was the same as the level in living organisms today. That is, we assume that the time t satisfies $P(t) = .20P_0$, which says that

$$P_0 e^{-.00012t} = .2P_0.$$

Solving for t,

$$\begin{aligned} e^{-.00012t} &= .2 \\ -.00012t &= \ln(.2) \\ t &= \frac{(\ln .2)}{(-.00012)} \\ &\approx 13{,}412 \text{ years.} \end{aligned}$$

That was the estimated age of the cave paintings more than 50 years ago. The estimated age now is about 13,500 years.

31. Read through the answers and note that in some cases such as (c) and (d), a function is evaluated at a specified value of t, that is, the time t is known but the value of the function must be computed. In other cases such as (a) and (f), the time t is unknown and an equation is solved to find t.

Next, note that some answers such as (b) and (c) involve the function $P(t)$, the amount of material present, while other answers such as (d) and (f) involve the derivative $P'(t)$, the rate at which the amount of material is changing. Finally, answers (g) and (h) relate to the specific type of function and differential equation that are associated with the model for radioactive decay.

Now, before you read further in this solution, go back to the text, read each question, and try to find the appropriate answer. After you have done this, read the solutions that follow.

A. The only differential equation is in (g).

B. The question "How fast?" indicates an answer involving the derivative (rate of change). The time is known, so the answer is (d): compute $P'(.5)$.

C. The general form of the function $P(t)$ is (h).

D. Half-life is the time at which $P(t)$ is half of the original amount, $P(0)$. The answer is (a).

E. "How many grams?" relates to $P(t)$, not $P'(t)$. The time is known, so the answer is (c): compute $P(.5)$.

F. The phrase "disintegrating at the rate" relates to $P'(t)$, and the question "When?" asks for a value of t. The answer is (f).

G. The question "When?" implies that the answer is either (a), (b) or (f). Rate of change is not involved, so (f) is ruled out. To choose between (a) and (b), you need to realize that the .5 in (b) is the value of $P(t)$ while the .5 in (a) only compares the value of $P(t)$ to $P(0)$. So the answer is (b).

H. "How much" relates to $P(t)$, and "present initially" actually gives the time, namely $t = 0$, although the word "year" is not mentioned. The answer is (e).

Helpful Hint: Exercise 31 provides a good review of basic concepts of exponential decay. Similar questions for exponential growth are in Exercise 25 of Section 5.2 and Exercise 13 of the Supplementary Exercises.

Helpful Hint: Use estimation to check whether your answer to a growth or decay problem is reasonable.

(a) Suppose a population doubles every 24 years and you compute that it will increase tenfold in 100 years. Is this reasonable? Round the doubling time to 25 years and notice that in 100 years, the initial population P_0 will double four times, from P_0 to $2P_0$, then to $4P_0$, $8P_0$ and finally $16P_0$, which is much more than just ten times P_0. Therefore, the answer is not reasonable.

(b) Suppose that a radioactive material has a half-life of 5 years and you compute that 1/10 of the material will remain after 22 years. Is this reasonable? Consider the following table which was constructed solely from the fact that the material will halve every five years.

Number of years	5	10	15	20
Fraction of original remaining	$\frac{1}{2}$	$\frac{1}{4}$	$\frac{1}{8}$	$\frac{1}{16}$

Since 1/10 is between 1/8 and 1/16, the time lies between 15 and 20 years. Therefore, the answer of 22 is not reasonable.

5.2 Compound Interest

The exercises in this section are similar to those in Section 5.1. A problem involving continuously compounded interest can be thought of as a problem about a population of money that is growing exponentially. The exercises in this section are of the same five types discussed in the notes for Section 5.1.

1. **(a)** The initial amount deposited in $A(0) = \$5000$.

(b) The interest rate is .04 or 4% interest per year.

(c) After 10 years, the amount in the account is

$$A(10) = 5000e^{.04(10)} \approx \$7459.12.$$

(d) $A(t)$ satisfies the differential equation

$$A'(t) = .04A(t)$$

which is also written as $y' = .04y$.

(e) From the differential equation and **(c)**,

$$\begin{aligned} A'(10) &= .04 \cdot A(10) = .04 \cdot (7459.12) \\ &\approx 298.36 \end{aligned}$$

After 10 years, the savings amount balance is growing at the rate of about \$298.36 per year.

(f) Let t_0 be the time at which $A'(t_0) = \$280$ per year. From the differential equation, $A(t_0)$ must satisfy

$$280 = .04 \cdot A(t_0)$$

so that $A(t_0) = 280/.04 = \$7000$. The account balance will be about \$7000 when it is growing at the rate of \$280 per year.

Helpful Hint: The phrase "after 10 years" in Exercise 1 means "after (exactly) 10 years of time have passed" This is a fairly common way of specifying $t = 10$. (The number 10, of course, could be replaced by any positive number.)

7. The initial amount and the interest rate are given. So $A(t) = 1000e^{.06t}$. To find when $A(t) = 2500$, set the formula for $A(t)$ equal to 2500, and solve for t:

$$1000e^{.06t} = 2500$$
$$e^{.06t} = 2.5$$

Apply the natural logarithm to both sides:

$$\ln e^{.06t} = \ln 2.5$$
$$.06t = \ln 2.5$$
$$t = \frac{\ln 2.5}{.06} \approx 15.27 \quad \text{(to two decimal places).}$$

About fifteen and one-quarter years are required for the account balance to reach $2500.

13. Since $A(t) = Pe^{rt}$, the fact that an initial investment P triples to $3P$ in 15 years means that the interest rate r satisfies

$$Pe^{r(15)} = 3P.$$

Divide by P and take the natural logarithm of each side:

$$e^{15r} = 3$$
$$15r = \ln 3$$
$$r = \frac{(\ln 3)}{15} \approx .07324.$$

Interest rates are usually reported to two or three significant figures. In this case, estimate the rate either as 7.3% or 7.32%.

Helpful Hint: Exercises 15–18 are two-step problems, similar to Exercise 19 in Section 5.1.

Step 1: Find the interest rate r. Keep at least two or preferably three significant figures in r. Otherwise, the answer to step 2 will have less accuracy.

Step 2: Write the formula for $A(t)$ with the value of r filled in, and use $A(t)$ to find the time required for the investment to grow to a certain amount.

19. The text supplies a formula for present value, but you may find it easier just to remember the equation $A = Pe^{rt}$ in the form:

$$[\text{future } \underline{\text{A}}\text{mount}] = [\underline{\text{P}}\text{resent value}] \cdot e^{rt}.$$

If P, r, and t are given, you can compute A. If A, r, and t are given, you can compute P (the "present value"). In Exercise 19 you are told the value of the investment at a future time (3 years), so that value is A. Thus $A = \$1000$, $t = 3$ years, r is .08, and

$$1000 = Pe^{.08(3)}$$
$$P = \frac{1000}{e^{.24}} \approx \$786.63 \quad \text{(to the nearest cent).}$$

25. This exercise is analogous to Exercise 31 in Section 5.1, with the function $A(t) = Pe^{rt}$ in place of $P(t)$. You might review that exercise before continuing here.

For most of the questions in this exercise, you need to decide (1) whether the question involves $A(t)$ or the rate of change of $A(t)$ and (2) whether the time t is known or unknown.

A. "How fast?" relates to $A'(t)$, the rate of growth of the balance $A(t)$. Since the time is known, the answer is (d): compute $A'(3)$.

B. Answer: (a), the general form of $A(t)$ is Pe^{rt}.

C. The question involves the amount $A(t)$ in the account and its relation to the initial amount $A(0)$. The question, "How long?" indicates you must solve to find t. Answer: (h), solve $A(t) = 3A(0)$ for t.

D. "The balance" refers to $A(t)$. The phrase "after 3 years" means that $t = 3$. Answer: (b), compute $A(3)$.

E. The question concerns $A(t)$. "When?" means that t must be found. Answer: (f), solve $A(t) = 3$ for t.

F. The phrase, "the balance . . . growing . . . rate" relates to $A'(t)$. "When?" means that t must be found. Answer: (e), solve $A'(t) = 3$ for t.

G. Answer: (c), principal amount means initial amount.

H. The differential equation is in (g): $y' = ry$.

Helpful Hint: Use estimation to check that the answer to a compound interest problem is reasonable.

(a) Suppose \$1000 is invested at 5% interest and you compute it will grow to about \$1103 in two years. Is this reasonable? Well, the deposit will earn .05(\$1000) or \$50 interest the first year and a little more than that (due to interest on the interest) during the second year. Therefore, it will earn a little more than \$100 in interest in two years, and so the result is reasonable.

(b) Suppose an investment doubles every eight years and you compute that it will increase tenfold in 20 years. Is this reasonable? Well, the investment will increase fourfold in 16 years and eightfold in 24 years. Therefore, the answer is not reasonable.

(c) Bankers have a Rule of 70 that can be used to estimate the doubling time of an investment (because 70 is approximately 100 times ln 2). The Rule of 70 says that an investment earning an interest rate of r% will double in about 70/r years. For instance, an investment earning 7 percent interest will double in about 70/7 or 10 years. Similarly, an investment doubling in d years has earned an interest rate of about 70/d percent per year. (In former times, bankers used a rule of 72, because they relied on mental arithmetic, and 72 is easily divisible by many common interest rates.)

31. Set $Y_1 = 1200 * e\hat{}(-3X) + 800 * e\hat{}(-4X) + 500 * e\hat{}(-5X)$. Then use the SOLVER command (on the TI-83/84, MATH 0) to solve the equation $Y_1 - 2000 = 0$. A good initial guess is $X = .10$ (a 10% interest rate). The answer to six decimal places is $X = .060276$, so the interest rate is about 6.0%.

5.3 Applications of the Natural Logarithm Function to Economics

The material in this section is not used in any other part of the book. Elasticity is one of the most important concepts of economics and provides insights into the pricing of goods and services. In addition to being able to compute the elasticity of demand, you should be able to interpret what it means for the demand to be, say, elastic. The box on Page 281 summarizes this. Here is another more informal way to remember it. When demand is elastic, an increase in price causes such a decline in sales that the total revenue falls. When demand is inelastic, an increase in price causes only a relatively small decline in sales, so that the total revenue still increases.

1. $f(t)=t^2$, so $f'(t)=2t$.

Thus at $t=10$, $\dfrac{f'(10)}{f(10)}=\dfrac{2(10)}{10^2}=\dfrac{20}{100}=.2=20\%$.

At $t=50$, $\dfrac{f'(50)}{f(50)}=\dfrac{2(50)}{50^2}=\dfrac{100}{2500}=.04=4\%$.

7. $f(p)=\dfrac{1}{p+2}$, thus $f'(p)=\dfrac{-1}{(p+2)^2}$.

So at $p=2$, $\dfrac{f'(2)}{f(2)}=\dfrac{-1/16}{1/4}=-\dfrac{4}{16}=-.25=-25\%$.

And at $p=8$, $\dfrac{f'(8)}{f(8)}=\dfrac{-1/100}{1/10}=-\dfrac{10}{100}=-.1=-10\%$.

13. $f(p)=q=700-5p$, so $f'(p)=-5$.

$$E(p)=\frac{-pf'(p)}{f(p)}=\frac{-p(-5)}{700-5p}=\frac{5p}{700-5p}=\frac{p}{140-p}.$$

So at $p=80$, $E(80)=\dfrac{80}{140-80}=\dfrac{80}{60}=\dfrac{4}{3}$.

Since $\frac{4}{3}>1$, demand is elastic.

19. **(a)** $f(p)=q=600(5-\sqrt{p})=3000-600p^{1/2}$, and $f'(p)=-300p^{-1/2}$.

$$E(p)=\frac{-pf'(p)}{f(p)}=\frac{-p(-300p^{-1/2})}{3000-600p^{1/2}}=\frac{300p^{1/2}}{3000-600p^{1/2}}=\frac{p^{1/2}}{10-2p^{1/2}}.$$

Thus at $p=4$, $E(4)=\dfrac{4^{1/2}}{10-2(4)^{1/2}}=\dfrac{2}{10-2(2)}=\dfrac{1}{3}$.

Since $\frac{1}{3}<1$, demand is inelastic.

(b) Since demand is inelastic, an increase in price will bring about an increase in revenue. Thus, the price of a ticket should be raised.

25. $E_c(x)=\dfrac{\frac{d}{dx}\ln C(x)}{\frac{d}{dx}\ln x}=\dfrac{\frac{C'(x)}{C(x)}}{\frac{1}{x}}=\dfrac{xC'(x)}{C(x)}$.

29. Use X for the variable p, and set $Y_1 = 60000 * e^\wedge(-.5 * X)$. If Y_4 is the first derivative of Y_1, then use Y_2 for the elasticity of demand function and set $Y_2 = -X*Y_4/Y_1$. To solve $Y_2 = 1$, apply the SOLVER program to the equation $0 = Y_2 - 1$.

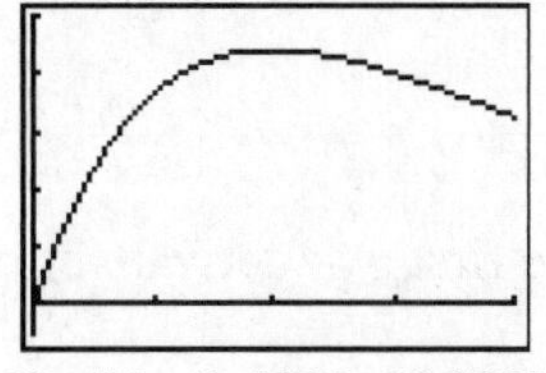

[0, 4] by [−5000, 50,000]

The graph is increasing for $x < 2$, so revenue is increasing for $p < 2$.

5.4 Further Exponential Models

This section illustrates the wide variety of applications in which exponential function appear. The material in this section is not needed for any other part of the book. The logistic curve will be studied further in Chapter 10. However, the discussion there is independent of the discussion in Section 5.4.

1. **(a)** $f(x) = 5(1-e^{-2x}) = 5 - 5e^{-2x}$, $x \geq 0$. Thus $f'(x) = 10e^{-2x}$. Since $10e^{-2x} > 0$ for every value of x, the derivative is positive for all x, in particular when $x \geq 0$. Thus $f(x)$ is increasing. $f''(x) = -20e^{-2x}$. Since $-20e^{-2x} < 0$, the second derivative is negative for all x. Thus $f(x)$ is concave down.

(b) $f(x) = 5(1-e^{-2x})$, $x \geq 0$.

Note that as x gets larger, $e^{-2x} = \frac{1}{e^{2x}}$ gets closer and closer to zero. Hence when x is very large the values of $f(x)$ are very close to 5. (The values of $f(x)$ are slightly less than 5 because $1-e^{-2x}$ is slightly less than 1.)

(c) y

$y = 5$

$y = 5(1 - e^{-2x})$

x

7. The number of people who have heard about the indictment by time t is given by $f(t) = P(1-e^{-kt})$, where P is the total population. Since after one hour, one quarter of the citizens had heard the news, $f(1) = P(1-e^{-k(1)}) = \frac{1}{4}P = .25P$. Hence

$$1 - e^{-k} = .25,$$
$$e^{-k} = .75,$$
$$-k = \ln .75 \approx -.29, \quad \text{or} \quad k = .29.$$

Before continuing with the solution, write out the formula for $f(t)$ for future reference.

$$f(t) = P(1-e^{-.29t}).$$

To find the time t when $f(t) = \frac{3}{4}P = .75P$, set

$$P(1-e^{-.29t}) = .75P.$$

Thus,

$$1 - e^{-.29t} = .75,$$
$$e^{-.29t} = .25,$$
$$-.29t = \ln .25 \approx -1.4,$$
$$t = \frac{-1.4}{-.29} = \frac{1.4}{.29} \approx 4.8 \text{ hours.}$$

13. Set $Y_1 = 122(e \wedge (-.2X) - e \wedge (-X))$. We assume that Y_4 and Y_5 are the first two derivatives of Y_1.

(a)

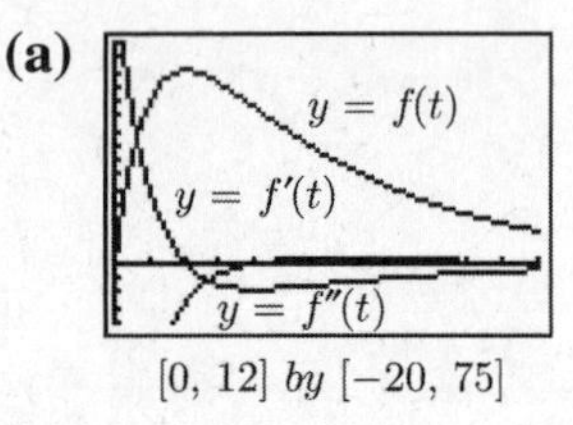

[0, 12] *by* [−20, 75]

(b) Evaluate $Y_1(7) \approx 29.97$. About 30 units are present.

(c) Evaluate $Y_4(1) \approx 24.90$. The drug level is increasing at the rate of about 25 units per hour.

(d) Solve $Y_1 - 20 = 0$. If you use the SOLVER command, you'll need to look first at the graph of Y_1 and use TRACE to find an approximate initial guess for X, on the part of the graph where the level is decreasing. Another method is to graph Y_1 and $Y_2 = 20$ in the same window and use the intersect command. You can choose an initial guess for X while in the graphing window. In any case, you should obtain $X \approx 9.038$. So the level of the drug is 20 units at about $t = 9$ hours.

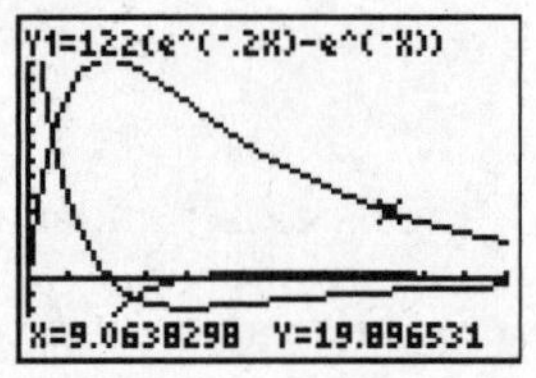

(e) You can use TRACE to estimate the maximum point on the graph of Y_1. On the TI-83, with the suggested window, you should find (2.04, 65.26). A more accurate method is to use SOLVER on the equation $Y_4 = 0$. You should obtain $X \approx 2.0118$. Evaluating $Y_1(X)$, with the exact value of X produced by SOLVER, yields $Y_1 \approx 65.269$. In either case, you might report that after about 2 hours the drug reaches its maximum level of about 65.3 units.

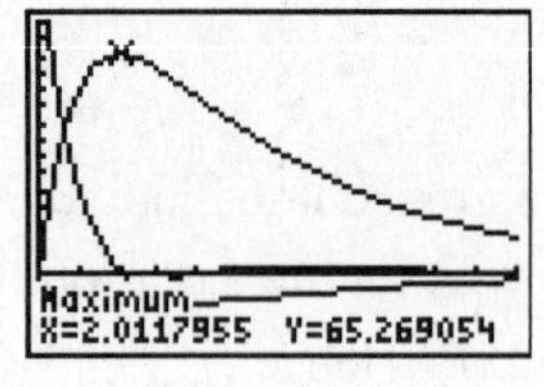

(f) The point of steepest decline on the graph of Y_1 is the inflection point. That's practically impossible to estimate on the graph of Y_1 using TRACE. The two reasonable choices are either to use TRACE on Y_4 and look for the minimum point, or to use SOLVER to find where the second derivative Y_5 is zero. SOLVER will work on Y_5, but only if your initial guess for X is small enough. (If you try X = 12 on the TI-83, the calculator will work for about 2 minutes and give the answer as $X = 9.99\ldots E98$, or about 10^{99}, which is the calculator's view of heaven.)

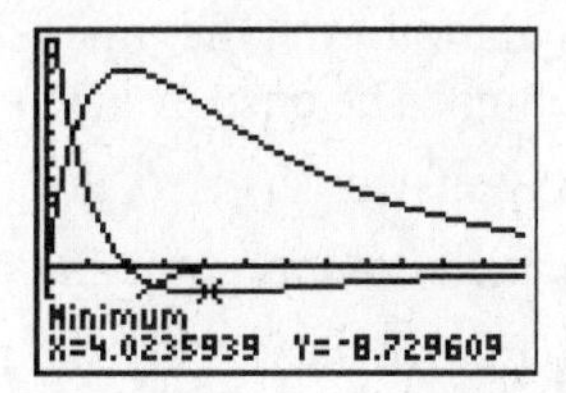

For this graph, the best method is to use TRACE on the graph of Y_4, and find that its minimum point occurs when $X \approx 4.08$. So the drug level is decreasing the fastest when $t = 4$ hours.

Chapter 5: Supplementary Exercises

1. $P'(x) = -.2P(x)$ implies that $P(x)$ is an exponential function of the form $P(x) = P_0 e^{-.2x}$, where $P_0 = P(0)$ is the atmospheric pressure at sea level. Thus $P_0 = 29.92$, and $P(x) = 29.92e^{-.2x}$.

7. **(a)** $P(t) = 17e^{kt}$, where $t = 0$ corresponds to January 1, 1990. To find the growth constant, set $P(7) = 19.3$ and solve for k:

$$17e^{k(7)} = 19.3$$

$$e^{7k} = \frac{19.3}{17}$$

$$7k = \ln\left(\frac{19.3}{17}\right)$$

$$k = \frac{1}{7}\ln\left(\frac{19.3}{17}\right) \approx .0181274 \quad \text{or} \quad .0181.$$

Thus $P(t) = 17e^{.0181t}$.

(b) The year 2000 corresponds to $t = 10$. At that time,

$$P(10) = 17e^{.0181(10)}$$

$$\approx 20.4 \text{ million.}$$

(c) The population reaches 25 million when t satisfies

$$17e^{.0181t} = 25$$

$$e^{.0181t} = \frac{25}{17}$$

$$.0181t = \ln\left(\frac{25}{17}\right)$$

$$t = \frac{1}{.0181}\ln\left(\frac{25}{17}\right) \approx 21.3.$$

Since $t = 0$ corresponds to 1990, the population will reach 25 million 21 years later, in 2011.

13. **(A)** "How fast" concerns $P'(t)$. The time is known, so the answer is (c).

(B) The general form of the function is in (g), P_0e^{kt}.

(C) The question involves the population $P(t)$ and the time is unknown, so the answer is either (a) or (f). The actual size of the population is not given. Rather, it is described as twice the current population. So the answer is (f), which involves $2P(0)$.

(D) The question involves the population $P(t)$, the time is known, and the answer to "what size" requires a value of $P(t)$, not a time. The answer is (b).

(E) The initial size is $P(0)$, which is answer (h).

(F) The exact value of $P(t)$ is specified, and the time is unknown, so the answer is (a).

(G) The question involves the rate of growth, $P'(t)$, and the time is unknown, so the answer is (d).

(H) The differential equation is in (e), $y' = ky$.

19. This question is based on the elasticity of demand, discussed at the end of Section 5.3. If the elasticity at $p = 8$ is 1.5, then the demand is elastic, because $E(p)$ is greater than 1. In this case, the change in revenue is in the opposite direction of the change in price, for prices close to $p = 8$. So if the price is increased to \$8.16, the revenue will decrease. Furthermore, from the discussion before Example 4 in Section 5.3,

$$\frac{[\text{relative rate of change of quantity}]}{[\text{relative rate of change of price}]} = -1.5.$$

The relative increase of price from \$8.00 to \$8.16 is .16/8.00 = .02. Therefore the relative change (decrease) of quantity demanded is $(.02)(-1.5) = -.03$. So the demand will fall by about 3%.

25. **(a)** The temperature "after 11 seconds" means the temperature at $t = 11$. This temperature is the y-coordinate of the point on the graph of $f(t)$ at which t is 11, namely, 400°F. (Each horizontal square represents 1 second, and each vertical square represents 100°.)

(b) The "rate of temperature" involves the graph of $f'(t)$. "After 6 seconds" means that $t = 6$. The corresponding y-coordinate is at about −100°per second.

(c) The temperature of the rod is 200 degrees at the time corresponding to a y-coordinate of 200 on the graph of $f(t)$, namely, at $t = 17$ seconds.

(d) The rod is cooling at the rate of 200°/sec when the y-coordinate of $f'(t)$ is −200, namely, when $t = 2$ sec.

Chapter 6
The Definite Integral

6.1 Antidifferentiation

To find an antiderivative, first make an educated guess and then differentiate the guess. (This may be done mentally if the differentiation is simple.) Your guess should be based on your experience with derivatives. In most cases, functions and their antiderivatives are the same basic type, such as power functions or exponential functions. If your first guess for the antiderivative is correct, you are finished (after you add " + *C*"). With some practice, your guess usually will be correct except for a constant factor, and you will only need to multiply the guess by a suitable constant. Exercises 25 to 36 show the correct form of the antiderivative. All you have to do is to adjust the constant *k* correctly.

1. By the constant multiple rule,

$$\frac{d}{dx}\frac{1}{2}x^2 = \frac{1}{2}\cdot\frac{d}{dx}x^2 = \frac{1}{2}\cdot 2x = x.$$

Hence one antiderivative of $f(x) = x$ is $F(x) = \frac{1}{2}x^2$, and all antiderivatives are of the form $F(x) = \frac{1}{2}x^2 + C$.

7. Since $\frac{d}{dx}x^4 = 4x^3$,

$$\int 4x^3 dx = x^4 + C.$$

13. Rewriting the integral, we have

$$\int\left(\frac{2}{x}+\frac{x}{2}\right)dx = \int\left(2x^{-1}+\frac{1}{2}x\right)dx.$$

Since $\frac{d}{dx}2\ln|x| = \frac{2}{x}$ and $\frac{d}{dx}\left(\frac{1}{4}x^2\right) = \frac{1}{2}x$,

$$\int\left(2x^{-1}+\frac{1}{2}x\right)dx = 2\ln|x| + \frac{1}{4}x^2 + C.$$

19. Since $\frac{d}{dx}\left(-\frac{3}{2}e^{-2x}\right) = 3e^{-2x}$,

$$\int 3e^{-2x}dx = -\frac{3}{2}e^{-2x} + C.$$

25. The equation $\int 5e^{-2t}dt = ke^{-2t} + C$ reminds you that e^{-2t} is the correct type of function to use as an antiderivative—you only have to adjust the constant multiple in front of it.

After some practice, you will be able to guess immediately what k should be. At first, however, just compute

$$\frac{d}{dt}ke^{-2t} = -2ke^{-2t}.$$

You want k to make $-2k$ equal to 5 (because there is a factor 5 inside the indefinite integral). To make $-2k = 5$, choose $k = 5/(-2)$, or $k = -5/2$. Check:

$$\begin{aligned}\frac{d}{dt}\left(\frac{5}{-2}\right)e^{-2t} &= \left(\frac{5}{-2}\right)\cdot\frac{d}{dt}e^{-2t}\\ &= \left(\frac{5}{-2}\right)\cdot(-2)e^{-2t}\\ &= 5e^{-2t}.\end{aligned}$$

Thus $k = -5/2$ is correct, and $(-5/2)e^{-2t}$ is an antiderivative of $5e^{-2t}$. The -2 in the denominator of $\frac{5}{-2}$ anticipates the -2 that will appear when e^{-2t} is differentiated. Once you have the antiderivative $(-5/2)e^{-2t}$, don't forget to add "+ C" to generate all the other antiderivatives.

31. The formula $\int(4-x)^{-1}dx = k\ln|4-x| + C$ reminds you of the type of antiderivative you need. If you guess that $k = 1$, then you have forgotten about the -1 that appears when $\ln|4-x|$ is differentiated:

$$\frac{d}{dx}\ln|4-x| = (4-x)^{-1}\cdot(-1).$$

The -1 is the derivative of the function $4-x$ inside the logarithm. To compensate for the -1, take $k = -1$. A check will show why that works:

$$\begin{aligned}\frac{d}{dx}(-1)\cdot\ln|4-x| &= (-1)\frac{d}{dx}\ln|4-x|\\ &= (-1)\cdot(4-x)^{-1}(-1)\\ &= (4-x)^{-1}.\end{aligned}$$

Thus $-\ln|4-x|$ is an antiderivative of $(4-x)^{-1}$, and the others are found by adding an arbitrary constant.

Helpful Hint: The antiderivatives in Exercises 27–29 and 31–34 are a little harder to find because you must remember how the chain rule works. (Later we will study a general method for handling such problems. For now, however, you may use the technique of simply guessing the general form of an antiderivative.) For instance, to handle Exercise 33, compute

$$\begin{aligned}\frac{d}{dx}k(3x+2)^5 &= 5k(3x+2)^4\cdot\frac{d}{dx}(3x+2)\\ &= 5k(3x+2)^4\cdot 3\\ &= 15k(3x+2)^4.\end{aligned}$$

You want $15k = 1$, so take $k = 1/15$. Perhaps you can do the calculation above mentally. In any case, once you "see" the 5 and the 3 appearing after the differentiation is performed, you can guess how to choose k so that

$$\frac{d}{dx}k(3x+2)^5 = (3x+2)^4.$$

Remember that you can check your final answer by differentiating what you think is the correct antiderivative.

37. Since $\frac{d}{dt}\left(\frac{2}{5}t^{5/2}\right) = t^{3/2}$, $f(t) = \frac{2}{5}t^{5/2} + C$.

Helpful Hint: In Exercises 43–46 and 57–63, each answer is a single function. This function is found by first antidifferentiating and then using additional information to find the proper value of C. On an exam or homework assignment, this value of C should be substituted for C in the antiderivative to obtain the answer.

43. $f(x)$ is an antiderivative of x. Since $\frac{d}{dx}\left(\frac{1}{2}x^2\right) = x$, $\frac{1}{2}x^2$ is an antiderivative of x and $f(x) = \frac{1}{2}x^2 + C$ for some C. To make $f(0) = 3$, we need

$$f(0) = \frac{1}{2}0^2 + C = 3.$$

Therefore, $C = 3$ and $f(x) = \frac{1}{2}x^2 + 3$.

49. **(a)** $\frac{d}{dx}\left(\frac{1}{x} + C\right) = \frac{d}{dx}(x^{-1} + C) = -x^{-2} \neq \ln x$, so (a) is not the answer.

(b) $\frac{d}{dx}(x\ln x - x + C) = x\left(\frac{1}{x}\right) + \ln x(1) - 1 = \ln x$, (b) is the answer.

(c) $\frac{d}{dx}\left(\frac{1}{2}(\ln x)^2\right) = \frac{1}{2}(2)(\ln x)\left(\frac{1}{x}\right) = \frac{\ln x}{x} \neq \ln x$, so the answer is not (c).

55. **(a)** Since velocity is the derivative of position, or in this case, height, to find $s(t)$, the height of the ball at time t, we find the antiderivative of the velocity:

$$s(t) = \int (96 - 32t)dt = 96t - 16t^2 + C.$$

We know the initial height of the ball was 256 feet, so

$$s(0) = 96(0) - 16(0)^2 + C = 256 \Rightarrow C = 256$$

and

$$s(t) = -16t^2 + 96t + 256.$$

(b) To find when the ball hits the ground, we solve $s(t)=0$:

$$-16t^2+96t+256=0$$
$$t^2-6t-16=0$$
$$(t+2)(t-8)=0.$$

Therefore, $t=-2$ or $t=8$. Since $t=8$ is the only sensible solution, we conclude the ball hits the ground after 8 seconds.

(c) To find how high the ball goes, we find when $s'(t)=0$:

$$s'(t)=96-32t=0$$
$$t=3.$$

Since $s''(t)=-32<0$, $t=3$ is a relative maximum and the ball reaches a height of

$$s(3)=-16(3)^2+96(3)+256$$
$$=400 \text{ feet.}$$

61. The marginal profit, *MP*, is the derivative of the profit function. Therefore, to find the profit function, we must antidifferentiate the marginal profit function:

$$P(x)=\int MPdx$$
$$=\int(1.30+.06x-.0018x^2)\,dx$$
$$=1.30x+.03x^2-.0006x^3+C.$$

At a sales level of $x=0$ per day, the shop will loose \$95. Thus $P(0)=C=-95$ and

$$P(x)=1.30x+.03x^2-.0006x^3-95.$$

67. We have

$$\int\left(e^{2x}+e^{-x}+\frac{1}{2}x^2\right)dx=\frac{1}{2}e^{2x}-e^{-x}+\frac{1}{6}x^3+C.$$

Let $C=0$, then $f(x)=\frac{1}{2}e^{2x}-e^{-x}+\frac{1}{6}x^3$.

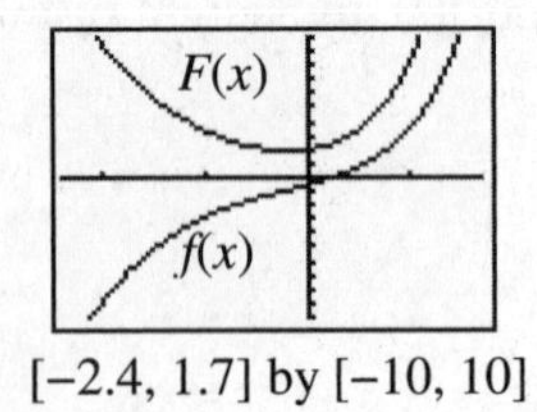

[−2.4, 1.7] by [−10, 10]

6.2 Areas and Riemann Sums

Riemann (pronounced "Reemahn") sums are introduced here in order to approximate the area under the graph of a nonnegative function, but their importance extends far beyond that. As you will see later, Riemann sums are used to derive some important formulas in applications. Computations with Riemann sums can be time-consuming, but some numerical homework is necessary for a proper understanding. Check with your instructor about how far to carry calculations on an exam. In some cases you may only have to write down the complete Riemann sum (using numbers and not symbols such as Δx), and omit the final arithmetic computation.

One goal of this section is to suggest how the change in a function over some interval is related to the area under the graph of the derivative of that function. See Example 4 and the table on Page 312. This concept will be made precise in Section 6.3.

1. $a = 0, b = 2, n = 4,$ so $\Delta x = \dfrac{b-a}{n} = \dfrac{2-0}{4} = .5.$

The first midpoint is a $a + \dfrac{\Delta x}{2} = 0 + \dfrac{.5}{2} = .25.$ Subsequent midpoints are found by adding Δx repeatedly. So the other midpoints are:

$$.25 + .5 = .75$$
$$.75 + .5 = 1.25$$
$$1.25 + .5 = 1.75.$$

Helpful Hint: You may use fractions, if you wish, and write 1/4 in place of .25, and so on. However, you can use a calculator more efficiently if you convert to decimals.

Helpful Hint: To draw a partition that has 5 subintervals, draw a horizontal line and, beginning at one end, mark off five equal subintervals. This is easier than trying to divide one larger interval into five equal pieces. (See Exercises 3, 4, and 7–10.)

7. $\Delta x = \dfrac{b-a}{n} = \dfrac{3-1}{5} = \dfrac{2}{5} = .4$ The first left endpoint is $x_1 = a = 1.$ Subsequent left endpoints are found by adding Δx repeatedly.

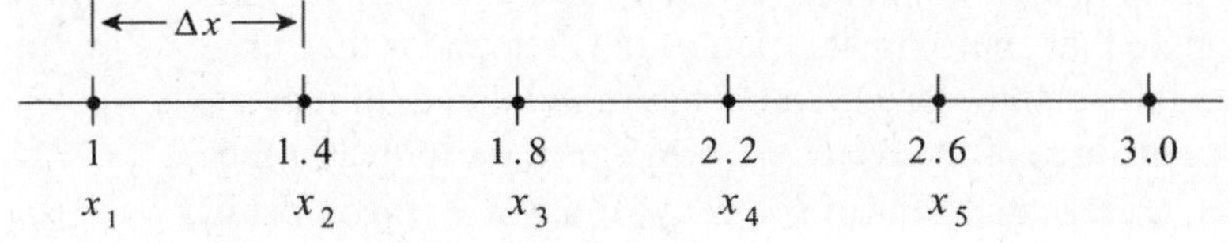

Thus the area is approximated by the Riemann sum:

$$\Delta x[f(1) + f(1.4) + f(1.8) + f(2.2) + f(2.6)]$$
$$= (.4)[(1)^3 + (1.4)^3 + (1.8)^3 + (2.2)^3 + 2.6)^3]$$
$$= (.4)[1 + 2.744 + 5.832 + 10.648 + 17.576]$$
$$= (.4)[37.8] = 15.12.$$

13. Since $\Delta x = \frac{b-a}{n} = \frac{9-4}{5} = 1$, the subintervals have length 1. The first right endpoint is at $4 + \Delta x = 5$. Focus your attention on the five right endpoints by placing small dots on the x-axis at 5, 6, 7, 8, and 9.

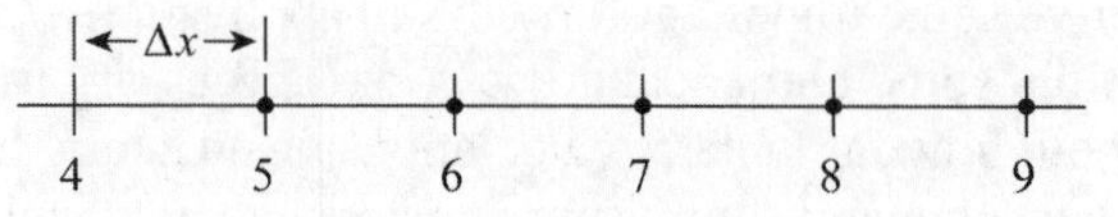

From the figure in the text, estimate the y-coordinates of the points on the curve where $x = 5, 6, \ldots, 9$, and compute

$$\begin{aligned} A &\approx [f(5) + \cdots + f(9)]\Delta x \\ &= (6+4+2+1+2]\cdot 1 = 15. \end{aligned}$$

19. For $0 \le x \le 2$ and $n = 5$, $\Delta x = \frac{b-a}{n} = \frac{2-0}{5} = \frac{2}{5} = .4$. The first midpoint is at $\frac{\Delta x}{2} = \frac{.4}{2} = .2$ The next midpoint is at $.2 + \Delta x = .2 + .4 = .6$, and so on.

In the figure for Exercise 19, the interval $0 \le x \le 1$ is divided into five segments, so each represents .2 second, which means that Δx is the length of two segments:

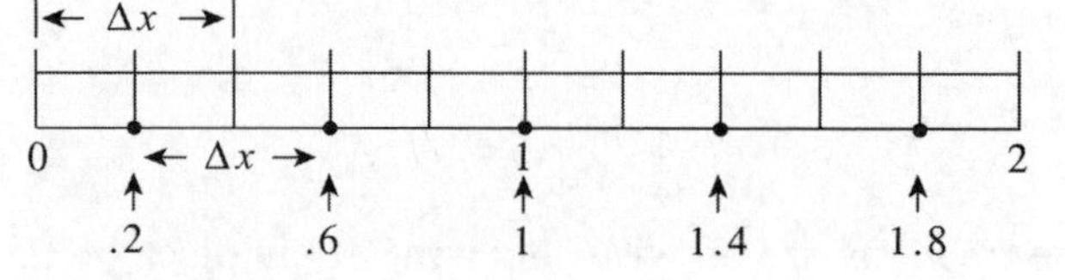

From the figure, read the heights of the curve above $.2, .6, \ldots, 1.8$. The Riemann sum is

$$\begin{aligned} [f(.2) + \cdots + f(1.8)]\Delta x &= [.2 + .7 + .8 + .7 + .3](.4) \\ &= (2.7)(.4) = 1.08. \end{aligned}$$

The heights of the rectangles represent flow rate, in liters/second; the widths of the rectangles represent time, in seconds; the product of the height and the width represents (liters/second) · (seconds) = liters. Thus the total volume of air during exhalation is approximately 1.08 liters.

25. To understand the answer, it might help to consider the units involved. Consider a rectangle that approximates part of the area under the curve. Its height represents rate of soil erosion, in tons/day, and its width represents time, in days. So, the area of the rectangle, which is the product of the height and width, represents [(tons of soil)/day) · [days] = [tons of soil].

Thus, the area under the rate curve equals the total number of tons of soil that has eroded during the five days from $t = 0$ to $t = 5$.

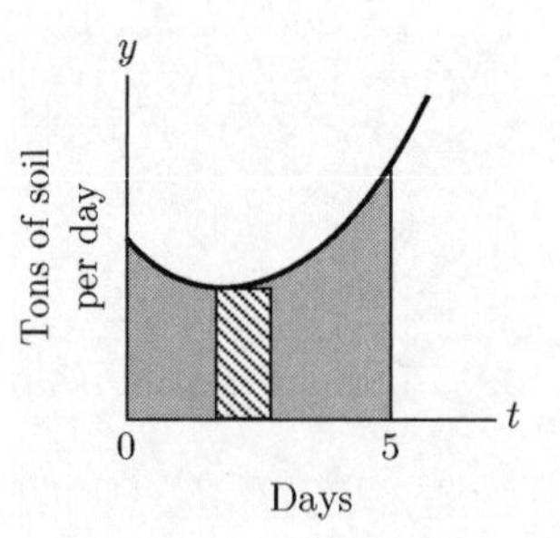

31. Here $\Delta x = [2 - (-2)]/100 = .04$. The left endpoint of the first subinterval is $x_1 = -2$. The left endpoint of the last subinterval is $x_{100} = 2 - \Delta x = 2 - .04 = 1.96$. Refer to the INCORPORATING TECHNOLOGY section on page 312 to see how to compute the Riemann sum in the form

$$[f(x_1) + f(x_2) + \cdots f(x_{100})](.04).$$

The **seq** command creates the sequence of terms

$$f(x_1), f(x_2), \ldots, f(x_{100}).$$

The **sum** command adds the terms and computes

$$f(x_1)+f(x_2)+\cdots+f(x_{100}).$$

After that, multiply the result by $\Delta x = .04$. You should find that

$$\text{sum(seq(e^(}-\text{X^2), X,}-2, 1.96, .04)\text{)*}.04 \approx 1.764143.$$

6.3 Definite Integrals and the Fundamental Theorem

Although definite integrals are defined by Riemann sums, they are usually calculated as the net change in an antiderivative. Since you have computed a few Riemann sums, you will probably appreciate the value of the Fundamental Theorem of Calculus.

1. The function is $f(x) = 1/x$. The interval is $.5 \le x \le 2$, which is determined by the x-coordinates of the points shown in the figure. The area is the value of

$$\int_{.5}^{2} \frac{1}{x}\,dx.$$

7. $\displaystyle \int_{-1}^{1} x\,dx = \frac{1}{2}x^2\Big|_{-1}^{1} = \frac{1}{2}(1)^2 - \left(\frac{1}{2}(-1)^2\right) = \frac{1}{2} - \frac{1}{2} = 0.$

13. $\displaystyle \int_0^1 4e^{-3x}dx = -\frac{4}{3}e^{-3x}\Big|_0^1 = \left(-\frac{4}{3}e^{-3}\right) - \left(-\frac{4}{3}\right) = \frac{4}{3} - \frac{4}{3}e^{-3}, \text{ or } \frac{4}{3}(1-e^{-3}).$

19. $\displaystyle \int_3^6 x^{-1}dx = \ln|x|\Big|_3^6 = \ln 6 - \ln 3 = \ln\frac{6}{3} = \ln 2.$

25. $\displaystyle \int_0^3 (x^3 + x - 7)dx = \left(\frac{1}{4}x^4 + \frac{1}{2}x^2 - 7x\right)\Big|_0^3 = \left(\frac{81}{4} + \frac{9}{2} - 21\right) - 0 = \frac{81}{4} + \frac{18}{4} - \frac{84}{4} = \frac{15}{4} = 3\frac{3}{4}.$

31. The desired area is

$$\int_0^1 e^{x/2}dx = 2e^{x/2}\Big|_0^1 = 2e^{1/2} - 2 = 2(e^{1/2} - 1).$$

37. Number of cigarettes sold from 1980 to 1998 is

$$\int_{20}^{38} (.1t + 2.4)dt = \left(\frac{.1t^2}{2} + 2.4t\right)\Bigg|_{20}^{38} \qquad \left(\textit{Note}: \ \frac{.1}{2} = .05\right)$$

$$= .05(38)^2 + 2.4(38) - [.05(20)^2 + 2.4(20)]$$

$$= 95.4 \text{ trillion.}$$

43. This represents the increase in profits generated by raising the production level from $x = 44$ to $x = 48$ units.

49. By Theorem III on Page 322, $A'(x) = f(x) = x^2 + 1$. Hence $A'(3) = f(3) = 3^2 + 1 = 10$.

51. The command fnInt is discussed in the INCORPORATING TECHNOLOGY section on page 323.

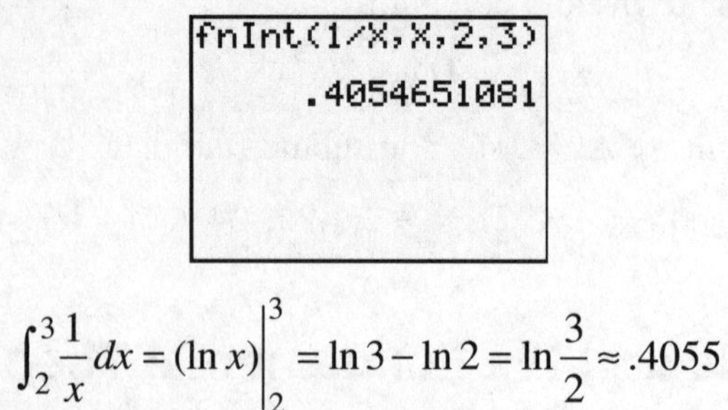

$$\int_2^3 \frac{1}{x}\,dx = (\ln x)\Big|_2^3 = \ln 3 - \ln 2 = \ln\frac{3}{2} \approx .4055$$

6.4 Areas in the *xy*-Plane

1.

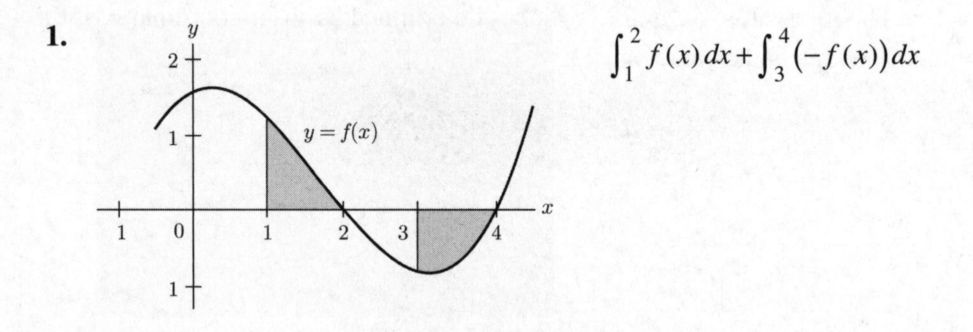

$$\int_1^2 f(x)\,dx + \int_3^4 (-f(x))\,dx$$

7. The graph of $y = x^2 - 6x + 12$ is a parabola that opens upward. Its minimum point occurs where $y' = 2x - 6 = 0$, that is, at $x = 3$. The minimum point is (3, 3). So the parabola lies entirely above the horizontal line $y = 1$.

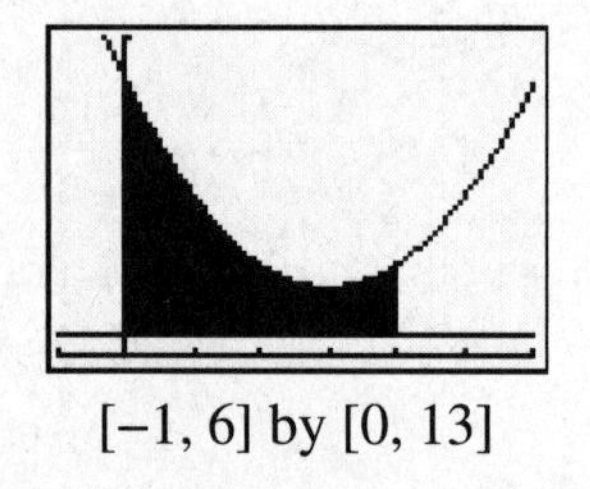

[−1, 6] by [0, 13]

So $f(x) = x^2 - 6x + 12$ and $g(x) = 1$. The area between the two curves from $x = 0$ to $x = 4$ is equal to,

$$\int_0^4 [(x^2 - 6x + 12) - (1)]dx = \int_0^4 (x^2 - 6x + 11)dx = \frac{x^3}{3} - 3x^2 + 11x\Big|_0^4$$

$$= \left(\frac{64}{3} - 48 + 44\right) - 0 = \frac{64}{3} = \frac{52}{3}.$$

13. Only a rough sketch is needed. The graph of $y = -x^2 + 6x - 5$ is a parabola that opens downward. (You can see this from the $-x^2$ term.) Also, the graph of $y = 2x - 5$ is a line with negative slope. Without plotting any points or using a coordinate system, you can visualize the two possible locations of the parabola and line:

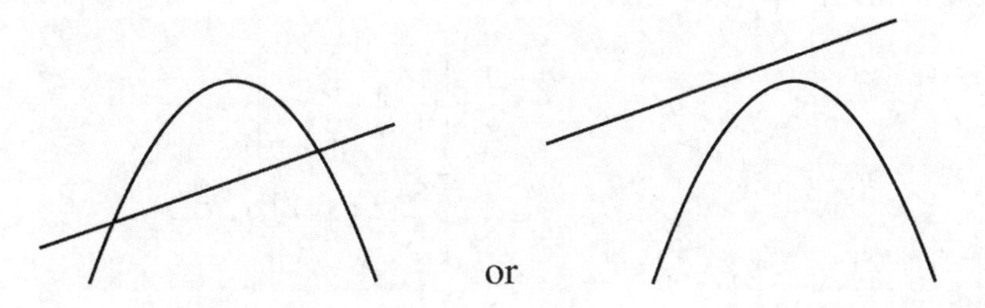

Hopefully, the line intersects the parabola. Otherwise there will be no region "bounded" by the graphs. To find the x-values of the intersection points, if any, equate the y-values of the two curves and solve for x:

$$-x^2 + 6x - 5 = 2x - 5$$
$$-x^2 + 4x = 0$$
$$-x(x-4) = 0$$
$$x = 0, \quad \text{or} \quad x = 4$$

The rough sketch shows that parabola is the top curve between $x = 0$ and $x = 4$. So the area bounded by the curves is

$$\int_0^4 [(-x^2 + 6x - 5) - (2x - 5)]dx = \int_0^4 (-x^2 + 4)dx = \left.\frac{-x^3}{3} + 2x^2\right|_0^4 = \frac{-64}{3} + 32 = \frac{32}{3}.$$

19. The graph of $y = x^2 - 3x$ is obviously a parabola that opens upward. It crosses the x-axis at values of x that satisfy $x^2 - 3x = 0$, i.e., $x(x-3) = 0$. Thus $x = 0$, or $x = 3$.

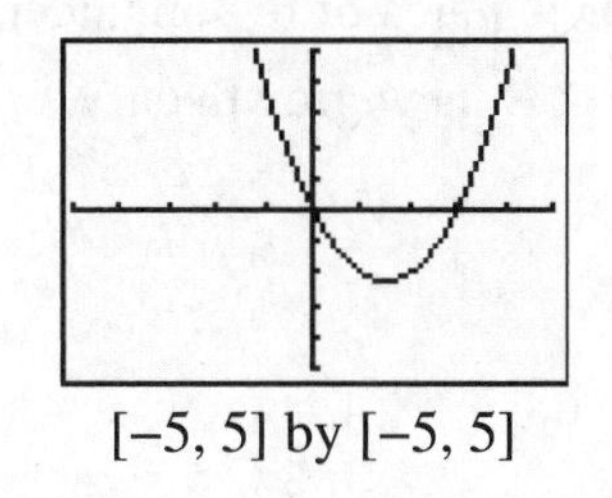

[−5, 5] by [−5, 5]

(a) The x-axis is the graph of the (constant) function $y = 0$. Since this graph is above the parabola from $x = 0$ to $x = 3$, the area of the region for Part (a) is

$$\int_0^3 [0 - (x^2 - 3x)]dx = \int_0^3 3x - x^2 dx$$
$$= \left.\frac{3}{2}x^2 - \frac{1}{3}x^3\right|_0^3 = \left[\frac{27}{2} - 9\right] - 0 = \frac{9}{2}.$$

(b) When x varies from $x = 0$ to $x = 4$, the two curves $y = 0$ and $y = 3x - x^2$ cross. We must consider two regions, one where $y = 0$ is on top and one where $y = 3x - x^2$ is on top. The first region was considered in Part (a). The total area is equal to

$$[\text{Area found in (a)}] + \int_3^4 [(x^2 - 3x) - 0]\,dx$$

$$= \frac{9}{2} + \left[\frac{1}{3}x^3 - \frac{3}{2}x^2\right]_3^4$$

$$= \frac{9}{2} + \left[\frac{64}{3} - 24\right] - \left[9 - \frac{27}{2}\right]$$

$$= \frac{27}{6} + \frac{128}{6} - 24 - 9 + \frac{81}{6}$$

$$= \frac{236}{6} - 33 = \frac{38}{6} = \frac{19}{3}.$$

(c) This is similar to Part (b). When x is between -2 and 0, the graph of $y = 3x - x^2$ is on top; when x is between 0 and 3, the x-axis is on top. Thus the total area is

$$\int_{-2}^{0} [(x^2 - 3x) - 0]\,dx + [\text{Area found in (a)}]$$

$$= \left(\frac{1}{3}x^3 - \frac{3}{2}x^2\right)\Bigg|_{-2}^{0} + \frac{9}{2} = 0 - \left[\frac{1}{3}(-2)^3 - \frac{3}{2}(-2)^2\right] + \frac{9}{2}$$

$$= -\left[-\frac{8}{3} - 6\right] + \frac{9}{2} = \frac{26}{3} + \frac{9}{2} = \frac{52}{6} + \frac{27}{6} = \frac{79}{6}.$$

25. $$\begin{bmatrix}\text{amount of}\\ \text{depletion}\end{bmatrix} = \begin{bmatrix}\text{amount of}\\ \text{consumption}\end{bmatrix} - \begin{bmatrix}\text{amount of}\\ \text{new growth}\end{bmatrix}$$

$$= \begin{bmatrix}\text{area of region under}\\ \text{consumption curve}\end{bmatrix} - \begin{bmatrix}\text{area of region under}\\ \text{new growth curve}\end{bmatrix}$$

$$= \begin{bmatrix}\text{area of region}\\ \text{between the curves}\end{bmatrix}$$

$$= \int_0^{20} [76.2e^{.03t} - (50 - 6.03e^{.09t})]\,dt$$

$$= \int_0^{20} (76.2e^{.03t} - 50 + 6.03e^{.09t})\,dt.$$

31. Set $Y_1 = 5 - (X - 2)\hat{}2$ and $Y_2 = e\hat{}X$. Deselect all functions except Y_1 and Y_2. You will need a window to the graph the functions. Start with one of the basic windows that are predefined on the ZOOM window, and then adjust the range of x-values and y-values as needed. Notice that between the two intersection points, the graph of Y_1 is above the graph of Y_2.

For high accuracy, find the values of a and b. You should be able to find intersection points at $x = 0$ and at approximately $x \approx 1.572$.

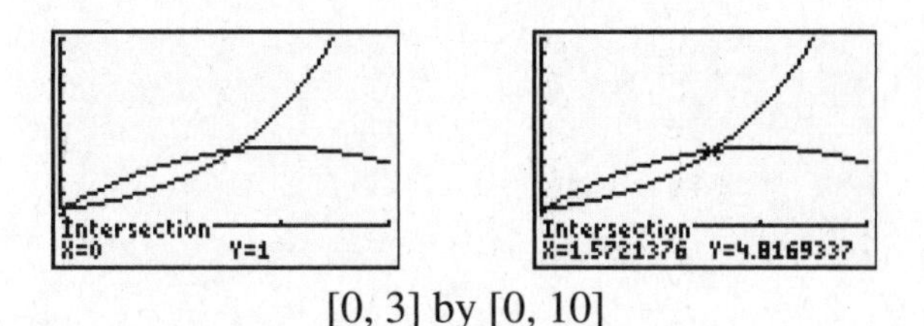

[0, 3] by [0, 10]

The desired area is the definite integral of $Y_1 - Y_2$ from a to b, where a and b are the x-coordinates of the intersection points. (Refer to the INCORPORATING TECHNOLOGY section on page 331 to see how to shade a region under the graph of a function.)

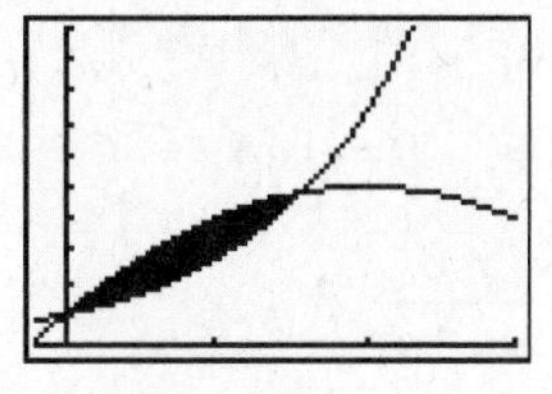

Then use the command fnInt to find the definite integral.

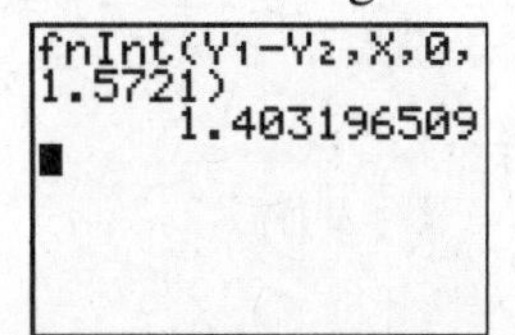

You can also get a fairly good estimate of the desired area without explicitly finding a and b. using the command $\int$f(x)dx on the CALC menu. Adjust the window to [−1, 2] by [−1, 2] and apply the command $\int$f(x)dx to the function $Y_3 = Y_1 - Y_2$. (Before you integrate Y_3, deselect Y_1 and Y_2.) The area should turn out to be approximately 1.403.

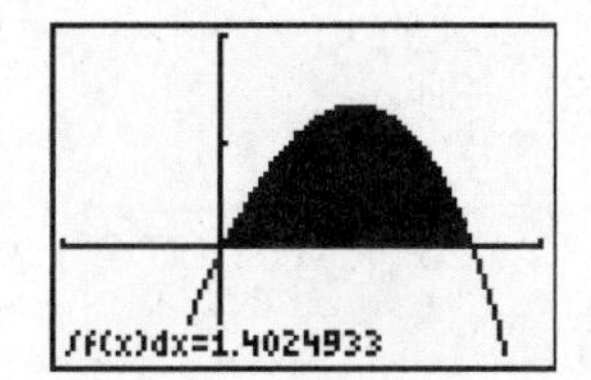

6.5 Applications of the Definite Integral

Although the derivations of the formulas presented in this section rely on Riemann sums, the exercises are worked by substitution into the formulas and do not require analysis in terms of Reimann sums. Example 1 is important.

1. $f(x) = x^2$. By definition, the average value of $f(x)$ between a and b is given by:

$$\text{Average value} = \frac{1}{b-a}\int_a^b f(x)dx = \frac{1}{3-0}\int_0^3 x^2 dx$$

$$= \frac{1}{3}\int_0^3 x^2 dx = \frac{1}{3}\left[\left.\frac{x^3}{3}\right|_0^3\right]$$

$$= \frac{1}{3}\left[\frac{3^3}{3}\right] = 3.$$

7. Average temperature $= \dfrac{1}{12-0}\displaystyle\int_0^{12}\left(47+4t-\frac{1}{3}t^2\right)dt$

$$= \frac{1}{12}\left[47t+2t^2-\frac{1}{9}t^3\Big|_0^{12}\right]$$

$$= \frac{1}{12}\left[47(12)+2(12)^2-\frac{1}{9}(12)^3\right]$$

$$= 47+2(12)-\frac{1}{9}(12)^2 = 55°.$$

13. Consumer's surplus $= \displaystyle\int_A^B [f(x)-B]\,dx, \quad B = f(A).$

$$f(x) = \frac{500}{x+10}-3$$

$$f(40) = \frac{500}{x+10}-3 = \frac{500}{50}-3 = 7$$

So the surplus $= \displaystyle\int_0^{40}\left(\frac{500}{x+10}-3-7\right)dx = 500\int_0^{40}\frac{dx}{x+10}-\int_0^{40}10\,dx$

$$= 500\ln(x+10)\Big|_0^{40} - 10x\Big|_0^{40} = 500\ln 50 - 500\ln 10 - 400$$

$$= 500\ln 5 - 400 \approx \$404.72.$$

19. To find (A, B) set

$$12-\frac{x}{50} = \frac{x}{20}+5$$

$$7 = \frac{x}{20}+\frac{x}{50} = \frac{7x}{100} \Rightarrow x = 100.$$

Thus $A = 100$, and $B = 12-\frac{100}{50} = 10.$

Consumer's surplus $= \displaystyle\int_0^{100}\left(12-\frac{x}{50}-10\right)dx$

$$= \int_0^{100}\left(2-\frac{x}{50}\right)dx = 2x-\frac{x^2}{100}\Big|_0^{100}$$

$$= 200-100 = \$100.$$

Producer's surplus $= \displaystyle\int_0^A (f(A)-f(x))\,dx$

$$= \int_0^{100}\left[10-\left(\frac{x}{20}+5\right)\right]dx$$

$$= \int_0^{100}\left(5-\frac{x}{20}\right)dx = 5x-\frac{x^2}{40}\Big|_0^{100}$$

$$= 500-\frac{10000}{40} = \$250.$$

25. Future amount $= \int_0^N Pe^{r(N-t)}dt$. We don't know *N*, and this is what we wish to evaluate. So

$$140,000 = \int_0^N 5000e^{.1(N-t)}dt$$
$$= 5000\int_0^N e^{.1N}e^{-.1t}dt$$

hence, $\frac{140,000}{5,000} = 28 = \int_0^N e^{.1N}e^{-.1t}dt$

$$= e^{.1N}\int_0^N e^{-.1t}dt = e^{.1N}\left[\frac{e^{-.1t}}{-.1}\Bigg|_0^N\right]$$
$$= -10e^{.1N}[e^{-.1N} - 1].$$

So, $28 = -10 + 10e^{.1N}$,

$38 = 10e^{.1N}$,

$3.8 = e^{.1N}$.

Taking the logarithm of both sides, we have $\ln 3.8 = .1N$. Then $10\ln 3.8 = N$ and $N \approx 13.35$. Thus, it will take about 13.35 years until the values of the investment reaches $140,000.

31. The volume is given by

$$\int_1^2 \pi(x^2)^2 dx = \int_1^2 \pi x^4 dx = \frac{\pi}{5}x^5\Bigg|_1^2 = \frac{31\pi}{5} \text{ cubic units.}$$

37. The Riemann sum tells us that $f(x) = x^3$, $n = 4$, and $\Delta x = .5$. Since $a = 8$ and $\Delta x = \frac{b-a}{n}$ we have

$$.5 = \frac{b-8}{4},$$

so $2 = b - 8, \quad \text{or} \quad b = 10.$

43. **(a)** At interest rate *r*, the average value of the amount in the account over three years is

$$\frac{1}{3}\int_0^3 1000e^{rt}dt = \frac{1}{3}\cdot\frac{1000}{r}e^{rt}\Bigg|_0^3 = \frac{1000}{3r}e^{3r} - \frac{1000}{3r}.$$

(b) For the average balance to be $1070.60, the rate *r* must satisfy the equation

$$\frac{1000}{3r}e^{3r} - \frac{1000}{3r} = 1070.60 \qquad \text{or} \qquad 1000e^{3r} - 1000 = 3211.8r$$

To solve for *r*, set Y_1 = 1000*e^(3X) – 1000 – 3211.8X. Now use the **SOLVER** command (MATH **0**). Press △ to display the the edit screen of the **EQUATION SOLVER**. Press CLEAR if necessary, then enter Y_1 (using the VARS menu) to the right of "**eqn: 0 =** ". Next press ENTER. The equation to be solved will be on the first line of the screen and the cursor will be just to the right of "**X =** ". Enter an initial guess of X = .10 (a 10% interest rate). Do not press the ENTER key! You can ignore the next line. Press ALPHA [SOLVE] (above the ENTER key.)The equation solver produces X = .04497… so the desired interest rate is about 4.45%. For more information about using the **EQUATION SOLVER**, refer to the user manual for your calculator.

An alternate way to find the solution is to set $Y_1 = 1000*e\text{^}(3X) - 1000$ and $Y_2 = 3211.8X$, and then find the intersection of the two graphs. The two graphs are close together, so you will need to zoom in.

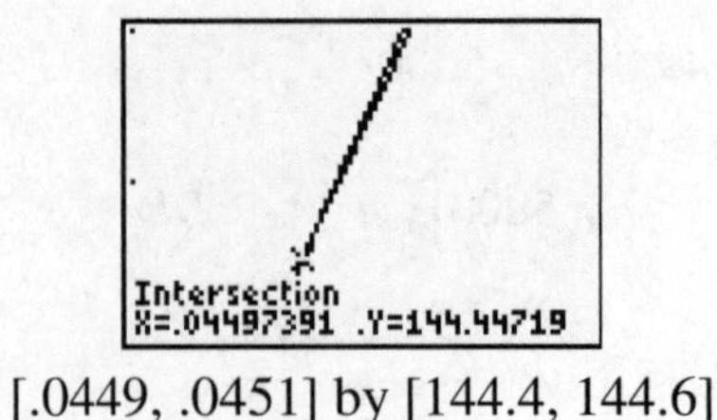

[.0449, .0451] by [144.4, 144.6]

Chapter 6: Supplementary Exercises

1. The integrand (3^2) is a constant which can be simplified:

$$\int 3^2\,dx = \int 9\,dx.$$

Since

$$\frac{d}{dx}(9x) = 9,$$

$$\int 9\,dx = 9x + C.$$

7. $\int e^{-x/2}\,dx$

$$\frac{d}{dx}(-2e^{-x/2}) = e^{-x/2},$$

$$\int e^{-x/2}dx = -2e^{-x/2} + C.$$

13. The integrand $(x+1)^2$ can be simplified:

$$\int_{-1}^{1}(x+1)^2\,dx = \int_{-1}^{1}(x^2+2x+1)\,dx.$$

Since

$$\frac{d}{dx}\left(\frac{x^3}{3}+x^2+x\right) = x^2+2x+1,$$

$$\int_{-1}^{1}(x^2+2x+1)dx = \left(\frac{x^3}{3}+x^2+x\right)\Bigg|_{-1}^{1} = \left(\frac{1}{3}+1+1\right)-\left(-\frac{1}{3}+1-1\right)$$

$$= \frac{7}{3} - \frac{-1}{3} = \frac{8}{3}.$$

19. $\int_1^4 \frac{1}{x^2}\,dx = \int_1^4 x^{-2}dx = -x^{-1}\big|_1^4 = -\frac{1}{4}+1 = \frac{3}{4}.$

Warning: A common error is to think that $\frac{1}{x^3}$ is an antiderivative of $\frac{1}{x^2}$. You can avoid this mistake when you switch to the negative exponent notation, $\frac{1}{x^2} = x^{-2}$.

Helpful Hint: Many integration problems, including Exercise 25, require finding an antiderivative of a function of the form e^{kx} for some nonzero constant k. For this reason it is helpful to have a method of finding an antiderivative of e^{kx} in your repertoire. Begin by differentiating e^{kx} (use the chain rule):

$$\frac{d}{dx}(e^{kx}) = ke^{kx}.$$

Thus:

$$\frac{d}{dx}\left(\frac{e^{kx}}{k}\right) = e^{kx},$$

$$\text{i.e.,} \quad \int e^{kx}\,dx = \frac{e^{kx}}{k} + C.$$

25. Begin by simplifying the integrand:

$$\int_0^{\ln 3}\left(\frac{e^x + e^{-x}}{e^{2x}}\right)dx = \int_0^{\ln 3}\left(\frac{e^x}{e^{2x}} + \frac{e^{-x}}{e^{2x}}\right)dx = \int_0^{\ln 3}(e^{-x} + e^{-3x})\,dx.$$

The formula from the helpful hint above tells us that $\frac{e^{-x}}{-1}$ is an antiderivative of e^{-x} and $\frac{e^{-3x}}{-3}$ is an antiderivative of e^{-3x}. Thus

$$\int_0^{\ln 3}(e^{-x} + e^{-3x})\,dx = \left(-e^{-x} - \frac{e^{-3x}}{3}\right)\Bigg|_0^{\ln 3} = \left(-e^{-\ln 3} - \frac{e^{-3\ln 3}}{3}\right) - \left(-e^0 - \frac{e^0}{3}\right)$$

$$= \left(-\frac{1}{3} - \frac{3^{-3}}{3}\right) - \left(-1 - \frac{1}{3}\right) = \left(\frac{1}{3} - \frac{1}{81}\right) + \frac{4}{3}$$

$$= -\frac{28}{81} + \frac{108}{81} = \frac{80}{81}$$

31. The shaded region of the graph represents the area between the upper curve $y = e^x$ and the lower curve $y = e^{-x}$ from $x = 0$ to $x = \ln 2$. This area is represented by the integral

$$\int_0^{\ln 2}(e^x - e^{-x})\,dx.$$

Evaluating the integral gives the area of the shaded region of the graph. The formula in the helpful hint preceding the solution to Exercise 25 can be used to find an antiderivative of e^{-x}.

$$\int_0^{\ln 2}(e^x - e^{-x})\,dx = \left[e^x - (-e^{-x})\right]\Big|_0^{\ln 2}$$

$$= (e^x + e^{-x})\Big|_0^{\ln 2}$$

$$= (e^{\ln 2} + e^{-\ln 2}) - (e^0 + e^0)$$

$$= \left(2 + \frac{1}{2}\right) - (1 + 1)$$

$$= \frac{1}{2}.$$

37. A sketch of the graph reveals three points of intersection.

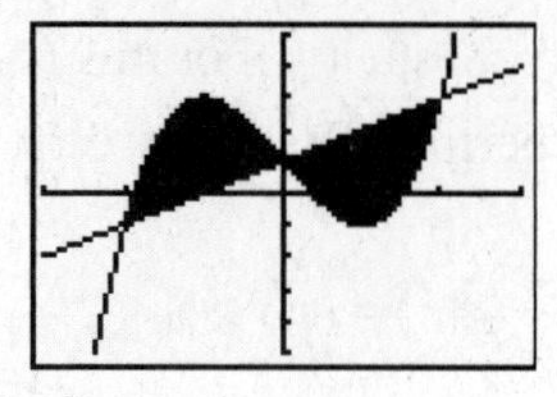

Equate the two expressions in x to obtain

$$x^3 - 3x + 1 = x + 1,$$
$$x^3 - 4x = 0,$$
$$x(x^2 - 4) = 0,$$
$$x(x-2)(x+2) = 0.$$

Thus, the graphs intersect at $x = 0, 2, -2$. The total area bounded by the curves is the sum of the areas of the two regions.

$$\begin{aligned}
\text{Area} &= \int_{-2}^{0} [(x^3 - 3x + 1) - (x+1)]\,dx + \int_{0}^{2} [(x+1) - (x^3 - 3x + 1)]\,dx \\
&= \int_{-2}^{0} (x^3 - 4x)\,dx + \int_{0}^{2} (-x^3 + 4x)\,dx \\
&= \left(\frac{1}{4}x^4 - 2x^2\right)\Bigg|_{-2}^{0} + \left(-\frac{1}{4}x^4 + 2x^2\right)\Bigg|_{0}^{2} \\
&= 0 - (4 - 8) + (-4 + 8) - 0 \\
&= 8.
\end{aligned}$$

43. The cost of producing x tires a day is an antiderivative of the marginal cost function $.04x + 150$. Hence, if the cost function is $C(x)$, then

$$\begin{aligned}
C(x) &= \int (.04x + 150)\,dx \\
&= .02x^2 + 150x + C.
\end{aligned}$$

If fixed costs are \$500 per day, then the cost of producing no tires is \$500 per day. That is,

$$C(0) = 0 + 0 + C = C = 500.$$

Hence $$C(x) = .02x^2 + 150x + 500.$$

49. Here $\Delta x = \frac{b-a}{n} = \frac{2-0}{2} = 1$, so the midpoints of the two subintervals are given by

$$x_1 = a + \frac{\Delta x}{2} = 0 + \frac{1}{2} = \frac{1}{2},$$
$$x_2 = \frac{1}{2} + \Delta x = \frac{1}{2} + 1 = \frac{3}{2}.$$

By the midpoint rule,

$$\begin{aligned}\text{Area} &\approx \left[f\left(\frac{1}{2}\right) + f\left(\frac{3}{2}\right)\right](1)\\ &= \left[\frac{1}{\frac{1}{2}+2} + \frac{1}{\frac{3}{2}+2}\right]\\ &= \frac{2}{5} + \frac{2}{7} = \frac{24}{35} \approx .68571.\end{aligned}$$

Computing the integral directly,

$$\begin{aligned}\int_0^2 \frac{1}{x+2}\,dx &= \ln|x+2|\Big|_0^2 = \ln 4 - \ln 2\\ &= \ln\frac{4}{2} = \ln 2 = .69315 \quad \text{(rounded)}.\end{aligned}$$

55. The region corresponding to

$$\int_a^c f(x)\,dx$$

contains one part above the x-axis, with area .68, and one part below the x-axis, with area .42. From Section 6.2,

$$\int_a^c f(x)\,dx = .68 - .42 = .26.$$

Similarly,

$$\int_a^d f(x)\,dx = .68 - .42 + 1.70 = 1.96,$$

because the part from b to c is below the x-axis.

61. Observe that the sum is a Riemann sum for the function $f(t) = 5000e^{-.1t}$ from $t = 0$ to $t = 3$. This sum is approximately equal to the value of the definite integral:

$$\begin{aligned}\int_0^3 5000e^{-.1t}\,dt &= -50{,}000e^{-.1t}\Big|_0^3\\ &= -50{,}000e^{-.3} + 50{,}000\\ &\approx 13{,}000.\end{aligned}$$

67. Amount of water used between 1960 and 1995:

$$= \int_0^{35} 860e^{.04t}\,dt$$

$$= 860\frac{e^{.04t}}{.04}\Bigg|_0^{35} = 21{,}500(e^{.04(35)} - e^0)$$

$$= 21{,}500(e^{1.4} - 1) \approx 65{,}000 \text{ km}^3.$$

73. $\int_0^1 (\sqrt{x} - x^2)\,dx = \int_0^1 (x^{1/2} - x^2)dx$

$$= \int_0^1 x^{1/2}dx - \int_0^1 x^2 dx$$

$$= \left(\frac{x^{3/2}}{\frac{3}{2}} - \frac{x^3}{3}\right)\Bigg|_0^1$$

$$= \left(\frac{2}{3}x^{3/2} - \frac{x^3}{3}\right)\Bigg|_0^1$$

$$= \left(\frac{2}{3}1^{3/2} - \frac{1^3}{3}\right) - \left(\frac{2}{3}0^{3/2} - \frac{0^3}{3}\right)$$

$$= \left(\frac{2}{3} - \frac{1}{3}\right) - 0 = \frac{1}{3}.$$

Chapter 7
Functions of Several Variables

7.1 Examples of Functions of Several Variables

You should be sure that you completely understand the meanings of the variables in Examples 2 and 3. These examples will be used to illustrate the concepts introduced in later sections.

1. $f(x, y) = x^2 - 3xy - y^2$. So

$$f(5,0) = (5)^2 - 3(5)(0) - (0)^2 = 25,$$
$$f(5,-2) = (5)^2 - 3(5)(-2) - (-2)^2 = 25 + 30 - 4 = 51,$$
$$f(a,b) = a^2 - 3ab - b^2.$$

7. [Cost] = [Cost of top and bottom] + [Cost of four sides]
= [Total area of top and bottom] · [Cost per sq. ft.] + [Total area of four sides] · [Cost per sq. ft.]
$= [xy + xy] \cdot 3 + [xz + yz + xz + yz] \cdot 5$
$= 6xy + 10xz + 10yz$
$= C(x, y, z)$.

13. $T = f(r, v, x) = \frac{r}{100}(.40v - x)$

(a) $v = 200{,}000$, $x = 5000$, $r = 2.5$,

$$T = \frac{r}{100}[.4v - x],$$
$$T = \frac{2.5}{100}[.4(200{,}000) - 5000] = \$1875.00.$$

(b) $v = 200{,}000$, $x = 5000$, $r = 3$,

$$T = \frac{3.00}{100}[.4(200{,}000) - 5000] = \$2250$$

The tax due also increases by 20% (or 1/5), since

$$1875.00 + \frac{1}{5}(1875.00) = \$2250.00.$$

19. A level curve has the form $f(x, y) = C$ (C is arbitrary). Rearrange the equation $y = 3x - 4$ so that all terms involving x and y are on the left-hand side:

$$y - 3x = -4.$$

So,

$$y - 3x = C, \text{ some constant.}$$

Thus $f(x, y) = y - 3x$.

25. The graph in Exercise 25 is matched with the level curves in (c). Imagine slicing "near the top" of the four humps. You should visualize a cross section of four circular-like figures and as you move further down the z-axis these circular-like figures become larger and larger. Similarly,
Exercise 23 is matched with (d),
Exercise 24 is matched with (b),
Exercise 26 is matched with (a).

7.2 Partial Derivatives

Any letter can be used as a variable of differentiation, even y. As a warm-up to this section, look over the following ordinary derivatives.

$$\text{If } f(y) = ky^2, \quad \text{then } \frac{df}{dy} = 2ky.$$

$$\text{If } g(y) = 3y^2x, \text{ then } \frac{dg}{dy} = 6yx.$$

Notice that x is a constant in the formula for $g(y)$ because the notation $g(y)$ means that y is the only variable in the function.

1. $\frac{\partial}{\partial x}5xy = \frac{\partial}{\partial x}[5y]x = 5y,$ [treat $5y$ as a constant]

$\frac{\partial}{\partial y}5xy = \frac{\partial}{\partial y}[5x]y = 5x.$ [treat $5x$ as a constant]

7.
$$\begin{aligned}\frac{\partial}{\partial x}(2x - y + 5)^2 &= 2(2x - y + 5)\cdot\frac{\partial}{\partial x}(2x - y + 5)\\ &= 2(2x - y + 5)\cdot(2 - 0 + 0) \quad \text{[treat } y \text{ as a constant]}\\ &= 4(2x - y + 5).\end{aligned}$$

$$\begin{aligned}\frac{\partial}{\partial y}(2x - y + 5)^2 &= 2(2x - y + 5)\cdot\frac{\partial}{\partial y}(2x - y + 5)\\ &= 2(2x - y + 5)\cdot(0 - 1 + 0) \quad \text{[treat } 2x \text{ as a constant]}\\ &= -2(2x - y + 5).\end{aligned}$$

13. $f(L,K)=3\sqrt{LK}=3(LK)^{1/2}=3L^{1/2}K^{1/2}$. Treat 3 and $K^{1/2}$ as constants:

$$\frac{\partial}{\partial L}3L^{1/2}K^{1/2}=\frac{\partial}{\partial L}[3K^{1/2}]L^{1/2}=[3K^{1/2}]\frac{1}{2}L^{-1/2}$$

$$=\frac{3}{2}K^{1/2}L^{-1/2}=\frac{3\sqrt{K}}{2\sqrt{L}}.$$

19. $\frac{\partial}{\partial x}(x^2+2xy+y^2+3x+5y)=2x+2y+0+3+0=2x+2y+3.$

Thus $\quad \frac{\partial f}{\partial x}(2,-3)=2(2)+2(-3)+3=4-6+3=1.$

$\frac{\partial}{\partial y}(x^2+2xy+y^2+3x+5y)=0+2x+2y+0+5=2x+2y+5.$

Thus $\quad \frac{\partial f}{\partial y}(2,-3)=2(2)+2(-3)+5=4-6+5=3.$

25. $f(x,\ y)=200\sqrt{6x^2+y^2}$

(a) Marginal productivity of labor is given by $\frac{\partial f}{\partial x}$.

$$f(x,\ y)=200\sqrt{6x^2+y^2}=200(6x^2+y^2)^{1/2}.$$

$$\frac{\partial f}{\partial x}=200\left(\frac{1}{2}\right)(6x^2+y^2)^{-1/2}\cdot\frac{\partial}{\partial x}(6x^2+y^2)$$

$$=100(6x^2+y^2)^{-1/2}(12x)=\frac{1200x}{\sqrt{6x^2+y^2}}.$$

$$\frac{\partial f}{\partial x}(10,5)=\frac{1200(10)}{\sqrt{16(10)^2+(5)^2}}=\frac{12{,}000}{\sqrt{625}}=\frac{12{,}000}{25}=480.$$

Here is what the value 480 of the marginal productivity of labor really means:

At $(x, y) = (10, 5)$, if the amount x of labor changes slightly from 10 units and y stays fixed at 5 units, production increases by 480 times the amount of change in the amount of labor.

See the discussion in Example 5.

Marginal productivity of capital is given by $\frac{\partial f}{\partial y}$.

$$\frac{\partial f}{\partial y}=200\left(\frac{1}{2}\right)(6x^2+y^2)^{-1/2}\cdot\frac{\partial f}{\partial y}(6x^2+y^2)$$

$$=100(6x^2+y^2)^{-1/2}\cdot(2y)=\frac{200y}{\sqrt{6x^2+y^2}}.$$

$$\frac{\partial f}{\partial y}(10,5)=\frac{200(5)}{\sqrt{6(10)^2+(5)^2}}=\frac{1000}{\sqrt{625}}=\frac{1000}{25}=40.$$

The value 40 for the marginal productivity of capital means that:

For $(x, y) = (10, 5)$, production changes at a rate 40 times the change in capital, when x stays fixed at 10 units and the amount y of capital changes slightly from 5 units.

This interpretation and the one above illustrate what is written in the box preceding Example 6. The next two parts of the exercise make this discussion more concrete.

(b) Let h represent a small number (positive or negative). If labor is changed by h units from 10 to $10 + h$ units and if capital is fixed at 5 units, then the quantity of goods produced will change by approximately $480 \cdot h$ units of goods.

(c) Suppose labor decreases from 10 to 9.8 units (while capital stays fixed at 5 units). Then the h from part (b) is $-.5$, and production will change by approximately $480(-.5)$ units. That is, production will *decrease* by about 240 units.

31.

$$V = \frac{.08T}{P} = .08TP^{-1},$$

$$\frac{\partial V}{\partial P} = -(.08)TP^{-2},$$

$$\frac{\partial V}{\partial P}(20, 300) = -(.08)(300)(20)^{-2} = \frac{-(.80)(300)}{400} = -.06,$$

$$\frac{\partial V}{\partial T} = .08P^{-1},$$

$$\frac{\partial V}{\partial T}(20, 300) = .08(20)^{-1} = \frac{.08}{20} = .004.$$

$\frac{\partial V}{\partial P}(20, 300)$ represents the rate of change of volume, when temperature is fixed at 300 and pressure is allowed to vary near 20. That is, if the pressure is increased by 1 small unit, volume will decrease by about .06 units.

$\frac{\partial V}{\partial T}(20, 300)$ represents the rate of change of volume when pressure is fixed at 20 and temperature is allowed to vary near 300. That is, if temperature is increased by 1 small unit, volume will increase by approximately .004 units.

37. $f(x, y) = 3x^2 + 2xy + 5y$

$$\begin{aligned} f(1+h, 4) - f(1, 4) &= 3(1+h)^2 + 2(1+h)(4) + 5(4) - [3(1)^2 + 2(1)(4) + 5(4)] \\ &= 3(1+h^2+2h) + 8 + 8h + 20 - 3 - 8 - 20 \\ &= 3 + 3h^2 + 6h + 8h - 3 \\ &= 3h^2 + 14h. \end{aligned}$$

7.3 Maxima and Minima of Functions of Several Variables

Finding relative extreme points for functions of several variables is accomplished in much the same way as for functions of one variable. In Chapter 2, you found relative maximum and minimum points by differentiating and then solving one equation for x. In this section, you will differentiate (with partials) and then solve a system of two equations in two unknowns.

Three types of systems of equations arise in this section.

First type: Each equation involves just one variable. An example is:

$$\begin{cases} x^2 + 1 = 5, \\ 3y - 4 = 11. \end{cases}$$

To solve this system, solve each equation separately for its variable. The solutions to the system consist of all combinations of these individual solutions. In this example above, the first equation has the two solutions $x = 2$ and $x = -2$. The second equation has the single solution $y = 5$. Therefore, the system has two solutions: $x = 2$, $y = 5$ and $x = -2$, $y = 5$. [*Note:* These two solutions can also be written as (2, 5) and (−2, 5).]

Second type: One equation contains both variables, the other equation contains just one variable. An example is

$$\begin{cases} 2x + 4y = 26, \\ 3y - 4 = 11. \end{cases}$$

To solve this system, solve the equation having just one variable, substitute this value into the other equation, and then solve that equation for the remaining variable. In the example above, the second equation has the solution $y = 5$. Substitute 5 for y into the first equation and obtain $2x + 4(5) = 26$ or $x = 3$. Therefore, the system has the solution $x = 3$, $y = 5$.

Third type: Both equations contain both variables. An example is

$$\begin{cases} -9x + y = 3, \\ \quad y + 2 = 4x. \end{cases}$$

Two methods for solving such systems are the EQUATE METHOD and the SOLVE-SUBSTITUTE METHOD.

With the EQUATE METHOD, both equations are solved for y, the two expressions for y are equated and solved for x, and the value of x is substituted into one of the equations to obtain the value of y. For the system above, begin the writing

$$\begin{cases} y = 3 + 9x, \\ y = 4x - 2. \end{cases}$$

Next, equate the two expressions for y and solve:

$$3 + 9x = 4x - 2,$$
$$5x = -5,$$
$$x = -1.$$

Finally, substitute $x = -1$ into the first equation: $y = 3 + 9(-1) = -6$. Therefore, the solution is $x = -1$, $y = -6$.

With the SOLVE-SUBSTITUTE METHOD, one equation is solved for y as an expression in x, this expression is substituted into the other equation, that equation is solved for x, and the value(s) of x is substituted into the expression for y to obtain the value(s) of y. For the system above, proceed as follows:

Solve first equation for y: $y = 3 + 9x.$

Substitute into second equation: $(3 + 9x) + 2 = 4x.$

Solve for x: $x = -1.$

Substitute the x value into the equation for y:

$$y = 3 + 9(-1) = -6.$$

1. $\frac{\partial}{\partial x}(x^2 - 3y^2 + 4x + 6y + 8) = 2x - 0 + 4 + 0 + 0 = 2x + 4,$

$\frac{\partial}{\partial y}(x^2 - 3y^2 + 4x + 6y + 8) = 0 - 6y + 0 + 6 + 0 = -6y + 6.$

To find possible relative maxima or minima, set the first partial derivatives equal to zero.

$$\frac{\partial f}{\partial x} = 2x + 4 = 0, \qquad \frac{\partial f}{\partial y} = -6y + 6 = 0,$$

$$2x = -4, \qquad -6y = -6,$$

$$x = \frac{-4}{2} = -2, \qquad y = 1.$$

Thus $f(x, y)$ has a possible relative maximum or minimum at $(-2, 1)$.

7. $\frac{\partial}{\partial x}\left(\frac{1}{3}x^3 - 2y^3 - 5x + 6y - 5\right) = x^2 - 0 - 5 + 0 - 0 = x^2 - 5,$

$\frac{\partial}{\partial y}\left(\frac{1}{3}x^3 - 2y^3 - 5x + 6y - 5\right) = 0 - 6y^2 - 0 + 6 - 0 = -6y^2 + 6.$

Again, set the first partial derivatives equal to zero.

$$\frac{\partial f}{\partial x} = x^2 - 5 = 0, \qquad \frac{\partial f}{\partial y} = -6y^2 + 6 = 0,$$

$$x^2 = 5, \qquad 6y^2 = 6,$$

$$x = \sqrt{5}, \quad x = -\sqrt{5}, \qquad y = 1, \quad y = -1.$$

There are four points (x, y) where there is a possible relative maximum or minimum:

$$(\sqrt{5}, 1), \quad (-\sqrt{5}, 1) \quad (\sqrt{5}, -1), \quad (-\sqrt{5}, -1).$$

13. $f(x, y) = 2x^2 - x^4 - y^2$,

$$\frac{\partial f}{\partial x} = 4x - 4x^3, \quad \frac{\partial^2 f}{\partial x^2} = 4 - 12x^2,$$

$$\frac{\partial f}{\partial y} = -2y, \quad \frac{\partial^2 f}{\partial y^2} = -2,$$

$$\frac{\partial^2 f}{\partial x \partial y} = 0.$$

$$\begin{aligned} D(x, y) &= \frac{\partial^2 f}{\partial x^2} \cdot \frac{\partial^2 f}{\partial y^2} - \left(\frac{\partial^2 f}{\partial x \partial y} \right)^2 \\ &= (4 - 12x^2) \cdot (-2) - 0 \\ &= -2(4 - 12x^2). \end{aligned}$$

Use the second-derivative test for functions of two variables. Look first at $(-1, 0)$:

$$D(-1, 0) = -2(4 - 12), \text{ which is positive; and}$$

$$\frac{\partial^2 f}{\partial x^2}(-1, 0) = 4 - 12(-1)^2, \text{ which is negative.}$$

Hence $f(x, y)$ has a relative maximum at $(-1, 0)$.

For $(0, 0)$: $D(0, 0) = -2(4 - 0) = -8$, which is negative, so $f(x, y)$ has neither a relative maximum nor a relative minimum at $(0, 0)$.

Finally at $(1, 0)$: $D(1, 0) = -2(4 - 12)$, which is positive, and

$$\frac{\partial^2 f}{\partial x^2}(1, 0) = 4 - 12 = -8, \text{ which is negative.}$$

Therefore again by the test, $f(x, y)$ has a relative maximum at $(1, 0)$.

19. $\frac{\partial}{\partial x}(-2x^2 + 2xy - y^2 + 4x - 6y + 5) = -4x + 2y + 4,$

$\frac{\partial}{\partial y}(-2x^2 + 2xy - y^2 + 4x - 6y + 5) = 2x - 2y - 6.$

Set these partials equal to zero and solve:

$$\frac{\partial f}{\partial x} = -4x + 2y + 4 = 0, \quad \frac{\partial f}{\partial y} = 2x - 2y - 6 = 0,$$

$$2y = 4x - 4, \quad 2y = 2x - 6,$$

$$y = 2x - 2, \quad y = x - 3.$$

Equate these two expressions for y:

$$2x - 2 = x - 3,$$

$$x = -1.$$

Substitute $x = -1$ into $y = x - 3$ and obtain $y = x - 3 = (-1)(-3) = -4$. So there is a possible relative maximum or minimum at $(-1, -4)$. Use the second derivative test to determine which, if either, is the case.

$$\frac{\partial^2 f}{\partial x^2} = \frac{\partial}{\partial x}(-4x+2y+4) = -4,$$

$$\frac{\partial^2 f}{\partial y^2} = \frac{\partial}{\partial y}(2x-2y-6) = -2,$$

$$\frac{\partial^2 f}{\partial x \partial y} = \frac{\partial}{\partial x}(2x-2y-6) = 2.$$

Therefore, $D(x, y) = (-4)(-2) - (2)^2 = 8 - 4 = 4$, and hence $D(-1, -4) = 4$. Since $4 > 0, f(x, y)$ indeed has a relative maximum or minimum at $(-1, -4)$. Since $\frac{\partial^2 f}{\partial x^2}(-1, -4) = -4 < 0$, $f(x, y)$ has a relative maximum at $(-1, -4)$.

25. $\frac{\partial f}{\partial x} = 4x - 1, \ \frac{\partial f}{\partial y} = 3y^2 - 12.$

Set the two partials equal to 0 and solve:

$$\begin{aligned} 4x - 1 &= 0, & 3y^2 - 12 &= 0, \\ 4x &= 1, & 3y^2 &= 12, \\ x &= \frac{1}{4}, & y^2 &= 4, \\ & & y &= 2, -2. \end{aligned}$$

So there are two possible relative maxima and/or minima: $\left(\frac{1}{4}, 2\right), \left(\frac{1}{4}, -2\right)$. For the second derivative test, compute

$$\frac{\partial^2 f}{\partial x^2} = \frac{\partial}{\partial x} = (4x-1) = 4,$$

$$\frac{\partial^2 f}{\partial y^2} = \frac{\partial}{\partial y}(3y^2-12) = 6y,$$

$$\frac{\partial^2 f}{\partial x \partial y} = \frac{\partial}{\partial x}(3y^2-12) = 0.$$

$$\begin{aligned} D(x, y) &= 4(6y) - 0^2 \\ &= 24y. \end{aligned}$$

Now $D\left(\frac{1}{4}, 2\right) = 24(2) = 48 > 0$, so $f(x, y)$ does indeed have a relative maximum or minimum at $\left(\frac{1}{4}, 2\right)$. Since

$$\frac{\partial^2 f}{\partial x^2}\left(\frac{1}{4}, 2\right) = 4 > 0,$$

there is a relative minimum at $\left(\frac{1}{4}, 2\right)$.

Finally, $D\left(\frac{1}{4}, -2\right) = 24(-2) = -48 < 0$, indicating that $\left(\frac{1}{4}, -2\right)$ is a saddle point; this is neither a relative maximum nor a relative minimum.

31. Note that the revenue can be expressed by $10x + 9y$, since the company sells x units of Product I for 10 dollars each, and y units of Product II for 9 dollars each. Since profit = (revenue) – (cost), the profit function is

$$P(x, y) = 10x + 9y - [400 + 2x + 3y + .01(3x^2 + xy + 3y^2)]$$
$$= 8x + 6y - .03x^2 - .01xy - .03y^2 - 400.$$

Now proceed as in Exercises 19 and 25 above.

$$\frac{\partial P}{\partial x} = 8 - .06x - .01y,$$
$$\frac{\partial P}{\partial y} = 6 - .01x - .06y.$$

$$8 - .06x - .01y = 0, \qquad 6 - .01x - .06y = 0,$$
$$.01y = 8 - .06x, \qquad .06y = 6 - .01x,$$
$$y = 800 - 6x, \qquad y = 100 - \frac{1}{6}x.$$

So,

$$800 - 6x = 100 - \frac{1}{6}x, \qquad 700 = \frac{35}{6}x, \qquad x = 120,$$

and

$$y = 800 - 6x = 800 - 6(120) = 800 - 720 = 80.$$

So (120, 80) is the only possible maximum. For the second derivative test, compute

$$\frac{\partial^2 P}{\partial x^2} = -.06, \qquad \frac{\partial^2 P}{\partial y^2} = -.06, \qquad \frac{\partial^2 P}{\partial x \partial y} = -.01.$$

Therefore, $D(x, y) = (-.06)(-.06) - (-.01)^2 = .0035$. Thus $D(120, 80) = .0035 > 0$ and (120, 80) is indeed a maximum or minimum. Since $\frac{\partial^2 P}{\partial x^2}(120, 80) = -.06 < 0$, (120, 80) is a maximum. Thus profit is maximized by manufacturing and selling 120 units of Product I and 80 units of Product II.

7.4 Lagrange Multipliers and Constrained Optimization

Constrained optimization problems are frequently encountered in economics, operations research, and science; the Greek letter λ commonly appears whenever Lagrange multipliers are discussed. Think of λ as just another (new) symbol for a variable. The following results illustrate derivatives using λ.

$$\text{If } f(\lambda) = 5\lambda, \qquad \text{then} \qquad \frac{\partial f(\lambda)}{\partial \lambda} = 5.$$

$$\text{If } f(\lambda) = \lambda k, \qquad \text{then} \qquad \frac{\partial f(\lambda)}{\partial \lambda} = k.$$

Partial derivatives of functions involving λ are not difficult to compute once the unfamiliarity of using a Greek letter is overcome.

The problems in this section require that systems of three and four variables be solved. The method following Example 1 applies to all equations in x, y, and λ. Systems with the additional variable z can usually be solved by mimicking the solution to Example 4.

1. Construct the function

$$F(x, y, \lambda) = x^2 + 3y^2 + 10 + \lambda(8 - x - y).$$

Set the first partial derivatives equal to zero and solve for λ to obtain

$$\frac{\partial F}{\partial x} = 2x - \lambda = 0, \quad \text{so} \quad \lambda = 2x,$$

$$\frac{\partial F}{\partial y} = 6y - \lambda = 0, \quad \text{so} \quad \lambda = 6y,$$

$$\frac{\partial F}{\partial \lambda} = 8 - x - y = 0.$$

Equate $\lambda = 2x$ and $\lambda = 6y$ to obtain $2x = 6y$, or $x = 3y$. Next, substitute $3y$ for x into the third equation and solve for y:

$$\begin{aligned} 8 - (3y) - y &= 0, \\ 8 - 4y &= 0, \\ 4y &= 8, \\ y &= 2. \end{aligned}$$

Finally, $x = 3y = 3(2) = 6$. So the minimum is at (6, 2), and the minimum value is

$$\begin{aligned} F(6, 2) &= 6^2 + 3(2^2) + 10 \\ &= 36 + 12 + 10 \\ &= 58. \end{aligned}$$

7. Minimize the function, $f(x, y) = x + y$ subject to the constraint

$$xy = 25, \quad \text{or} \quad xy - 25 = 0.$$

$F(x, y, \lambda) = x + y + \lambda(xy - 25)$ is the Lagrange function. For a relative extremum, set the first partial derivatives equal to zero:

$$\left.\begin{aligned} \frac{\partial F}{\partial x} &= 1 + \lambda y = 0 \\ \frac{\partial F}{\partial y} &= 1 + \lambda x = 0 \end{aligned}\right\} \Rightarrow \lambda y = \lambda x, \quad \text{or} \quad y = x, \qquad (1)$$

$$\frac{\partial F}{\partial \lambda} = xy - 25 = 0 \qquad \text{(substitute in equation (1))},$$

$$x^2 - 25 = 0, \quad \text{or} \quad x = \pm 5.$$

If $x = 5$ (since x must be positive): $y = 5$ by equation (1). Therefore, $x = 5$, $y = 5$.

13. Maximize $P(x, y) = 3x + 4y$ subject to the constraint

$$9x^2 + 4y^2 - 18{,}000 = 0, \; x \geq 0, \; y \geq 0.$$

$F(x, y, \lambda) = 3x + 4y + \lambda(9x^2 + 4y^2 - 18{,}000)$. Set the first partial derivatives equal to zero:

$$\frac{\partial F}{\partial x} = 3 + 18\lambda x = 0,$$
$$\frac{\partial F}{\partial y} = 4 + 8\lambda y = 0,$$
$$\frac{\partial F}{\partial \lambda} = 9x^2 + 4y^2 - 18{,}000 = 0. \quad (1)$$

Thus, $3 + 18\lambda x = 0$ and $4 + 8\lambda y = 0$, which give

$$1 + 6\lambda x = 0, \quad \text{and} \quad 1 + 2\lambda y = 0,$$

respectively. Therefore,

$$1 + 6\lambda x = 1 + 2\lambda y,$$
$$6\lambda x = 2\lambda y,$$
$$3x = y.$$

Substitute $3x$ for y in equation (1):

$$9x^2 + 4(3x)^2 - 18{,}000 = 0,$$
$$45x^2 = 18{,}000,$$
$$x^2 = 400, \quad \text{or} \quad x = \pm 20.$$

Since $y = 3x$, $y = \pm 60$. But $x,\ y \geq 0$ so,

$$x = 20, \quad \text{and} \quad y = 60.$$

19. Let $F(x, y, z, \lambda) = 3x + 5y + z - x^2 - y^2 - z^2 + \lambda(6 - x - y - z)$, and set the first partial derivatives equal to zero.

$$\frac{\partial F}{\partial x} = 3 - 2x - \lambda = 0,$$
$$\frac{\partial F}{\partial y} = 5 - 2y - \lambda = 0,$$
$$\frac{\partial F}{\partial z} = 1 - 2z - \lambda = 0.$$

This provides three expressions for λ:

$$\lambda = 3 - 2x,$$
$$\lambda = 5 - 2y,$$
$$\lambda = 1 - 2z.$$

Equate the first two equations and solve for x in terms of y:

$$3 - 2x = 5 - 2y,$$
$$-2x = 2 - 2y,$$
$$x = -1 + y.$$

Next, equate the second two equations and solve for z in terms of y:

$$5 - 2y = 1 - 2z,$$
$$2z = 2y - 4,$$
$$z = y - 2.$$

Finally, substitute these expressions for x and for z into the equation

$$\frac{\partial F}{\partial \lambda} = 6 - x - y - z = 0.$$

$$6 - (-1 + y) - y - (y - 2) = 0$$
$$6 + 1 - y - y - y + 2 = 0$$
$$9 - 3y = 0$$
$$9 = 3y, \quad \text{or} \quad y = 3.$$

Thus, $x = -1 + y = -1 + 3 = 2$ and $z = y - 2 = 3 - 2 = 1$. The function is maximized when $x = 2$, $y = 3$, and $z = 1$.

25. Throughout the problem use the general expression $f(x, y)$ for the production function, which we wish to maximize. The constraint is the fact that the cost of labor plus the cost of capital is c dollars, that is:

$$ax + by = c, \quad \text{or} \quad ax + by - c = 0.$$

Construct the function

$$F(x, y, \lambda) = f(x, y) + \lambda(ax + by - c).$$

Take partial derivatives and set them equal to zero.

$$\frac{\partial F}{\partial x} = \frac{\partial f}{\partial x} + \lambda a = 0,$$

$$\frac{\partial F}{\partial y} = \frac{\partial f}{\partial y} + \lambda b = 0.$$

Solve each equation for λ and get

$$\lambda = -\frac{1}{a}\frac{\partial f}{\partial x} \quad \text{and} \quad \lambda = -\frac{1}{b}\frac{\partial f}{\partial y}.$$

And equating these expressions for λ, we conclude that

$$-\frac{1}{a}\frac{\partial f}{\partial x} = -\frac{1}{b}\frac{\partial f}{\partial y}$$

$$b\frac{\partial f}{\partial x} = a\frac{\partial f}{\partial y}$$

$$\frac{\frac{\partial f}{\partial x}}{\frac{\partial f}{\partial y}} = \frac{a}{b}.$$

7.5 The Method of Least Squares

Each problem in this section requires the solution of a system of two equations in two unknowns. These systems differ from those previously studied in three ways.

1. The variables (that is, the unknowns) will always be A and B, rather then x and y.
2. The equations are linear. That is, they have the form $aA + bB = c$, where a, b, and c are constants.
3. The systems have exactly one solution. This solution is best found by multiplying one equation by a constant and subtracting it from the other equation to eliminate one of the variables. The value of the other variable is then easily found. By substituting this value into one of the equations, the value of the first variable is easily found.

1. The given points are (1, 3), (2, 6), (3, 8), and (4, 6) with straight line $y = 1.1x + 3$. When $x = 1, 2, 3, 4$ the corresponding y-coordinates are $1.1 + 3$, $2(1.1) + 3$, $3(1.1) + 3$, $4(1.1) + 3$ or 4.1, 5.2, 6.3, and 7.4, respectively. Then

$$E = E_1^2 + \cdots + E_4^2 = (A+B-3)^2 + (2A+B-6)^2 + (3A+B-8)^2 + (4A+B-6)^2$$
$$= (1.1+3-3)^2 + (2(1.1)+3-6)^2 + (3(1.1)+3-8)^2 + (4(1.1)+3-6)^2 = 6.7.$$

7. Let the straight line be $y = Ax + B$. Then

$$E = (A+B-9)^2 + (2A+B-8)^2 + (3A+B-6)^2 + (4A+B-3)^2$$

To minimize the error E, take partial derivatives with respect to A and B, and set them equal to zero.

$$\frac{\partial E}{\partial A} = 2(A+B-9) + 2(2A+B-8)\cdot 2 + 2(3A+B-6)\cdot 3 + 2(4A+B-3)\cdot 4$$
$$= (2+B+18+32)A + (2+4+6+8)B - (18+32+36+24)$$
$$= 60A + 20B - 110 = 0.$$

$$\frac{\partial E}{\partial B} = 2(A+B-9) + 2(2A+B-8) + 2(3A+B-6) + 2(4A+B-3)$$
$$= (2+4+6+8)A + (2+2+2+2)B - (18+16+12+6)$$
$$= 20A + 8B - 52 = 0.$$

To solve the system of simultaneous linear equations

$$60A + 20B = 110$$
$$20A + 8B = 52,$$

multiply the second equation by -3, and add it to the first equation:

$$60A + 20B = 110$$
$$-60A - 24B = -156$$
$$-4B = -46, \quad B = 11.5.$$

Then find A by substituting this value of B into the first equation:

$$60A + 20(11.5) = 110, \qquad 60A = 110 - 230, \qquad A = -2.$$

The least-squares line is $y = -2x + 11.5$.

13. **(a)** The data points are (0, 10.7), (5, 13.9), (10, 16.2), (15, 19.4), (20, 21.3), and (25, 23.0). The table for these data is

x	y	xy	x^2
0	10.7	0	0
5	13.9	69.5	25
10	16.2	162	100
15	19.4	291	225
20	21.3	426	400
25	23.0	575	625

$$\sum x = 75 \qquad \sum y = 104.5 \qquad \sum xy = 1523.5 \qquad \sum x^2 = 1375$$

The formulas for A and B are:

$$A = \frac{6 \cdot 1523.5 - 75 \cdot 104.5}{6 \cdot 1375 - (75)^2} = \frac{1303.5}{2625} \approx .4966$$

$$B = \frac{104.5 - .4966 \cdot 75}{6} \approx 11.2092.$$

The least-squares line is $y = .4966x + 11.2092$. The number of decimal places to use in the equation of the line should be determined by the accuracy of the original data. In this problem, the data have three significant figures, so a reasonable answer for the least-squares line might be $y = .497x + 11.2$, or even $y = .50x + 11.2$.

Alternatively, we can use a graphing calculator to solve the problem. Enter the data, then use the linear regression function to determine the equation of the line of best fit.

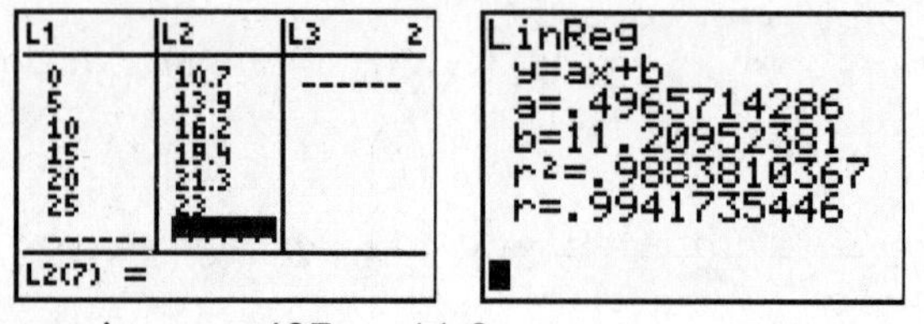

Thus, we obtain the equation $y = .497x + 11.2$.

(b) Since 1993 corresponds to $x = 23$, the best estimate for the percent completion in 1993 is $.497(23) + 11.2 = 22.631$, which you probably should report as 22.6%. The same 22.6% answer would come from the line $y = .4966x + 11.2092$. The line $y = .5x + 11.2$ produces 22.7%.

(c) Solve $.497x + 11.2 = 27.1$ and obtain $x = 31.99$, that is, 32 years after 1970. The completion rate should reach 27.1% in 2002.

7.6 Double Integrals

The problems in this section amount to performing two integrations, once with respect to y as the variable and once with respect to x. Antidifferentiation with respect to y takes a little getting used to. It is just the reverse operation to taking partial derivatives with respect to y.

1. First evaluate the inner integral.

$$\int_0^1 e^{x+y}dy = e^{x+y}\Big|_0^1 = e^{x+1} - e^x.$$

[The dy in the integral says that y is the variable of integration when evaluating e^{x+y} at 0 and 1.] Now evaluate the outer integral.

$$\int_0^1 (e^{x+1} - e^x)dx = (e^{x+1} - e^x)\Big|_0^1 = (e^2 - e) - (e - 1) = e^2 - 2e + 1.$$

7. First evaluate the inner integral.

$$\int_x^{2x} (x+y)dy = \left(xy + \frac{1}{2}y^2\right)\Bigg|_x^{2x} = \left[x(2x) + \frac{1}{2}(2x)^2\right] - \left[x \cdot x + \frac{1}{2}x^2\right]$$
$$= \left[2x^2 + 2x^2\right] - x^2 - \frac{1}{2}x^2 = \frac{5}{2}x^2.$$

Now evaluate the outer integral.

$$\int_{-1}^1 \frac{5}{2}x^2 dx = \frac{5}{6}x^3\Bigg|_{-1}^1 = \frac{5}{6} - \left(-\frac{5}{6}\right) = \frac{5}{3}.$$

13. The desired volume is given by the double integral:

$$\iint_R (x^2 + y^2)\,dx\,dy,$$

which is equivalent to the iterated integral:

$$\int_1^3 \left(\int_0^1 (x^2 + y^2)\,dy \right) dx.$$

First evaluate the inner integral:

$$\int_0^1 (x^2 + y^2)dy = \left(x^2 y + \frac{1}{3} y^3 \right)\Bigg|_0^1 = \left(x^2 + \frac{1}{3} \right) - (0+0) = x^2 + \frac{1}{3}.$$

Then evaluate the outer integral:

$$\int_1^3 \left(x^2 + \frac{1}{3} \right) dx = \left(\frac{1}{3} x^3 + \frac{1}{3} x \right)\Bigg|_1^3 = (9+1) - \left(\frac{1}{3} + \frac{1}{3} \right) = \frac{28}{3} = 9\frac{1}{3}.$$

Chapter 7: Supplementary Exercises

1. $f(2,9) = \dfrac{2\sqrt{9}}{1+2} = \dfrac{6}{3} = 2, \quad f(5,1) = \dfrac{5\sqrt{1}}{1+5} = \dfrac{5}{6}, \quad f(0,0) = \dfrac{0\sqrt{0}}{1+0} = \dfrac{0}{1} = 0.$

7. $f(x, y) = e^{x/y} = e^{xy^{-1}},$

$$\frac{\partial f}{\partial x} = y^{-1} e^{xy^{-1}} = \frac{1}{y} e^{x/y}, \qquad \text{(treat } y \text{ as a constant)}$$

$$\frac{\partial f}{\partial y} = e^{xy^{-1}} \left[\frac{\partial}{\partial y} xy^{-1} \right] = -xy^{-2} e^{xy^{-1}} = -\frac{x}{y^2} e^{x/y}, \quad \text{(treat } x \text{ as a constant).}$$

13. $f(x, y) = x^5 - 2x^3 y + \dfrac{1}{2} y^4.$

In order to find the second partial derivatives, first determine the first partial derivatives with respect to x and y.

$$\frac{\partial f}{\partial x} = 5x^4 - 6x^2 y, \quad \frac{\partial f}{\partial y} = -2x^3 + 2y^3.$$

Differentiating each of these with respect to x and y, we have

$$\frac{\partial^2 f}{\partial x^2} = \frac{\partial}{\partial x}\left(\frac{\partial f}{\partial x} \right) = \frac{\partial}{\partial x}(5x^4 - 6x^2 y) = 20x^3 - 12xy,$$

$$\frac{\partial^2 f}{\partial y^2} = \frac{\partial}{\partial y}\left(\frac{\partial f}{\partial y} \right) = \frac{\partial}{\partial y}(-2x^3 + 2y^3) = 6y^2,$$

$$\frac{\partial^2 f}{\partial x \partial y} = \frac{\partial}{\partial x}\left(\frac{\partial f}{\partial y} \right) = \frac{\partial}{\partial x}(-2x^3 + 2y^3) = -6x^2,$$

$$\frac{\partial^2 f}{\partial y \partial x} = \frac{\partial}{\partial y}\left(\frac{\partial f}{\partial x} \right) = \frac{\partial}{\partial y}(5x^4 - 6x^2 y) = -6x^2.$$

19. $f(x, y) = x^3 + 3x^2 + 3y^2 - 6y + 7$

$$\frac{\partial}{\partial x}(x^3 + 3x^2 + 3y^2 - 6y + 7) = 3x^2 + 6x,$$

$$\frac{\partial}{\partial y}(x^3 + 3x^2 + 3y^2 - 6y + 7) = 6y - 6.$$

To find possible relative maxima or minima, set the first derivatives equal to zero.

$$\frac{\partial f}{\partial x} = 3x^2 + 6x = 0, \qquad \frac{\partial f}{\partial y} = 6y - 6 = 0,$$

$$3x(x+2) = 0, \qquad 6y = 6,$$

$$x = 0, \quad x = -2, \qquad y = 1.$$

Thus $f(x, y)$ has a possible relative maximum or minimum at (0, 1) and at (−2, 1).

25. Construct the function

$$F(x, y, \lambda) = 3x^2 + 2xy - y^2 + \lambda(5 - 2x - y).$$

Set the first partial derivatives equal to zero and solve for λ:

$$\frac{\partial F}{\partial x} = 6x + 2y - 2\lambda = 0,$$

$$2\lambda = 6x + 2y,$$

$$\lambda = 3x + y.$$

$$\frac{\partial F}{\partial y} = 2x - 2y - \lambda = 0,$$

$$\lambda = 2x - 2y.$$

Next, equate the two expressions for λ:

$$3x + y = 2x - 2y \quad \text{or} \quad x = -3y.$$

The equation $\frac{\partial F}{\partial \lambda} = 5 - 2x - y = 0$ is the constraint. Substitute $-3y$ for x and obtain:

$$5 - 2(-3y) - y = 0,$$

$$5 + 5y = 0,$$

$$5y = -5, \quad \text{or} \quad y = -1.$$

Hence, $x = -3y = -3(-1) = 3$, so the maximum is at (3, −1), and the maximum value is $f(3, -1) = 3(9) + 2(3)(-1) - 1 = 27 - 6 - 1 = 20$.

31. Let the straight line by $y = Ax + B$. When $x = 1, 2, 3$, the corresponding y-coordinates of the points of the line are $A + B$, $2A + B$, $3A + B$, respectively. Therefore, the squares of the vertical distances from the line to the points (1, 1), (2, 3), and (3, 6) are

$$E_1^2 = (A+B-1)^2,$$
$$E_2^2 = (2A+B-3)^2,$$
$$E_3^2 = (3A+B-6)^2.$$

Thus the least-squares error is

$$f(A, B) = E_1^2 + E_2^2 + E_3^2 = (A-B-1)^2 + (2A+B-3)^2 + (3A+B-6)^2.$$

To minimize $f(A, B)$, take partial derivatives with respect to A and B, and set them equal to zero.

$$\begin{aligned}\frac{\partial f}{\partial A} &= 2(A+B-1) + 2(2A+B-3)\cdot 2 + 2(3A+B-6)\cdot 3\\ &= (2A+2B-2) + (8A+4B-12) + (18A+6B-36)\\ &= 28A+12B-50 = 0.\end{aligned}$$

$$\begin{aligned}\frac{\partial f}{\partial B} &= 2(A+B-1) + 2(2A+B-3)\cdot 2 + 2(3A+B-6)\\ &= (2A+2B-2) + (4A+2B-6) + (6A+2B-12)\\ &= 12A+6B-20 = 0.\end{aligned}$$

To find A and B, you must solve the system of simultaneous linear equations

$$28A+12B = 50,$$
$$12A+6B = 20.$$

Subtract 2 times the second equation from the first equation and obtain

$$4A+10, \quad \text{and} \quad A = \frac{5}{2}.$$

Hence

$$6B = 20 - 12\left(\frac{5}{2}\right) = -10, \quad \text{and} \quad B = -\frac{5}{3}.$$

Therefore, the straight line that minimizes the least-squares error is $y = \frac{5}{2}x - \frac{5}{3}$.

37. $\iint_R 5dxdy$ represents the volume of a box with dimensions $(4-0)\times(3-1)\times 5$.
So, $\iint_R 5dxdy = 4\cdot 2\cdot 5 = 40$.

Chapter 8
The Trigonometric Functions

Trigonometric functions are useful for describing physical processes that are cyclical or periodic. The goal of this short chapter is to study basic properties of these functions. We realize that many of our readers have never studied trigonometry, and so we mention a few simple connections with right triangles. However, our main interest is in calculus, not in trigonometry.

8.1 Radian Measure of Angles

All of the exercises in this text use radian measure in connection with trigonometric functions.

1. $30° = 30 \times \frac{\pi}{180}$ radians $= \frac{\pi}{6}$ radians

$120° = 120 \times \frac{\pi}{180}$ radians $= \frac{2\pi}{3}$ radians

$315° = 315 \times \frac{\pi}{180}$ radians $= \frac{7\pi}{4}$ radians.

7. The angle described by this figure consists of one full revolution plus three quarters of a revolution. That is

$$t = 2\pi + \frac{3}{4}(2\pi) = 2\pi + \frac{3}{2}\pi = \frac{7\pi}{2}.$$

13. See the figures below. Here is the reasoning that leads to those figures. First, since $\pi/2$ is one quarter-revolution of the circle, $3\pi/2$ is three quarter-revolutions. Next, since $\pi/4$ is one eighth-revolution of the circle, $3\pi/4$ is three eighth-revolutions. Notice also that $3\pi/4$ is exactly one half the angle described in the first case. Finally, since π is a one half-revolution of the circle, 5π is five half-revolutions or two-and-one-half revolutions.

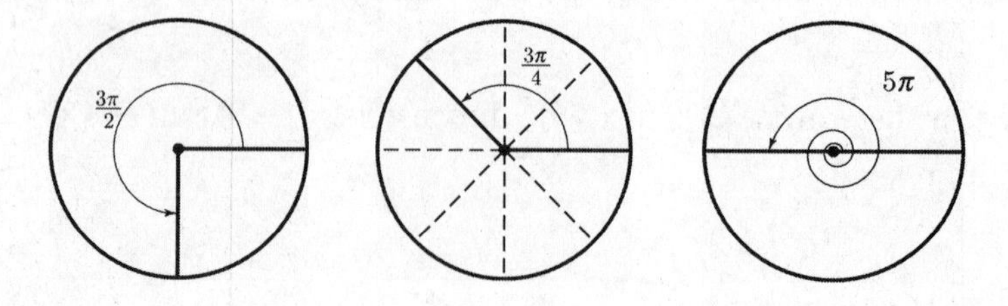

8.2 The Sine and the Cosine

The most important pictures in this section are on page 400. Years from now if someone asks you what the sine function is, you should think of the graphs in figures 12 and 13, not the triangles in figure 3, or even the circle in figure 7. It is the regular, fluctuating shape of the graphs of cos t and sin t that make them useful in mathematical models. Nevertheless, your instructor will want you to have a basic understanding of how cos t and sin t are defined, and that is the purpose of the other figures in this section.

Exercises 1–20 help you learn the basic definitions:

$$\cos t = \frac{x}{r}, \quad \text{and} \quad \sin t = \frac{y}{r},$$

where x and y are the coordinates of a point that determines an angle of t radians, and r is the distance of the point from the origin. Exercises 21–38 help you learn the alternative definitions of cos t and sin t as the x- and y-coordinates of an appropriate point on the unit circle. The alternative definitions enable you to understand some important properties of the sine and cosine. Be sure to find out which of these properties you must memorize. [Formula (7), for example, is only used later for proofs of derivative formulas, so you may not need to learn it.]

1. $\sin t = \dfrac{\text{opposite}}{\text{hypotenuse}} = \dfrac{1}{2}, \quad \cos t = \dfrac{\text{adjacent}}{\text{hypotenuse}} = \dfrac{\sqrt{3}}{2}.$

7. First compute r.

$$r = \sqrt{x^2 + y^2} = \sqrt{(-2)^2 + 1^2} = \sqrt{4+1} = \sqrt{5}.$$

Then

$$\sin t = \frac{y}{r} = \frac{1}{\sqrt{5}}, \quad \cos t = \frac{x}{r} = \frac{-2}{\sqrt{5}}.$$

13. First compute

$$\sin t = \frac{b}{c} = \frac{5}{13} = .385.$$

One way to find t is to look in the "sin t" column of a table of trigonometric functions. A number in this column close to .385 is .38942, in the row corresponding to $t = .4$. From this you can conclude that $t \approx .4$. For more accuracy, use a scientific calculator. Check to make sure that angle measurement is set for radians, and use the "inverse sine" function to compute $\sin^{-1} .385 = .39521$.

19. Since "a" is given and it is the side *adjacent* to the given angle t, use the formula

$$\cos t = \frac{\text{adjacent}}{\text{hypotenuse}}.$$

In this exercise

$$\cos .5 = \frac{2.4}{c}.$$

From a calculator, $\cos .5 = .87758$, so

$$.87758 = \frac{2.4}{c},$$
$$.87758c = 2.4,$$
$$c = \frac{2.4}{.87758} = 2.7348.$$

One way to find b, now that you know c, is to use the formula

$$\sin t = \frac{b}{c}.$$

That is,

$$\sin .5 = \frac{b}{2.73},$$
$$.47943 = \frac{b}{2.73},$$
$$2.73(.47943) = b,$$
$$b = 1.31.$$

Alternatively, using the Pythagorean theorem, $(2.4)^2 + b^2 = (2.73)^2$,

$$5.76 + b^2 = 7.45,$$
$$b^2 = 1.69,$$
$$b = 1.30.$$

The discrepancy between the two values for b is due to the round-off error in using 2.73 in place of 2.7348. Either value of b is a suitable answer.

25. On the unit circle, locate the point P that is determined by an angle of $-\frac{5\pi}{8}$ radians. The x-coordinate of P is $\cos\left(-\frac{5\pi}{8}\right)$. There is another point Q on the unit circle with the same x-coordinate. (See the figure.) Let t be the radian measure of the angle determined by Q. Then $\cos t = \cos\left(-\frac{5\pi}{8}\right)$ because Q and P have the same x-coordinate. Also, $0 \le t \le \pi$. From the symmetry of the diagram, it is clear that $t = \frac{5\pi}{8}$. Also, since $\cos(-t) = \cos t$ and $0 \le \frac{5\pi}{8} \le \pi$, we have $t = \frac{5\pi}{8}$.

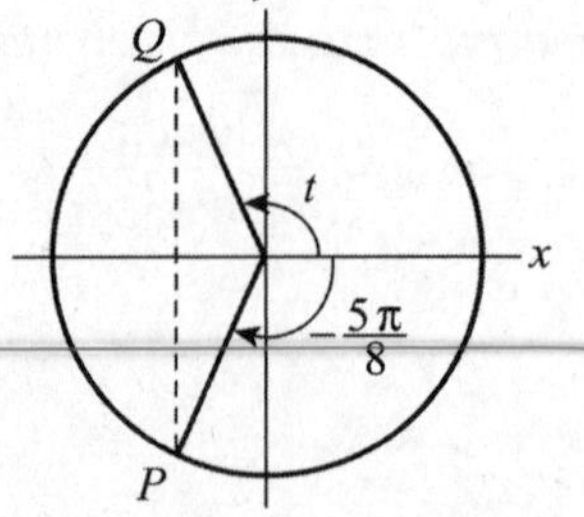

31. The equation $\sin t = -\sin(\pi/6)$ relates the y-coordinates of two points on the unit circle. One point P corresponds to an angle of $\pi/6$ radians. (See the figure.) The other point Q must be on the right half of the unit circle, since $-\frac{\pi}{2} \le t \le \frac{\pi}{2}$. But its y-coordinate must be the negative of the y-coordinate of P. From the symmetry of the figure, $t = -\frac{\pi}{6}$. Another method is to recall that $-\sin t = \sin(-t)$. In particular,

$$-\sin\left(\frac{\pi}{6}\right) = \sin\left(\frac{-\pi}{6}\right).$$

Thus $-\pi/6$ works for t. Since $-\pi/6$ is between $-\pi/2$ and $-\pi/2$, $t = -\pi/6$.

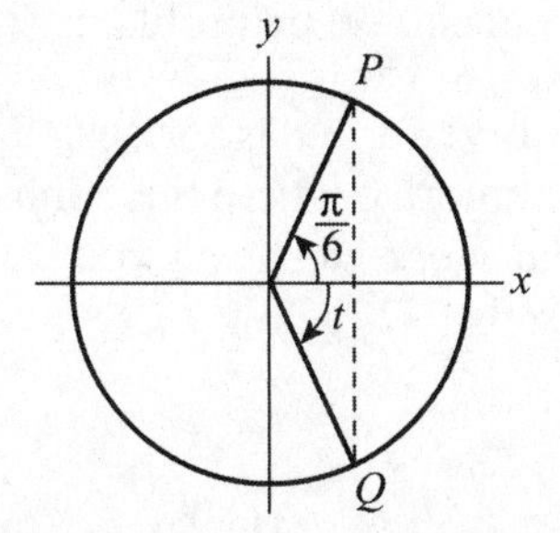

37. Think of $\sin t$ as the y-coordinate of the point P on the unit circle that is determined by an angle of t radians. From the figures below conclude that $\sin 5\pi = 0$ and $\sin(-2\pi) = 0$.

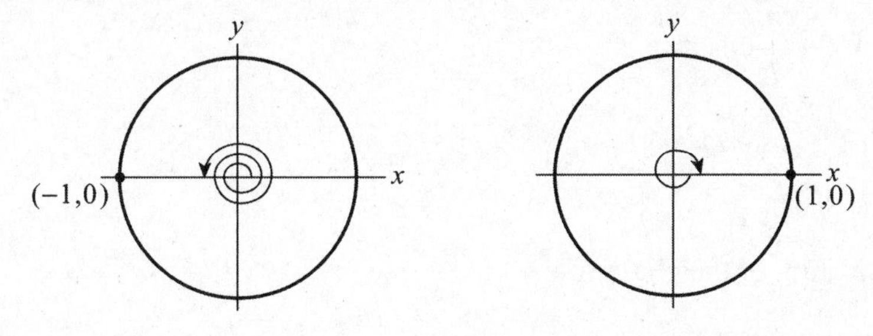

Since $\frac{17\pi}{2} = 8\pi + \frac{\pi}{2} = 4(2\pi) + \frac{\pi}{2}$, the number $\frac{17\pi}{2}$ corresponds to four complete revolutions of the circle plus one quarter-revolution. So P is the point (0, 1) and $\sin\frac{17\pi}{2} = 1$.

Since $-\frac{13\pi}{2} = -6\pi - \frac{\pi}{2} = 3(-2\pi) - \frac{\pi}{2}$, the number $-\frac{13\pi}{2}$ corresponds to three complete revolutions of the circle in the negative direction plus one plus one quarter-revolution in the negative direction. So P is the point (0, –1) and $\sin\left(-\frac{13\pi}{2}\right) = -1$.

8.3 Differentiation and Integration of sin *t* and cos *t*

The significance of the derivative formula

$$\frac{d}{dt}\sin t = \cos t$$

is revealed by figures 1 and 2 in the text. That is, for each number t, the value of $\cos t$ gives the slope of the sine curve at the specified value of t. This fundamental property of the derivative as a slope function is worth contemplating because it will be crucial for your work in Chapter 10.

Most of the exercises in this section are routine drill problems. Exercises 35–46 are just a "warm-up" to remind you of antiderivatives. We will concentrate on problems such as these in Section 9.1.

Helpful Hint: *Don't* memorize antiderivative formulas for $\cos t$ and $\sin t$, because you might confuse them with the derivative formulas. Just memorize where the minus sign goes in the derivative formulas, and then use this knowledge to check your work whenever you need to "guess" an antiderivative involving a sine or cosine. (See our solution to Exercise 37, below.)

1. By the chain rule,

$$\frac{d}{dt}(\sin 4t) = (\cos 4t)\cdot\frac{d}{dt}4t = 4\cos 4t.$$

7. $$\begin{aligned}\frac{d}{dt}(t+\cos \pi t) &= \frac{d}{dt}t+\frac{d}{dt}\cos \pi t\\ &= 1+(-\sin \pi t)\frac{d}{dt}\pi t\\ &= 1-\pi\sin \pi t.\end{aligned}$$

13. By repeated application of the chain rule,

$$\begin{aligned}\frac{d}{dx}\sin\sqrt{x-1} &= \cos\sqrt{x-1}\cdot\frac{d}{dx}\sqrt{x-1}\\ &= \cos\sqrt{x-1}\cdot\frac{d}{dx}(x-1)^{1/2}\\ &= \cos\sqrt{x-1}\cdot\left[\frac{1}{2}(x-1)^{-1/2}\right]\frac{d}{dx}(x-1)\\ &= \cos\sqrt{x-1}\cdot\left[\frac{1}{2}(x-1)^{-1/2}\right]\\ &= \frac{\cos\sqrt{x-1}}{2\sqrt{x-1}}.\end{aligned}$$

Helpful Hint: Instructors like problems such as Exercise 19 because they force students to understand the notation. To avoid mistakes, be sure to include the first step shown in the solution above.

19. By repeated application of the chain rule,

$$\begin{aligned}\frac{d}{dx}\cos^2 x^3 &= \frac{d}{dx}(\cos x^3)^2 \\ &= 2\cos x^3 \cdot \frac{d}{dx}\cos x^3 \\ &= 2\cos x^3(-\sin x^3)\cdot\frac{d}{dx}x^3 \\ &= 2\cos x^3(-\sin x^3)3x^2 \\ &= -6x^2\cos x^3 \sin x^3.\end{aligned}$$

25. By the quotient rules,

$$\begin{aligned}\frac{d}{dt}\left(\frac{\sin t}{\cos t}\right) &= \frac{\cos t(\cos t) - \sin t(-\sin t)}{\cos^2 t} \\ &= \frac{\cos^2 t + \sin^2 t}{\cos^2 t} = \frac{1}{\cos^2 t}\end{aligned}$$

31. In order to find the slope of the line tangent to the graph of $y = \cos 3x$ at $x = 13\pi/6$, first find the derivative of the equation.

$$\frac{d}{dx}(\cos 3x) = -(\sin 3x)(3) = -3\sin 3x$$

Now evaluate the derivative for $x = 13\pi/6$ to determine the slope.

$$\text{slope} = -3\sin 3\left(\frac{13\pi}{6}\right) = -3(1) = -3.$$

37. Our first guess is that an antiderivative of $\cos\frac{x}{7}$ would involve the sine function, namely $\sin\frac{x}{7}$. We may differentiate our guess (not the original function) to check our reasoning:

$$\frac{d}{dx}\sin\frac{x}{7} = \cos\frac{x}{7}\cdot\left(\frac{1}{7}\right) = \frac{1}{7}\cos\frac{x}{7}.$$

We need to adjust our guess so the constant in front of our derivative becomes $-\frac{1}{2}$. We adjust our guess by multiplying it by $-\frac{7}{2}$. Now

$$\frac{d}{dx}\left(-\frac{7}{2}\sin\frac{x}{7}\right) = -\frac{7}{2}\cos\frac{x}{7}\cdot() = -\frac{1}{2}\cos\frac{x}{7}.$$

Thus $-\frac{7}{2}\sin\frac{x}{7}$ is an antiderivative of $-\frac{1}{2}\cos\frac{x}{7}$. Therefore,

$$\int -\frac{1}{2}\cos\frac{x}{7} = -\frac{7}{2}\sin\frac{x}{7} + C.$$

43. Our first guess is that an antiderivative of sin(4*x* + 1) should involve the cosine function, namely, cos(4 + 1). We may differentiate our guess (*not* the original function) to check our reasoning:

$$\frac{d}{dx}\cos(4x+1) = -\sin(4x+1)\cdot 4 = -4\sin(4x+1).$$

This derivative is −4 times what it should be. So, we adjust our guess by dividing it by −4. Now

$$\begin{aligned}\frac{d}{dx}\left[-\frac{1}{4}\cos(4x+1)\right] &= -\frac{1}{4}\cdot\frac{d}{dx}\cos(4x+1)\\ &= -\frac{1}{4}[-\sin(4x+1)\cdot 4]\\ &= \sin(4x+1).\end{aligned}$$

Thus $-\frac{1}{4}\cos(4x+1)$ *is* an antiderivative of sin(4*x* + 1). Hence,

$$\int \sin(4x+1)dx = -\frac{1}{4}\cos(4x+1)+C.$$

49. We are looking for

$$\lim_{h\to 0}\frac{\sin\left(\frac{\pi}{2}+h\right)-1}{h},$$

or equivalently

$$\lim_{h\to 0}\frac{\sin\left(\frac{\pi}{2}+h\right)-\sin\left(\frac{\pi}{2}\right)}{h}.$$

Since

$$f'(x) = \lim_{h\to 0}\frac{f(a+h)-f(a)}{h},$$

we see that $f(x)=\sin x$ and $a=\frac{\pi}{2}$. Therefore,

$$\begin{aligned}\frac{d}{dx}\sin x\Big|_{x=\pi/2} = \cos x\Big|_{x=\pi/2} &= \cos\frac{\pi}{2} = 0\\ \Rightarrow \lim_{h\to 0}\frac{\sin\left(\frac{\pi}{2}+h\right)-1}{h} &= 0.\end{aligned}$$

8.4 The Tangent and Other Trigonometric Functions

You may need to memorize the discussion of how to get a derivative formula for tan *t*. (See the bottom of page 413 or the *Manual* solution to Exercise 25 in Section 8.2.) This is a nice application of the quotient rule. Of course you should also memorize the formula:

$$\frac{d}{dt}\tan t = \sec^2 t.$$

1. Since

$$\cos t = \frac{\text{adjacent}}{\text{hypotenuse}} \quad \text{and} \quad \sec t = \frac{1}{\cos t},$$

we have

$$\sec t = \frac{\text{hypotenuse}}{\text{adjacent}}.$$

7. $\tan t = \frac{y}{x} = \frac{2}{-2} = -1$. To find $\sec t$ you need r.

$$\begin{aligned} r &= \sqrt{x^2 + y^2} = \sqrt{(-2)^2 + 2^2} = \sqrt{4+4} \\ &= \sqrt{8} = 2\sqrt{2}, \\ \sec t &= \frac{r}{x} = \frac{2\sqrt{2}}{-2} = -\sqrt{2}. \end{aligned}$$

13. Note that $\sec t = 1/\cos t = (\cos t)^{-1}$. Hence,

$$\begin{aligned} \frac{d}{dt}\sec t &= \frac{d}{dt}(\cos t)^{-1} = (-1)(\cos t)^{-2} \cdot \frac{d}{dt}\cos t \\ &= (-1)(\cos t)^{-2}(-\sin t) = \frac{\sin t}{\cos^2 t}. \end{aligned}$$

This answer is acceptable. It may be written in other equivalent forms, such as $\sin t \sec^2 t$, or

$$\frac{\sin t}{\cos t} \cdot \frac{1}{\cos t} = \tan t \sec t.$$

19. By the chain rule,

$$\begin{aligned} f'(x) &= \frac{d}{dx} 3\tan(\pi - x) \\ &= 3\sec^2(\pi - x) \cdot \frac{d}{dx}(\pi - x) \\ &= -3\sec^2(\pi - x). \end{aligned}$$

25. By the product rule,

$$\begin{aligned} y' &= \frac{d}{dx}(x \tan x) \\ &= x \cdot \frac{d}{dx}(\tan x) + (\tan x) \cdot \frac{d}{dx} x \\ &= x\sec^2 x + \tan x. \end{aligned}$$

31. From the chain rule and the solution to Exercise 13,

$$y' = \frac{d}{dt}\ln(\tan t + \sec t) = \frac{1}{\tan t + \sec t}\cdot\frac{d}{dt}(\tan t + \sec t)$$

$$= \frac{1}{\tan t + \sec t}\left(\frac{d}{dt}\tan t + \frac{d}{dt}\sec t\right) = \frac{1}{\tan t + \sec t}(\sec^2 t + \tan t \sec t)$$

$$= \frac{(\sec t + \tan t)\sec t}{\tan t + \sec t} = \sec t.$$

37. In this section of the text it is shown that the derivative of *tan*(x) is *sec*^{2} (x). Hence *tan*(x) is an antiderivative of *sec*^{2}(x), so the integration can proceed as follows:

$$\int_{-\pi/4}^{\pi/4} \sec^2 x dx = (\tan x)\Big|_{-\pi/4}^{\pi/4} = \tan\left(\frac{\pi}{4}\right) - \tan\left(\frac{-\pi}{4}\right)$$

$$= 1 - (-1) = 2.$$

Chapter 8: Supplementary Exercises

Use the chapter checklist to review the main definitions and facts. You may also need to memorize some of the following facts: (1) the graphs of $\sin t$ and $\cos t$ (knowing the values of $\sin t$ and $\cos t$ when $t = 0$, $\pi/2$, π, $3\pi/2$, and 2π, for example); (2) selected trigonometric identities from Section 8.2; and (3) the definitions of $\sec t$, $\cot t$, and $\csc t$. Check with your instructor.

1. The angle described is three quarter-revolutions of the circle or $3\left(\frac{\pi}{2}\text{ radians}\right) = \frac{3\pi}{2}$ radians.

7. First find r.

$$r = \sqrt{x^2 + y^2} = \sqrt{3^2 + 4^2} = \sqrt{25} = 5.$$

Therefore,

$$\sin t = \frac{y}{r} = \frac{4}{5} = .8, \quad \cos t = \frac{x}{r} = \frac{3}{5} = .6, \quad \tan t = \frac{y}{x} = \frac{4}{3}.$$

13. $\sin t = \cos t$ whenever the point on the unit circle defined by the angle t has identical x and y coordinates (i.e., the point lies on the line $x = y$). The points on the unit circle with this property are $(1/\sqrt{2}, 1/\sqrt{2})$ and $(-1/\sqrt{2}, -1/\sqrt{2})$. The four angles between -2π and 2π that correspond to these points are $-\frac{3\pi}{4}, -\frac{7\pi}{4}, \frac{\pi}{4}, \frac{5\pi}{4}$.

19. $f'(t) = \frac{d}{dt}3\sin t = 3\cos t.$

25. By the quotient rule,

$$\begin{aligned}
f'(x) &= \frac{d}{dx}\left(\frac{\cos 2x}{\sin 3x}\right) \\
&= \frac{\sin(3x)\cdot \frac{d}{dx}\cos 2x - \cos(2x)\cdot \frac{d}{dx}\sin 3x}{(\sin 3x)^2} \\
&= \frac{\sin(3x)(-2\sin 2x) - 3\cos(2x)\cos(3x)}{\sin^2(3x)} \\
&= -\frac{2\sin(3x)\sin(2x) + 3\cos(2x)\cos(3x)}{\sin^2(3x)}
\end{aligned}$$

31. By the chain rule,

$$\begin{aligned}
y' &= \frac{d}{dx}\sin(\tan x) = \cos(\tan x)\frac{d}{dx}\tan x \\
&= \cos(\tan x)\sec^2 x.
\end{aligned}$$

37. By the product and chain rules,

$$\begin{aligned}
y' &= \frac{d}{dx}e^{3x}\sin^4 x \\
&= e^{3x}(4\sin^3 x)\cdot \frac{d}{dx}\sin x + (\sin^4 x)\cdot 3e^{3x} \\
&= 4e^{3x}\sin^3 x\cos x + 3e^{3x}\sin^4 x.
\end{aligned}$$

43. $f(t) = \sin^2 t.$

$$\begin{aligned}
f'(t) &= \frac{d}{dt}(\sin t)^2 \\
&= (2\sin t)\cdot \frac{d}{dt}\sin t \\
&= 2\sin t\cos t. \\
f''(t) &= \frac{d}{dt}(2\sin t\cdot \cos t) \\
&= 2\left[(\sin t)\cdot \frac{d}{dt}\cos t + (\cos t)\cdot \frac{d}{dt}\sin t\right] \\
&= 2\left[(\sin t)(-\sin t) + \cos t\cdot \cos t\right] \\
&= 2(\cos^2 t - \sin^2 t).
\end{aligned}$$

49. $y' = \frac{d}{dt}\tan t = \sec^2 t.$

The slope of the tangent line when $t = \frac{\pi}{4}$ is

$$\sec^2\left(\frac{\pi}{4}\right) = \left(\frac{1}{\cos\frac{\pi}{4}}\right)^2 = \left(\frac{1}{\frac{\sqrt{2}}{2}}\right)^2 = 2.$$

At $t = \frac{\pi}{4}$, $y = \tan\frac{\pi}{4} = 1$, so $\left(\frac{\pi}{4}, 1\right)$ is a point on the line. Hence, the equation of the line is

$$y - 1 = 2\left(t - \frac{\pi}{4}\right).$$

55. It is easy to check that the line $y = x$ is tangent to the graph of $y = \sin x$ at $x = 0$. From the graph of $y = \sin x$, it is clear that $y = \sin x$ lies below $y = x$ for $x \geq 0$. So, the area between these two curves from $x = 0$ to $x = \pi$ is given by

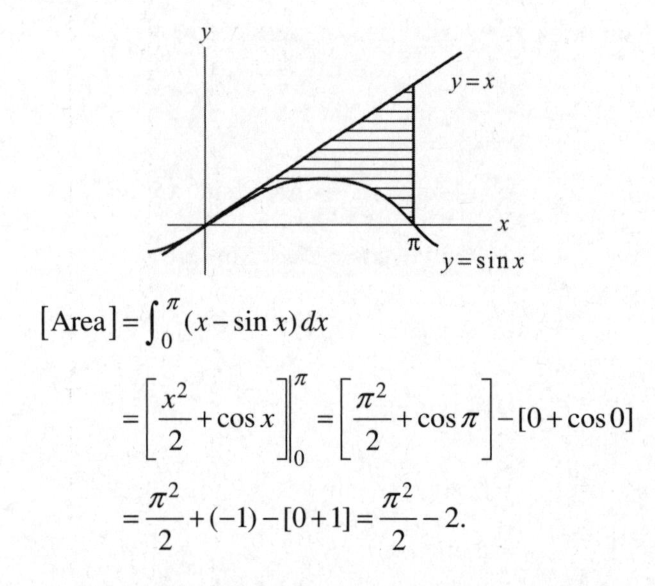

$$[\text{Area}] = \int_0^{\pi} (x - \sin x)\,dx$$

$$= \left[\frac{x^2}{2} + \cos x\right]_0^{\pi} = \left[\frac{\pi^2}{2} + \cos\pi\right] - [0 + \cos 0]$$

$$= \frac{\pi^2}{2} + (-1) - [0 + 1] = \frac{\pi^2}{2} - 2.$$

61. It may be helpful to review Example 5(b) of Section 8.3 before attempting this problem. To integrate, we will need to find an antiderivative of *cos*(6*x*). As a first guess, we could try *sin*(6*x*). Differentiating shows

$$\frac{d}{dx}(\sin 6x) = 6\cos 6x,$$

which is 6 times too much. So we divide our original guess by 6 to find that $\frac{1}{6}\sin(6x)$ is a correct antiderivative of $\cos(6x)$. (Check!) The integration proceeds as follows:

$$\int_0^{\pi/2} \cos(6x)dx = \left(\frac{1}{6}\sin(6x)\right)\Bigg|_0^{\pi/2} = \left(\frac{1}{6}\sin(3\pi)\right) - \left(\frac{1}{6}\sin(0)\right)$$

$$= 0 - 0 = 0.$$

67. The shaded region A_1 represents the region between the upper curve ($y = \cos x$) and the lower curve ($y = \sin x$) from $x = 0$ to $x = \pi/4$. Thus the area of A_1 can be computed by evaluating the integral

$$\int_0^{\pi/4} (\cos x - \sin x)dx.$$

Use the facts that sin(x) is an antiderivative of cos(x) and cos(x) is an antiderivative of –sin(x) to integrate:

$$\begin{aligned}\int_0^{\pi/4} (\cos x - \sin x)dx &= (\sin x + \cos x)\Big|_0^{\pi/4}\\ &= \left(\sin\frac{\pi}{4} + \cos\frac{\pi}{4}\right) - (\sin 0 + \cos 0)\\ &= \left(\frac{\sqrt{2}}{2} + \frac{\sqrt{2}}{2}\right) - (0+1)\\ &= \sqrt{2} - 1.\end{aligned}$$

73. Using the formula for the average value of a function on an interval given in Section 6.5, we see that the answer to this problem can be computed by evaluating

$$\frac{1}{\frac{3\pi}{4} - 0}\int_0^{3\pi/4}\left[1000 + 200\sin\left(2\left(t - \frac{\pi}{4}\right)\right)\right]dt.$$

This can be partially evaluated as follows:

$$\begin{aligned}&\frac{1}{3\pi/4 - 0}\int_0^{3\pi/4}\left[1000 + 200\sin\left(2\left(t - \frac{\pi}{4}\right)\right)\right]dt\\ &= \frac{1}{3\pi/4}\left[\int_0^{3\pi/4} 1000\,dt + 200\left(\int_0^{3\pi/4}\sin\left(2\left(t-\frac{\pi}{4}\right)\right)dt\right)\right]\\ &= \frac{4}{3\pi}\left[\left(1000t\Big|_0^{3\pi/4}\right) + 200\left(\int_0^{3\pi/4}\sin\left(2\left(t-\frac{\pi}{4}\right)\right)dt\right)\right]\\ &= \frac{4}{3\pi}\left[750\pi + 200\left(\int_0^{3\pi/4}\sin\left(2\left(t-\frac{\pi}{4}\right)\right)dt\right)\right]\\ &= 1000 + \frac{800}{3\pi}\left(\int_0^{3\pi/4}\sin\left(2\left(t-\frac{\pi}{4}\right)\right)dt\right).\end{aligned}$$

All that remains is to evaluate the integral

$$\int_0^{3\pi/4}\sin\left(2\left(t-\frac{\pi}{4}\right)\right)dt$$

and simplify. To find an antiderivative of

$$\sin\left(2\left(t-\frac{\pi}{4}\right)\right),$$

it is helpful to refer to the method used in the solution to Exercise 61 and Example 5(b) of Section 8.3 to observe that

$$-\frac{1}{2}\cos(2t)$$

is an antiderivative of sin($2t$).

Use the chain rule to verify that

$$\cos\left(t-\frac{\pi}{4}\right)$$

is an antiderivative of $\sin\left(t-\frac{\pi}{4}\right)$. Thus

$$-\frac{1}{2}\cos\left(2\left(t-\frac{\pi}{4}\right)\right)$$

is a reasonable (and correct) guess for an antiderivative of $\sin\left(2\left(t-\frac{\pi}{4}\right)\right)$. Thus the remaining integral can be evaluated as follows:

$$\begin{aligned}\int_0^{3\pi/4}\sin\left(2\left(t-\frac{\pi}{4}\right)\right)dt &= \left(-\frac{1}{2}\cos\left(2\left(t-\frac{\pi}{4}\right)\right)\right)\Bigg|_0^{3\pi/4}\\ &= \left(-\frac{1}{2}\cos\left(2\left(\frac{\pi}{2}\right)\right)\right)-\left(-\frac{1}{2}\cos\left(\frac{-2\pi}{4}\right)\right)\\ &= \left(-\frac{1}{2}(-1)\right)-0=\frac{1}{2}.\end{aligned}$$

Finally, the average value can now be computed:

$$\begin{aligned}1000+\frac{800}{3\pi}\left(\int_0^{3\pi/4}\sin\left(2\left(t-\frac{\pi}{4}\right)\right)dt\right) &= 1000+\frac{800}{3\pi}\left(\frac{1}{2}\right)\\ &= 1000+\frac{400}{3\pi}\approx 1042.44\end{aligned}$$

79. First, use identity (1) of Section 8.4 to transform the integral before evaluating it, as suggested in the hint:

$$\int_{-\pi/8}^{\pi/8}\tan^2(2x)\,dx=\int_{-\pi/8}^{\pi/8}(\sec^2(2x)-1)\,dx.$$

Next, use the fact that $\frac{1}{2}\tan 2x$ is an antiderivative of $\sec^2(2x)$ to evaluate the new integral:

$$\begin{aligned}\int_{-\pi/8}^{\pi/8}(\sec^2(2x)-1)\,dx &= \left(\frac{1}{2}\tan(2x)-x\right)\Bigg|_{-\pi/8}^{\pi/8}\\ &= \left(\frac{1}{2}\tan\left(\frac{\pi}{4}\right)-\frac{\pi}{8}\right)-\left(\frac{1}{2}\tan\left(\frac{-\pi}{4}\right)-\frac{-\pi}{8}\right)\\ &= \left(\frac{1}{2}(1)-\frac{\pi}{8}\right)-\left(\frac{1}{2}(-1)+\frac{\pi}{8}\right)\\ &= \frac{1}{2}-\frac{\pi}{8}+\frac{1}{2}-\frac{\pi}{8}\\ &= 1-\frac{\pi}{4}\approx .2146.\end{aligned}$$

Chapter 9
Techniques of Integration

9.1 Integration by Substitution

The technique of integration by substitution is best learned by trial and error. The more problems you work the fewer errors you will make in your first guess at a substitution. The basic strategy is to arrange the integrand in the form $f(g(x))g'(x)$. (This is possible for every problem in this section.) Look for the most complicated part of the integrand that may be expressed as a composite function, and let u be the "inside" function $g(x)$. Then proceed as discussed in the text.

Remember that you can check your answers to an antidifferentiation problem by differentiating your answer. (You ought to do this at least mentally for every indefinite integral you find.) We urge you to work *all* the problems (odd and even) in Section 9.1. Use the solutions in this *Guide* to check your work, *not* to show you how to start a problem.

1. Let $u = x^2 + 4$. Then $du = \dfrac{d}{dx}(x^2 + 4)dx = 2x\,dx$. So

$$\begin{aligned}\int 2x(x^2+4)^5\,dx &= \int u^5 du \\ &= \frac{1}{6}u^6 + C \\ &= \frac{1}{6}(x^2+4)^6 + C.\end{aligned}$$

7. Let $u = 4 - x^2$. Then $du = \frac{d}{dx}(4 - x^2)dx = -2x\,dx$. So,

$$\begin{aligned}\int x\sqrt{4-x^2}\,dx &= \int\left(-\frac{1}{2}\right)(-2)x\sqrt{4-x^2}\,dx \\ &= -\frac{1}{2}\int\sqrt{4-x^2}\,\underbrace{(-2x)dx}_{du} = -\frac{1}{2}\int\sqrt{u}\,du \\ &= -\frac{1}{2}\int u^{1/2}du = -\frac{1}{2}\cdot\frac{2}{3}u^{3/2} + C \\ &= -\frac{1}{3}u^{3/2} + C = -\frac{1}{3}(4-x^2)^{3/2} + C.\end{aligned}$$

13. Let $u = \ln(2x)$. Then $du = \frac{d}{dx}[\ln(2x)]dx = \frac{1}{2x} \cdot 2dx = \frac{1}{x}dx$. Thus

$$\int \frac{\ln 2x}{x} dx = \int (\ln 2x) \cdot \frac{1}{x} dx = \int u\, du$$
$$= \frac{u^2}{2} + C = \frac{(\ln 2x)^2}{2} + C.$$

19. $\ln \sqrt{x} = \ln x^{1/2} = \frac{1}{2} \ln x$. This gives us the new form:

$$\int \frac{\ln \sqrt{x}}{x} dx = \int \frac{\frac{1}{2}\ln x}{x} dx = \frac{1}{2} \int \frac{\ln x}{x} dx.$$

So we first need to find $\int \frac{\ln x}{x} dx$.

Let $u = \ln x$. Then

$$du = \frac{d}{dx} \ln x\, dx = \frac{1}{x} dx.$$

Hence,

$$\int \frac{\ln x}{x} dx = \int u\, du = \frac{u^2}{2} + C = \frac{(\ln x)^2}{2} + C.$$

Finally,

$$\int \frac{\ln \sqrt{x}}{x} dx = \frac{1}{2} \frac{\ln x}{x} dx = \frac{1}{2}\left[\frac{(\ln x)^2}{2} + C\right] = \frac{(\ln x)^2}{4} + C.$$

25. A good strategy to use when faced with an unfamiliar integral is to ask yourself, "Can I rewrite the integrand in some way to make the problem simpler?" You should recall that $\ln x^2 = 2 \ln x$. That fact is helpful here:

$$\int \frac{1}{x \ln x^2} dx = \int \frac{1}{2x \ln x} dx = \frac{1}{2} \int \frac{1}{x \ln x} dx.$$

Let $u = \ln x$. Then $du = \frac{1}{x} dx$, and

$$\int \frac{1}{x \ln x^2} dx = \frac{1}{2} \int \frac{1}{x} \cdot \frac{1}{\ln x} dx = \frac{1}{2} \int \frac{1}{u} du = \frac{1}{2} \ln|u| + C = \frac{1}{2} \ln|\ln x| + C.$$

If you do not happen to notice the initial simplification of the integral shown, you can make the substitution $u = \ln x^2$. In this case,

$$du = \frac{d}{dx}(\ln x^2)\, dx = \frac{1}{x^2} \cdot 2x\, dx = \frac{2}{x} dx.$$

Then

$$\int \frac{1}{x \ln x^2} dx = \frac{1}{2} \int \frac{2}{x} \cdot \frac{1}{\ln x^2} dx = \frac{1}{2} \int \frac{1}{u} du = \frac{1}{2} \ln|\ln x^2| + C.$$

You may find it strange that this antiderivative is similar to the first one but is not quite the same. (Both antiderivatives are correct.) How can that be possible? How must two antiderivatives of the same function be related?

31. Let $u = 1 + 2e^x$. Then $du = \frac{d}{dx}(1 + 2e^x)dx = 2e^x dx$. Then

$$\begin{aligned}\int \frac{e^x}{1+2e^x}dx &= \frac{1}{2}\int \frac{1}{1+2e^x}\cdot 2e^x dx = \frac{1}{2}\int \frac{1}{u}du \\ &= \frac{1}{2}\ln|u| + C \\ &= \frac{1}{2}\ln|1+2e^x| + C.\end{aligned}$$

37. If $f'(x) = \frac{x}{\sqrt{x^2+9}}$, then $f(x) = \int \frac{x}{\sqrt{x^2+9}}dx$. Let $u = x^2 + 9$. Then $du = 2x\,dx$, so

$$\begin{aligned}\int \frac{x}{\sqrt{x^2+9}}dx &= \int \frac{\left(\frac{1}{2}\right)2x}{\sqrt{x^2+9}}dx = \frac{1}{2}\int \frac{du}{\sqrt{u}} = \frac{1}{2}\int u^{-1/2}du \\ &= \frac{1}{2}2u^{1/2} + C = u^{1/2} + C = \sqrt{x^2+9} + C.\end{aligned}$$

But $f(4) = 8$, so $8 = \sqrt{4^2+9} + C$ which implies that $C = 3$. Therefore, $f(x) = \sqrt{x^2+9} + 3$.

43. Let $u = \sin x$. Thus $du = \frac{d}{dx}(\sin x)dx = \cos x\,dx$. So,

$$\begin{aligned}\int \sin x \cos x\,dx &= \int u\,du \\ &= \frac{u^2}{2} + C = \frac{(\sin x)^2}{2} + C.\end{aligned}$$

Note: Another substitution is $u = \cos x$. Then $du = -\sin x\,dx$, and

$$\begin{aligned}\int \sin x \cos x\,dx &= \int -\cos x(-\sin x)dx = \int -u\,du \\ &= -\frac{u^2}{2} + C = -\frac{(\cos x)^2}{2} + C.\end{aligned}$$

How can the two functions $\frac{(\sin x)^2}{2}$ and $-\frac{(\cos x)^2}{2}$ both be antiderivatives of $\sin x \cos x$?

49. Let $u = 2 - \sin 3x$. Then

$$du = \frac{d}{dx}(2 - \sin 3x)dx = -3\cos 3x\,dx.$$

So,

$$\begin{aligned}\int \frac{\cos 3x}{\sqrt{2-\sin 3x}}dx &= \int (2-\sin 3x)^{-1/2}\left(-\frac{1}{3}\right)(-3\cos 3x)\,dx \\ &= -\frac{1}{3}\int u^{-1/2}du = -\frac{1}{3}(2u^{1/2}) + C \\ &= -\frac{2}{3}(2-\sin 3x)^{1/2} + C.\end{aligned}$$

9.2 Integration by Parts

Exercises 1–24 are all solved using integration by parts. This technique, like the method of substitution, is learned by working many problems. There is another method for integration by parts that some students may have seen elsewhere. It uses u in place of $f(x)$ and dv in place of $g(x)dx$. Years ago we tried teaching both methods and we found that the one we now present in the text is easier to learn and use. Once you have mastered integration by parts and by substitution, you are ready to try Exercises 25–36. These problems are more difficult because you have to decide which method to use!

(a) Always consider the method of substitution first. This will work if the integrand is the product of two functions and one of the functions is the derivative of the "inside" of the other function (except maybe for a constant multiple).

(b) If the integrand is the product of two unrelated functions, try integration by parts.

(c) Some antiderivatives cannot be found by either of the methods we have discussed. We will never give you such a problem to solve, but you should be careful if you look in another text for more problems to work.

1. Let $f(x) = x, \quad g(x) = e^{5x},$

$f'(x) = 1, \quad G(x) = \frac{1}{5}e^{5x}.$ [Stop and check that $G'(x) = g(x)$.]

Then

$$\begin{aligned}\int xe^{5x}dx &= \frac{1}{5}xe^{5x} - \int \frac{1}{5}e^{5x}dx \\ &= \frac{1}{5}xe^{5x} - \frac{1}{5}\int e^{5x}dx \\ &= \frac{1}{5}xe^{5x} - \frac{1}{5}\cdot\frac{1}{5}e^{5x} + C \\ &= \frac{1}{5}xe^{5x} - \frac{1}{25}e^{5x} + C.\end{aligned}$$

Helpful Hint: The various steps in your calculation of $\int xe^{5x}dx$ should be connected by equals signs, showing how you move from one step to the next. It requires a little more effort to rewrite the first term $\frac{1}{5}xe^{5x}$ on each line, but the solution is not correct otherwise. A careless student might write

$$\begin{aligned}\int xe^{5x}dx &= \frac{1}{5}xe^{5x} - \int \frac{1}{5}e^{5x}dx \\ &\boxed{=} \quad -\frac{1}{5}\int e^{5x}dx \\ \boxed{\text{errors!}} \quad &\boxed{=} \quad -\frac{1}{5}\cdot\frac{1}{5}e^{5x} + C \\ &\boxed{=} \quad -\frac{1}{25}e^{5x} + C.\end{aligned}$$

Avoid such "shortcuts" in your *homework* as well as on tests. Develop the habit of writing proper mathematical solutions on your homework. This will help to generate correct patterns of thought when you take an exam.

7. Let $f(x)=x,\qquad g(x)=\dfrac{1}{\sqrt{x+1}}=(x+1)^{-1/2},$

$f'(x)=1,\qquad G(x)=2(x+1)^{1/2}.$ [Stop and check that $G'(x)=g(x)$.]

Then

$$\begin{aligned}\int\frac{x}{\sqrt{x+1}}\,dx &= 2x(x+1)^{1/2}-\int 2(x+1)^{1/2}\,dx\\ &= 2x(x+1)^{1/2}-2\int(x+1)^{1/2}\,dx\\ &= 2x(x+1)^{1/2}-2\cdot\frac{2}{3}(x+1)^{3/2}+C\\ &= 2x(x+1)^{1/2}-\frac{4}{3}(x+1)^{3/2}+C.\end{aligned}$$

13. Let $f(x)=x,\qquad g(x)\sqrt{x+1}=(x+1)^{1/2},$

$f'(x)=1,\qquad G(x)=\dfrac{2}{3}(x+1)^{3/2}.$

Then

$$\begin{aligned}\int x\sqrt{x+1}\,dx &= \frac{2}{3}x(x+1)^{3/2}-\int\frac{2}{3}(x+1)^{3/2}\,dx\\ &= \frac{2}{3}x(x+1)^{3/2}-\frac{2}{3}\int(x+1)^{3/2}\,dx\\ &= \frac{2}{3}x(x+1)^{3/2}-\frac{2}{3}\cdot\frac{2}{5}(x+1)^{5/2}+C\\ &= \frac{2}{3}x(x+1)^{3/2}-\frac{4}{15}(x+1)^{5/2}+C.\end{aligned}$$

19. Let $f(x)=\ln 5x,\qquad g(x)=x,$

$f'(x)=\dfrac{5}{5x}=\dfrac{1}{x},\qquad G(x)=\dfrac{x^2}{2}.$ See Example 7 for a similar problem.

Then

$$\begin{aligned}\int x\ln 5x\,dx &= \frac{x^2}{2}\ln 5x-\int\frac{1}{x}\cdot\frac{x^2}{2}\,dx\\ &= \frac{x^2}{2}\ln 5x-\frac{1}{2}\int x\,dx\\ &= \frac{x^2}{2}\ln 5x-\frac{1}{2}\cdot\frac{1}{2}x^2+C\\ &= \frac{x^2}{2}\ln 5x-\frac{1}{4}x^2+C.\end{aligned}$$

25. Since $x(x+5)^4$ is the product of two functions, neither of which is a multiple of the derivative of the other, try integration by parts.

Set $f(x)=x, \quad g(x)=(x+5)^4,$

$f'(x)=1, \quad G(x)=\frac{1}{5}(x+5)^5.$

Then

$$\begin{aligned}\int x(x+5)^4\,dx &= \frac{1}{5}x(x+5)^5 - \int \frac{1}{5}(x+5)^5\,dx \\ &= \frac{1}{5}x(x+5)^5 - \frac{1}{5}\cdot\frac{1}{6}(x+5)^6 + C \\ &= \frac{1}{5}x(x+5)^5 - \frac{1}{30}(x+5)^6 + C.\end{aligned}$$

31. Since x^2+1 has $2x$ as a derivative, and $2x$ is a multiple of x, use substitution to evaluate the integral. Let $u=x^2+1$. Then $du=2x\,dx$. So,

$$\begin{aligned}\int x\sec^2(x^2+1)dx &= \frac{1}{2}\int 2x\sec^2(x^2+1)\,dx = \frac{1}{2}\int \sec^2 u\,du \\ &= \frac{1}{2}\tan u + C = \frac{1}{2}\tan(x^2+1)+C.\end{aligned}$$

37. The slope is $\frac{x}{\sqrt{x+9}}$, so

$$f'(x)=\frac{x}{\sqrt{x+9}} \quad \text{and} \quad f(x)=\int\frac{x}{\sqrt{x+9}}\,dx.$$

We can integrate by parts. Since $f(x)$ already denotes a function, write $h(x)$ in place of the usual $f(x)$ in the formula for integration by parts. That is, let

$$h(x)=x, \quad g(x)=(x+9)^{-1/9},$$
$$h'(x)=1, \quad G(x)=2(x+9)^{1/2}.$$

Then

$$\begin{aligned}\int x(x+9)^{-1/9}\,dx &= x\cdot 2(x+9)^{1/2} - \int 1\cdot 2(x+9)^{1/2}\,dx \\ &= 2x(x+9)^{1/2} - \int 2(x+9)^{1/2}\,dx \\ &= 2x(x+9)^{1/2} - 2\cdot\frac{2}{3}(x+9)^{3/2} + C.\end{aligned}$$

Now, (0, 2) is on the graph of $f(x)$, so $2=f(0)$. That is,

$$2 = 2(0)(0+9)^{1/2} - \frac{4}{3}(0+9)^{3/2} + C$$

$$2 = -\frac{4}{3}(9)^{3/2} + C = -\frac{4}{3}(27) + C = -36 + C.$$

Thus $C=38$, and $f(x)=2x(x+9)^{1/2}-\frac{4}{3}(x+9)^{3/2}+38.$

9.3 Evaluation of Definite Integrals

One purpose of this section is to give you lots of practice choosing between integration by parts and by substitution. Another purpose is to show how these procedures may be simplified somewhat when working with definite integrals.

A common mistake in this section is to make a change of variable in a definite integral and not to change the limits of integration. For example, here is an incorrect evaluation of the integral

$$\int_0^1 2x(x^2+1)^5\,dx.$$

Let $u = x^2 + 1$ and $du = 2x\,dx$.

Then

$$\int_0^1 (x^2+1)^5 \cdot 2x\,dx \boxed{=} \int_0^1 u^5 du = \frac{1}{6}u^6\Big|_0^1$$

$$\boxed{\text{error!}} \qquad = \frac{1}{6}(1)^6 - \frac{1}{6}(0)^6 = \frac{1}{6}.$$

The error in this calculation was made on the first line. The integral on the left equals 21/2, not 1/6. See Example 1.

Here is another incorrect evaluation whose final answer is correct. However, the work contains two errors and therefore does not justify the final answer. As before, let $u = x^2 + 1$, $du = 2x\,dx$.

Then

$$\int_0^1 (x^2+1)^5 \cdot 2x\,dx \boxed{=} \int_0^1 u^5 du = \frac{1}{6}u^6\Big|_0^1$$

$$\boxed{\text{errors!}} \qquad \boxed{=} \frac{1}{6}(x^2+1)^6\Big|_0^1 = \frac{1}{6}(2)^6 - \frac{1}{6}(1)^6$$

$$= \frac{63}{6} = \frac{21}{2}.$$

The symbol $\frac{1}{6}u^6\Big|_0^1$ represents the net change in $\frac{1}{6}u^6$ over the interval $0 \le u \le 1$. This number is not the same as the net change in $\frac{1}{6}(x^2+1)^6$ over the interval $0 \le x \le 1$.

The second incorrect evaluation above may be corrected by omitting the limits of integration at first, that is, by considering an antiderivative instead of a definite integral. Thus,

$$\int (x^2+1)^5 \cdot 2x\,dx = \int u^5 du = \frac{1}{6}u^6 + C = \frac{1}{6}(x^2+1)^6 + C,$$

and hence,

$$\int_0^1 (x^2+1)^5 2x\,dx = \frac{1}{6}(x^2+1)^6\Big|_0^1 = \frac{1}{6}(2)^6 - \frac{1}{6}(1)^6 = \frac{21}{2}.$$

This is the first solution method discussed in Example 1 in the text.

1. Let $u = 2x - 5$. Then $du = \frac{d}{dx}(2x-5)\,dx = 2\,dx$.

$$\text{If } x = \frac{5}{2}, \text{ then } u = 2\left(\frac{5}{2}\right) - 5 = 0.$$

$$\text{If } x = 3, \text{ then } u = 2(3) - 5 = 1.$$

Therefore

$$\int_{5/2}^{3} 2(2x-5)^{14}\,dx = \int_0^1 u^{14}\,du = \frac{1}{15}u^{15}\Big|_0^1$$
$$= \frac{1}{15}(1)^{15} - 0 = \frac{1}{15}.$$

7. Since $(x^2 - 9)$ has $2x$ as a derivative, and since this is a multiple of x, use the substitution $u = x^2 - 9$. Then $du = 2x\,dx$. If $x = 3$, then $u = 3^2 - 9 = 0$; if $x = 5$, then $u = 5^2 - 9 = 16$.

$$\int_3^5 x\sqrt{x^2-9}\,dx = \frac{1}{2}\int_3^5 2x\sqrt{x^2-9}\,dx = \frac{1}{2}\int_0^{16}\sqrt{u}\,du$$
$$= \frac{1}{2}\cdot\frac{2}{3}u^{3/2}\Big|_0^{16} = \frac{1}{3}\cdot 16^{3/2} - 0 = \frac{64}{3}.$$

13. Since x^3 has $3x^2$ as a derivative, and since this is a multiple of x^2, use the substitution $u = x^3$. Then $du = 3x^2 dx$. If $x = 1$, then $u = 1^3 = 1$, and if $x = 3$, then $u = 3^3 = 27$.

$$\int_1^3 x^2 e^{x^3}\,dx = \frac{1}{3}\int_1^3 3x^2 e^{x^3}\,dx = \frac{1}{3}\int_1^{27} e^u\,du$$
$$= \frac{1}{3}e^u\Big|_1^{27} = \frac{1}{3}e^{27} - \frac{1}{3}e = \frac{1}{3}(e^{27} - e).$$

19. Since the integrand is the product of two unrelated functions, use the technique of integration by parts. Let $f(x) = x$, and $g(x) = \sin \pi x$. Then

$$f'(x) = 1, \quad G(x) = -\frac{1}{\pi}\cos \pi x. \quad [\text{Check that } G'(x) = g(x).]$$

Then,

$$\int x \sin \pi x\,dx = -\frac{1}{\pi}x\cos \pi x - \int -\frac{1}{\pi}\cos \pi x\,dx$$
$$= -\frac{1}{\pi}x\cos \pi x + \frac{1}{\pi}\int \cos \pi x\,dx$$
$$= -\frac{1}{\pi}x\cos \pi x + \frac{1}{\pi}\left(\frac{1}{\pi}\sin \pi x\right) + C$$
$$= -\frac{1}{\pi}x\cos \pi x + \frac{1}{\pi^2}\sin \pi x + C.$$

So,

$$\begin{aligned}\int_0^1 x\sin\pi x\,dx &= \left(-\frac{1}{\pi}x\cos\pi x+\frac{1}{\pi^2}\sin\pi x\right)\Big|_0^1\\ &= \left[-\frac{1}{\pi}\cos\pi+\frac{1}{\pi^2}\sin\pi\right]-\left[-\frac{1}{\pi}(0)+\frac{1}{\pi^2}\sin 0\right]\\ &= \left[-\frac{1}{\pi}(-1)+0\right]\\ &= \frac{1}{\pi}.\end{aligned}$$

25. First, find where $y = x\sqrt{4-x^2}$ cuts the x-axis ($y = 0$):

$$\begin{aligned}0 &= x\sqrt{4-x^2},\\ x &= 0, \quad\text{or}\quad \sqrt{4-x^2}=0 \Rightarrow 4-x^2=0,\\ x^2 &= 4, \quad\text{or}\quad x=\pm 2.\end{aligned}$$

The area of the portion from $x = 0$ to $x = 2$ is given by

$$\int_0^2 x\sqrt{4-x^2}\,dx.$$

To find $\int x\sqrt{4-x^2}\,dx$, let

$$\begin{aligned}u &= 4-x^2,\\ du &= \frac{d}{dx}(4-x^2)dx = -2x\,dx.\end{aligned}$$

So,

$$\begin{aligned}\int x\sqrt{4-x^2}\,dx &= \int\left(-\frac{1}{2}\right)(-2x)\sqrt{4-x^2}\,dx\\ &= -\frac{1}{2}\int\sqrt{u}\,du = -\frac{1}{2}\int u^{1/2}du\\ &= -\frac{1}{2}\frac{u^{3/2}}{\frac{3}{2}} = -\frac{1}{3}u^{3/2}+C\\ &= -\frac{1}{3}(4-x^2)^{3/2}+C.\end{aligned}$$

Therefore, the area from $x = 0$ to $x = 2$ is

$$\begin{aligned}\int_0^2 x\sqrt{4-x^2}\,dx &= -\frac{1}{3}(4-x^2)^{3/2}\Big|_0^2\\ &= -\frac{1}{3}(4-4)^{3/2}+\frac{1}{3}(4)^{3/2} = \frac{8}{3}.\end{aligned}$$

By symmetry, the area from $x = -2$ to $x = 0$ is also $\frac{8}{3}$. The total area is $\frac{16}{3}$.

9.4 Approximation of Definite Integrals

It will be helpful for you to review Section 6.2 before you study Section 9.4. When you work the exercises in Section 9.4, do not be concerned if your answers differ slightly (in, say, the fourth or fifth significant figure) from the answers in the text. When numerical calculations are made that involve several steps, any decision (ours or yours) to "round off" intermediate answers can affect the final answer. Check with your instructor about how many significant figures you should retain in your calculations.

1. The interval has length $5-3=2$. When this interval is divided into 5 subintervals of equal length, each subinterval will have length $\Delta x = \frac{2}{5} = .4$. Therefore, the end points of the subintervals are:

$$\begin{aligned} a_0 &= 3, \\ a_1 &= 3+.4 = 3.4, \\ a_2 &= 3.4+.4 = 3.8, \\ a_3 &= 3.8+.4 = 4.2, \\ a_4 &= 4.2+.4 = 4.6, \\ a_5 &= 4.6+.4 = 5. \end{aligned}$$

7. If $n=2$, then $\Delta x = \dfrac{(4-0)}{2} = 2$. The first midpoint is $x_1 = a + \dfrac{\Delta x}{2} = 0 + \dfrac{2}{2} = 1$. The second midpoint is $x_2 = 1 + \Delta x = 1 + 2 = 3$. If $f(x) = x^2 + 5$, then $f(1) = 6$, $f(3) = 14$. By the midpoint rule,

$$\begin{aligned} \int_0^4 (x^2+5)dx &\approx [f(1)+f(3)]\cdot 2 \\ &= (6+14)\cdot 2 \\ &= 40. \end{aligned}$$

If $n=4$, then $\Delta x = \dfrac{(4-0)}{4} = 1$. The first midpoint is $x_1 = a + \dfrac{\Delta x}{2} = 0 + \dfrac{1}{2} = .5$. Since the other midpoints are spaced 1 unit apart, $x_2 = 1.5$, $x_3 = 2.5$, and $x_4 = 3.5$. Hence,

$$\begin{aligned} \int_0^4 (x^2+5)dx &\approx [f(x_1)+f(x_2)+f(x_3)+f(x_4)]\Delta x \\ &= \{[(.5)^2+5]+[(1.5)^2+5]+[(2.5)^2+5]+[(3.5)^2+5]\}(1). \end{aligned}$$

On an exam, some instructors may permit you to stop at this point, because the expression above shows that you know how to use the midpoint rule. (Check with your instructor.) Only arithmetic remains. For homework, of course, you should complete the calculation, and obtain

$$\begin{aligned} \int_0^4 (x^2+5)dx &\approx \{5.25+7.25+11.25+17.25\}(1) \\ &= 41. \end{aligned}$$

Evaluating the integral directly,

$$\begin{aligned} \int_0^4 (x^2+5)dx &= \left(\frac{1}{3}x^3+5x\right)\Bigg|_0^4 \\ &= \frac{64}{3}+20 = 41\frac{1}{3}. \end{aligned}$$

13. For $n = 3$, $\Delta x = \frac{5-1}{3} = \frac{4}{3}$. The endpoints of the three subintervals begin at $a + 0 = 1$ and are spaced $\frac{4}{3}$ units apart. Thus $a_1 = 1 + \frac{4}{3} = \frac{7}{3}$, $a_2 = \frac{7}{3} + \frac{4}{3} = \frac{11}{3}$, and $a_3 = \frac{11}{3} + \frac{4}{3} = \frac{15}{3} = 5$. By the trapezoidal rule,

$$\int_1^5 \frac{1}{x^2}dx \approx \left[f(1) + 2f\left(\frac{7}{3}\right) + 2f\left(\frac{11}{3}\right) + f(5) \right]\left(\frac{4}{3}\right)\cdot\frac{1}{2}$$

$$= \left[1 + 2\cdot\frac{1}{\left(\frac{7}{3}\right)^2} + 2\cdot\frac{1}{\left(\frac{11}{3}\right)^2} + \frac{1}{5^2} \right]\left(\frac{4}{3}\right)\cdot\frac{1}{2}.$$

At this point, all substitutions into the trapezoidal rule are finished; only arithmetic remains. (Check to see if you can stop here on an exam.) The fractions above become quite messy, and it is best to use decimal approximations. We'll use five decimal places. (Check to see how many you should use on an exam.)

$$\int_1^5 \frac{1}{x^2}dx \approx [1 + .36735 + .14876 + .04](.66667)$$

$$= 1.03741.$$

Evaluating the integral directly produces

$$\int_1^5 \frac{1}{x^2}dx = \int_1^5 x^{-2}dx = -x^{-1}\Big|_1^5$$

$$= -\frac{1}{5} - (-1) = \frac{4}{5} = .8.$$

19. When $n = 5$, we recommend that you use decimals. In this exercise,

$$\Delta x = \frac{(5-2)}{5} = \frac{3}{5} = .6.$$

It is often helpful to draw a picture. This is particularly true here because you need both the endpoints and midpoints of the subintervals, and the numbers involved are not as simple as in some exercises. Draw a line segment and, beginning at one end, mark of five equal subintervals. (This is easier than trying to divide one large interval into five equal parts.) Label the left endpoint $a_0 = 2$, and repeatedly add $\Delta x = .6$ to get the other endpoints.

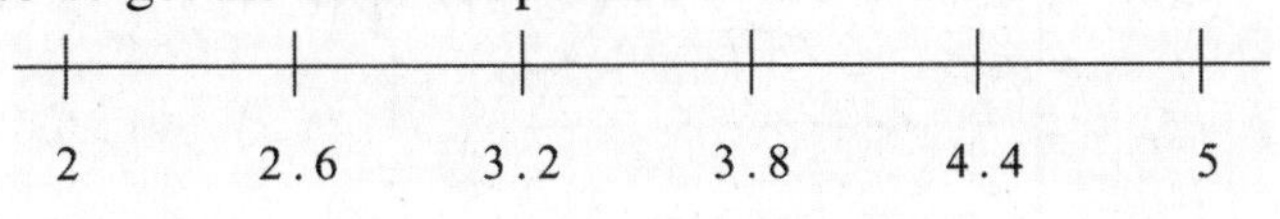

The first midpoints is $2 + \frac{\Delta x}{2} = 2 + .3 = 2.3$. Repeatedly add $\Delta x = .6$ to get the other midpoints.

2.3 2.9 3.5 4.1 4.7

2 2.6 3.2 3.8 4.4 5

Use the midpoint rule for $f(x) = xe^x$ and obtain

$$\int_2^5 xe^x dx \approx [2.3e^{2.3} + 2.9e^{2.9} + 3.5e^{3.5} + 4.1e^{4.1} + 4.7e^{4.7}](.6)$$

$$\approx (955.69661)(.6) = 573.41797 = M.$$

Using the trapezoidal rule,

$$\int_2^5 xe^x dx \approx [2e^2 + (2)2.6e^{2.6} + (2)3.2e^{3.2} + (2)3.8e^{3.8}$$

$$+(2)4.4e^{4.4} + 5e^5](.6)\cdot\frac{1}{2}$$

$$\approx (2040.36019)(.6)(.5)$$

$$= 612.10806 = T.$$

Use the values of M and T in Simpson's rule:

$$\int_2^5 xe^x dx \approx \frac{2M+T}{3}$$

$$= \frac{2(573.41797)+612.10806}{3}$$

$$= 586.31467 = S.$$

To evaluate the integral directly, use the technique of integration by parts. Set

$$f(x) = x, \qquad g(x) = e^x,$$

$$f'(x) = 1, \qquad G(x) = e^x.$$

Then,

$$\int_2^5 xe^x dx = xe^x\Big|_2^5 - \int_2^5 e^x dx$$

$$= (5e^5 - 2e^2) - e^x\Big|_2^5$$

$$= 5e^5 - 2e^2 - (e^5 - e^2)$$

$$= 4e^5 - e^2$$

$$\approx 586.26358.$$

25. Let $f(x)$ represent the distance (in feet) from the shore to the property line as x runs from the top to the bottom of the diagram, and is measured in feet.

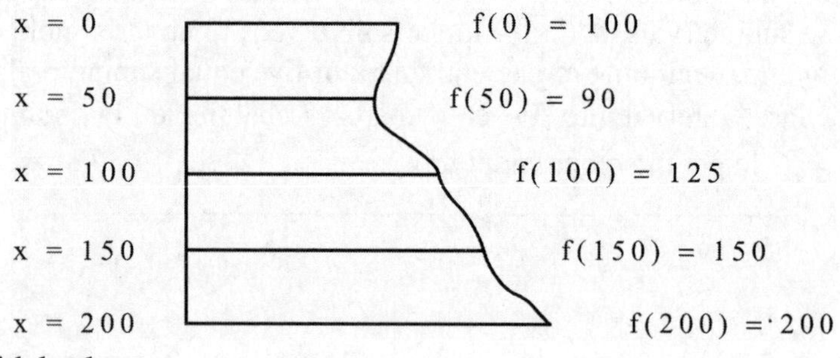

So by the trapezoidal rule,

$$[\text{Area of Property}] = \int_0^{200} f(x)dx$$

$$\approx [f(0) + 2f(50) + 2f(100) + 2f(150) + (200)]\left(\frac{50}{2}\right)$$

$$= [100 + 180 + 250 + 300 + 200]25$$

$$= (1030)25$$

$$= 25{,}750 \text{ square feet.}$$

31. **(a)** Note from figure 13(a) in the text that the height of the triangle is $k-h$. The area of the triangle is $\frac{1}{2}(k-h)\ell$, and the area of the rectangle is $h\ell$. Therefore, the area of the trapezoid is given by

$$A=\frac{1}{2}(k-h)\ell+h\ell=\left[\frac{1}{2}k-\frac{1}{2}h+h\right]\ell$$
$$=\left[\frac{1}{2}h+\frac{1}{2}k\right]\ell=\frac{1}{2}(h+k)\ell.$$

(b) Note that $h=f(a_0)$, $k=f(a_1)$, and $\ell=\Delta x$. Therefore,

$$A=\frac{1}{2}[(a_0)+f(a_1)]\,\Delta x.$$

(c) The area of the first trapezoid is:

$$A=\frac{1}{2}[f(a_0)+f(a_1)]\Delta x=[f(a_0)+f(a_1)]\frac{\Delta x}{2}.$$

Similarly, the respective areas of the second, third, and fourth rectangles are

$$A_2=[f(a_1)+f(a_2)]\frac{\Delta x}{2},$$
$$A_3=[f(a_2)+f(a_3)]\frac{\Delta x}{2},$$
$$A_4=[f(a_3)+f(a_4)]\frac{\Delta x}{2}.$$

So, the sum of the areas is given by

$$A_1+A_2+A_3+A_4=\left([f(a_0)+f(a_1)]+[(f(a_1)+f(a_2)]+[f(a_2)+f(a_3)]+[f(a_3)+f(a_4)]\right)\frac{\Delta x}{2}$$
$$=\left(f(a_0)+2f(a_1)+2f(a_2)+2f(a_3)+f(a_4)\right)\frac{\Delta x}{2}.$$

This quantity is the value of the trapezoidal rule with $n=4$ for

$$\int_a^b f(x)\,dx.$$

The combined area of the trapezoids is approximately equal to the area under the graph of $f(x)$ from $x=a$ to $x=b$.

37. In this problem, $f(x)=1/x$, $a=1$, $b=11$, $n=10$, and $\Delta x=1$. The formula for $f(x)$ is so simple, there is no need to use it to define Y_1. Just use it directly in the **sum (seq(...))** command. When you compute the midpoint and trapezoidal estimates, store the results to variables, which you might as well call M and T.

Midpoint rule: $M=\left[\frac{1}{1.5}+\frac{1}{2.5}+\cdots+\frac{1}{10.5}\right]\cdot 1.$

```
sum(seq(1/X,X,1.
5,10.5,1))*1
         2.361749156
Ans→M
         2.361749156
```

Trapezoidal rule: $T = \left[\frac{1}{1} + 2\left(\frac{1}{2}\right) + \cdots + 2\left(\frac{1}{10}\right) + \frac{1}{11}\right] \cdot \frac{1}{2}.$

```
(1+sum(seq(2/X,X
,2,10,1))+1/11)/
2
         2.474422799
Ans→T
         2.474422799
```

```
(2M+T)/3
         2.399307037
```

The exact answer is

$$\int_1^{11} \frac{1}{x}\,dx = \ln x\Big|_1^{11}$$
$$= \ln 11 - \ln 1$$
$$= 2.397895273.$$

If you store the exact answer (ln 11) in a variable, say E, then the following error calculations are easily made:

Error using the midpoint rule: $|M - E| = 0.036146117,$

Error using the trapezoidal rule: $|T - E| = 0.076527526,$

Error using Simpson's rule: $|(2M + T)/3 - E| = 0.001411764,$

Notice that the exact answer lies between M and T, and the error for T is about twice the error for M, as usually happens.

9.5 Some Applications of the Integral

The purpose of this section is to show how fairly difficult integration problems can arise easily in applications. The Riemann sum argument in Example 3 is well worth studying, but your instructor may not expect you to reproduce it every time you work an exercise. (Check with your instructor.)

1. Use the text's formula for present value, with $K(t) = 35,000, T_1 = 0, T_2 = 5,$ and $r = .07.$

$$[\text{present value}] = \int_0^5 35000e^{-.07t}\,dt = \frac{35000e^{-.07t}}{-.07}\Bigg|_0^5$$
$$= -500,000[e^{-.07(5)} - e^0]$$
$$= -500,000[e^{-.35} - 1]$$
$$\approx \$147,656.$$

7. **(a)** [present value] $= \int_0^2 (30+5t)e^{-.1t}\,dt.$

(b) First, find the appropriate antiderivative, using integration by parts:

$$f(t) = 30+5t \qquad g(t) = e^{-.1t}$$
$$f'(t) = 5 \qquad G(t) = -10e^{-.1t}$$

$$\begin{aligned}\int (30+5t)e^{-.1t}\,dt &= (30+5t)(-10e^{-.1t}) - \int 5(-10e^{-.1t})\,dt \\ &= (-300-50t)e^{-.1t} + 50\int e^{-.1t}\,dt \\ &= (-300-50t)e^{-.1t} + 50(-10e^{-.1t}) + C \\ &= (-800-50t)e^{-.1t} + C.\end{aligned}$$

Then use this to evaluate the definite integral.

$$\begin{aligned}\int_0^2 (30+5t)e^{-.1t}\,dt &= ((-800-50t)e^{-.1t})\Big|_0^2 \\ &= -900e^{-.2} - (-800e^0) \\ &\approx 63.14.\end{aligned}$$

The present value of the stream of earnings for the next two years is about 63.1 million dollars.

13. **(a)** Area of ring is $2\pi t(\Delta t)$. Population density is $40e^{-.5t}$ thousand per square mile. Thus the population living in the ring is:

$$2\pi t(\Delta t)40e^{-.5t} = 80\pi t(\Delta t)e^{-.5t} \text{ thousand people.}$$

(b) $\dfrac{dP}{dt}$, or $P'(t)$.

(c) It represents the number of people who live between 5 miles from the city center and $(5+\Delta t)$ miles from the city center.

(d) $P(t+\Delta t) - P(t) = 80\pi t(\Delta t)e^{-.5t}$ from (a), so

$$\begin{aligned}\frac{P(t+\Delta t) - P(t)}{\Delta t} &\approx P'(t) \\ &= 80\pi t e^{-.5t}.\end{aligned}$$

(e) $P(b) - P(a) = \int_a^b P'(t)\,dt$, (by the Fundamental Theorem of Calculus)

$$= \int_a^b 80\pi t e^{-.5t}\,dt.$$

9.6 Improper Integrals

Although some instructors may disagree, we feel that the evaluation of an improper integral such as

$$\int_0^\infty e^{-x}dx$$

should *not* use notation such as $-e^{-x}\Big|_0^\infty$ or $e^{-\infty}+e^{-0}$. In the context of Section 9.6, the value of a function "at infinity" is not well-defined. The proper notation should involve a limit, such as $\lim_{b\to\infty}(1-e^{-b})$. You should use two steps to evaluate an improper integral:

i. Compute the net change in the antidervиative over a finite interval, such as from $x=1$ to $x=b$.
ii. Find the limit of the result in (i) as $b\to\infty$.

1. As b gets large, $\frac{5}{b}$ approaches zero. That is, $\lim_{b\to\infty}\frac{5}{b}=0$.

7. As b gets large, $\sqrt{b+1}$ gets large, without bound. Hence $\frac{1}{\sqrt{b+1}}=(b+1)^{-1/2}$ approaches zero, and thus $2-(b+1)^{-1/2}$ approaches 2.

13. The area is $\int_2^\infty x^{-2}dx$. First, compute

$$\int_2^b x^{-2}dx=-x^{-1}\Big|_2^b$$

$$=-b^{-1}-(-2^{-1})=\frac{1}{2}-\frac{1}{b}.$$

Then, take the limit as $b\to\infty$, namely,

$$\int_2^\infty x^{-2}dx=\lim_{b\to\infty}\left(\frac{1}{2}-\frac{1}{b}\right)=\frac{1}{2}.$$

19. $\int_1^b(14x+18)^{-4/5}dx=\frac{1}{14}(5)(14x+18)^{1/5}\Big|_1^b$ (Check the antiderivative.)

$$=\frac{5}{14}(14b+18)^{1/5}-\frac{5}{14}(32)^{1/5}.$$

As $b\to\infty$, the quantity $\frac{5}{14}(14b+18)^{1/5}$ grows without bound, so the region under the graph of $(14x+18)^{-4/5}$ cannot be assigned any finite number as its area.

25. Take $b > 0$ and compute

$$\int_0^b e^{2x}dx = \frac{1}{2}e^{2x}\Big|_0^b = \frac{1}{2}e^{2b} - \frac{1}{2}.$$

Now consider the limit as $b \to \infty$. As $b \to \infty$, the number $\frac{1}{2}e^{2b} - \frac{1}{2}$ can be made larger than any specified number. Therefore,

$$\int_0^b e^{2x}dx$$

has no limit as $b \to \infty$, so $\int_0^\infty e^{2x}dx$ is divergent.

31. $\int_0^b 6e^{1-3x}dx = \frac{6}{-3}e^{1-3x}\Big|_0^b$

$$= -2e^{1-3b} - (-2e^1) = 2e - 2e^{1-3b}.$$

As $b \to \infty$, $1-3b$ becomes extremely negative, so $2e^{1-3b} \to 0$.

Thus,

$$\int_0^\infty 6e^{1-3x}dx = \lim_{b\to\infty}(2e - 2e^{1-3b}) = 2e.$$

37. First consider the indefinite integral $\int 2x(x^2+1)^{-3/2}dx$ and let $u = x^2+1$, $du = 2x\,dx$.

Thus

$$\int 2x(x^2+1)^{-3/2}dx = \int u^{-3/2}du = -2u^{-1/2} + C$$
$$= -2(x^2+1)^{-1/2} + C.$$

Using this antiderivative of $2x(x^2+1)^{-3/2}$, compute

$$\int_0^b 2x(x^2+1)^{-3/2}dx = -2(x^2+1)^{-1/2}\Big|_0^b$$
$$= -2(b^2+1)^{-1/2} - (-2\cdot 1).$$

As $b \to \infty$, $(b^2+1)^{-1/2} \to 0$, so

$$\int_0^\infty 2x(x^2+1)^{-3/2}dx = \lim_{b\to\infty}(2 - 2(b^2+1)^{-1/2}) = 2.$$

43. Let $u = e^{-x} + 2$ and $du = -e^{x}dx$, so that

$$\int \frac{e^{-x}}{(e^{-x}+2)^2}dx = \int \frac{-1}{u^2}du = u^{-1} + C$$
$$= (e^{-x}+2)^{-1} + C.$$

As $b \to \infty, e^{-b} \to 0,$ and $\frac{1}{e^{-b}+2} \to \frac{1}{2}$. Thus,

$$\int_0^{\infty} \frac{e^{-x}}{(e^{-x}+2)^2}dx = \lim_{b\to\infty} (e^{-x}+2)^{-1}\Big|_0^b$$
$$= \lim_{b\to\infty} [(e^{-b}+2)^{-1} - 3^{-1}]$$
$$= \frac{1}{2} - \frac{1}{3} = \frac{1}{6}.$$

49. When the rate of income is a constant K dollars per year, the antiderivative of Ke^{-rt} is easy to find:

$$[\text{capital value}] = \lim_{b\to\infty} \int_0^b Ke^{-rt}dt = \lim_{b\to\infty} -\frac{1}{r}Ke^{-rt}\Big|_0^b$$
$$= \lim_{b\to\infty} \left(-\frac{K}{r}e^{-rb} + \frac{K}{r}e^{0}\right)$$
$$= -\frac{K}{r}\left(\lim_{b\to\infty} e^{-rb}\right) + \frac{K}{r}$$
$$= \frac{K}{r}.$$

Chapter 9: Supplementary Exercises

Sections 9.1–9.3 and 9.6 provide the tools for solving many of the problems in the next three chapters. It is essential that you master the techniques in Sections 9.1 and 9.2. Supplementary Exercises 19–36 will help you learn to recognize which technique to use.

1. Since $3x^2$ has $6x$ as a derivative, and this is a multiple of x, use the technique of substitution. Let $u = 3x^2$, then $du = 6x\,dx$. Then

$$\int x\sin 3x^2 dx = \frac{1}{6}\int 6x\sin 3x^2 dx$$
$$= \frac{1}{6}\int \sin u\,du$$
$$= -\frac{1}{6}\cos u + C$$
$$= -\frac{1}{6}\cos 3x^2 + C.$$

7. The derivative of $4-x^2$ is a multiple of x (the factor next to the square root). So make the substitution

$$u = 4 - x^2,$$
$$du = -2x dx.$$

Then

$$\int x\sqrt{4-x^2}\,dx = -\frac{1}{2}\int(-2x)\sqrt{4-x^2}\,dx$$
$$= -\frac{1}{2}\int\sqrt{u}\,du$$
$$= -\frac{1}{2}\cdot\frac{2}{3}u^{3/2} + C$$
$$= -\frac{1}{3}(4-x^2)^{3/2} + C.$$

13. First simplify the integrand, using the algebraic property $\ln x^2 = 2\ln x$:

$$\int \ln x^2 dx = 2\int \ln x dx.$$

Then recall that $f \ln x\,dx$ is easily found by integration by parts. Use $f(x) = \ln x$ and $g(x) = 1$. The details are in Example 7 of Section 9.2. Using the results of that example,

$$\ln x^2 dx = 2\int \ln x dx = 2(x\ln x - x) + C.$$

Helpful Hint: Integrals of the form $\int x^k \ln x dx$ are found by integration by parts, with $f(x) = \ln x$ and $g(x) = x^k$.

19. Since the integrand is the product of two functions, x and e^{2x}, neither of which is a multiple of the derivative of the other, use integration by parts. Set

$$f(x) = x, \quad g(x) = e^{2x}.$$

25. Since the integrand is the product of two functions, e^{-x} and $(3x-1)^2$, and neither is a multiple of the derivative of the other, use integration by parts. Set

$$f(x) = (3x-1)^2, \qquad g(x) = e^{-x}.$$

In order to complete the problem, you must integrate by parts a second time. Note that if you interchange $f(x)$ and $g(x)$ above, integration by parts will not yield a solution.

31. Observe that the derivative of x^2+6x is $2x+6$, which is a multiple of $x+3$. Use the technique of substitution, setting $u = x^2+6x$.

37. Since x^2+1 has $2x$ as a derivative, use the technique of substitution. Let $u = x^2+1$. Then $du = 2x\,dx$. If $x=0$, then $u=1$, and if $x=1$, then $u=2$. So,

$$\int_0^1 \frac{2x}{(x^2+1)^3}dx = \int_1^2 \frac{1}{u^3}du = \int_1^2 u^{-3}du$$
$$= -\frac{1}{2}u^{-2}\Big|_1^2 = -\frac{1}{2}\left(\frac{1}{4}\right) + \frac{1}{2} = \frac{3}{8}.$$

43. The interval has length $9-1=8$ and $n=4$, so $\Delta x=\frac{8}{4}=2.$ The subintervals have endpoints $a_0=1,$ $a_1=3, a_2=5, a_3=7,$ and $a_4=9,$ and midpoints $x_1=2, x_2=4, x_3=6,$ and $x_4=8.$

1 2 3 4 5 5 7 8 9

Using the midpoint rule, with $f(x)=\frac{1}{\sqrt{x}}=x^{-1/2},$

$$\begin{aligned}\int_1^9 f(x)dx \approx M &= [f(2)+f(4)+f(6)+f(8)]\cdot 2\\ &= \left[\frac{1}{\sqrt{2}}+\frac{1}{\sqrt{4}}+\frac{1}{\sqrt{6}}+\frac{1}{\sqrt{8}}\right](2)\\ &= [.70711+.50000+.40825+.35355]\cdot 2\\ &= 3.93782.\end{aligned}$$

Using the trapezoidal rule,

$$\begin{aligned}\int_1^9 f(x)dx \approx T &= [f(1)+2f(3)+2f(5)+2f(7)+f(9)]\cdot\frac{2}{2}\\ &= \left[\frac{1}{\sqrt{1}}+\frac{2}{\sqrt{3}}+\frac{2}{\sqrt{5}}+\frac{2}{\sqrt{7}}+\frac{1}{\sqrt{9}}\right]\frac{2}{2}\\ &= [1+1.15470+.89443+.75593+.33333]\\ &= 4.13839.\end{aligned}$$

Using Simpson's rule,

$$\int_1^9 \frac{1}{\sqrt{x}}dx \approx \frac{2M+T}{3}=4.00468.$$

49. The derivative of x^2+4x-2 is $2x+4,$ which is a multiple of $x+2.$ Use the technique of substitution. Let $u=x^2+4x-2$ and $du=(2x+4)dx.$

$$\begin{aligned}\int\frac{x+2}{x^2+4x-2}dx &= \frac{1}{2}\int\frac{2x+4}{x^2+4x-2}dx=\frac{1}{2}\int\frac{1}{u}du\\ &= \frac{1}{2}\ln|u|=\frac{1}{2}\ln|x^2+4x-2|+C.\end{aligned}$$

Hence,

$$\begin{aligned}\int_1^b\frac{x+2}{x^2+4x-2}dx &= \frac{1}{2}\ln|x^2+4x-2|\Big|_1^b\\ &= \frac{1}{2}\ln|b^2+4b-2|-\frac{1}{2}\ln 3.\end{aligned}$$

As $b\to\infty,$ the number $|b^2+4b-2|$ gets large, without bound, and so $\ln|b^2+4b-2|$ gets large, without bound. Therefore,

$$\frac{1}{2}\ln|b^2+4b-2|-\frac{1}{2}\ln 3$$

can be made larger than any specific number.

Therefore,

$$\int_1^b \frac{x+2}{x^2+4x-2}\,dx$$

has no limit as $b \to \infty$ and

$$\int_1^\infty \frac{x+2}{x^2+4x-2}\,dx \text{ is divergent.}$$

55. Use the text's formula for present value given in Section 9.5, with $K(t) = 50e^{-.08t}$, $r = .12$, $T_1 = 0$, and $T_2 = 4$.

$$[\text{present value}] = \int_0^4 50e^{-.08t} \cdot e^{-.12t}\,dt.$$

Integration by parts is not necessary this time because the exponential functions can be combined:

$$[\text{present value}] = \int_0^4 50e^{-.20t}\,dt = -250e^{-.2t}\Big|_0^4$$

$$= -112.332 + 250 = 137.668.$$

The present value of the continuous stream of income over the next four years is $137,668 (to the nearest dollar).

Chapter 10
Differential Equations

10.1 Solutions of Differential Equations

This section lays the foundation for the chapter. You should read the text and the examples several times. The high point of the chapter is Section 10.6, which presents a wide variety of applications. Our students find this material extremely interesting, and we know that they can master it if they are given adequate preparation. So each section includes a few problems that teach you how to read and understand a verbal problem about a differential equation. Be sure to try exercise 19 or 20.

1. Replace y by $f(t)=\frac{3}{2}e^{t^2}-\frac{1}{2}$, and y' by $f'(t)=3te^{t^2}$.

Then

$$y'-2ty=3te^{t^2}-2t\left[\frac{3}{2}e^{t^2}-\frac{1}{2}\right]$$
$$=3te^{t^2}-[3te^{t^2}-t]=t.$$

This is true for all t, so $f(t)=\frac{3}{2}e^{t^2}-\frac{1}{2}$ is a solution.

7. The derivative of the constant function $f(t)=3$ is the constant function $f'(t)=0$. If you replace y by $f(t)=3$ and y' by $f'(t)=0$, then

$$y'=6-2y,$$
$$0=6-2(3).$$

Both sides are zero for all t, so $f(t)=3$ is a solution.

13. The differential equation $y'=.2(160-y)$ describes a relationship between the acceleration (the derivative of the downward velocity) and the downward velocity, at each moment during the free fall. At a time when the velocity is 60 ft/sec, the acceleration is

$$y'=.2(160-60)=.2(100)=20 \text{ ft/sec per second.}$$

19. Let $f(t)$ be the amount of capital invested. Then $f'(t)$ is the rate of net investment. We are told that

[rate of net investment] is proportional to $(C-[\text{capital investment}])$.

So there is a constant of proportionality, k, such that

$$f'(t)=k\{C-f(t)\}.$$

To determine the sign of k, suppose that at some time $f(t)$ is larger than C. Then $C - f(t)$ is negative. Should $f'(t)$ be positive or negative? That is, should $f(t)$ be increasing or decreasing? Well, if the amount of $f(t)$ is larger than the optimum level C, then it is reasonable to want $f(t)$ to decrease down towards C. So when $C - f(t)$ is negative, we want $f'(t)$ to be negative, too. This means that k must be positive.

$$f'(t) = k\{C - f(t)\},$$
$$[\text{neg}] = [\text{pos}] \cdot [\text{neg}].$$

You would arrive at the same conclusion that k is positive if you supposed that $f(t)$ is less than C. The differential equation is simplified by writing y for $f(t)$ and y' for $f'(t)$, namely,

$$y' = k(C - y), \qquad k > 0.$$

25. The initial condition $f(0) = 1500$ tells us that the point (0, 1500) is on the graph of the solution curve $y = f(t)$. Starting at this point in the ty-plane of Figure 5(a) of the text and tracing a curve tangent to the line segments in the slope field, we obtain a solution curve through the point (0, 1500) that appears to be similar to the solution through (0, 1000), shifted vertically up 500 units.

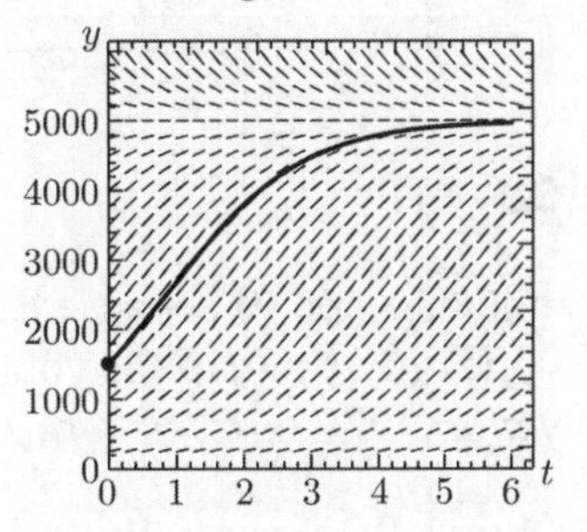

The key point to notice about the slope field in Figure 5(a) is that above the line $y = 5000$, the tangent lines pictured are negative. This can also be verified in general by observing that y' is always negative whenever a value of y greater than 5000 is plugged into the equation $y' = .0002y(5000 - y)$. For this reason, any solution curve to the differential equation $y' = .0002y(5000 - y)$ that contains a point (t, y) with $y < 5000$ will never go above the line $y = 5000$. Thus, in this scenario, $f(t)$ (the number of infected people) will never exceed 5000.

31. **(a)** Set Y_1 = 10 + 500*e^(-.2X) and deselect all other functions. Set Xmin = 0, Xmax = 30, Ymin = −75, and Ymax = 550, then graph the function. The graph shows a decreasing function in the first quadrant with a vertical asymptote at the y-axis, and a horizontal asymptote at the x-axis.

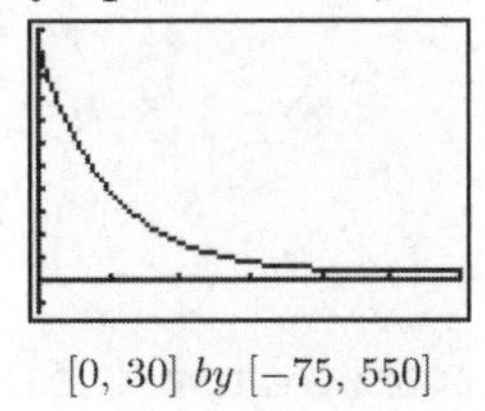

[0, 30] *by* [−75, 550]

(b) We are given that $y = 10 + 500e^{-.2t}$ is a solution to the differential equation $y' = .2(10 - y)$. This exercise is asking us to use the calculator to check that this is indeed the case when $t = 5$. Note that, based on the solution to part (a), we can expect the value of $f'(5)$ to be a negative number, since the curve $y = f(t)$ has a negative derivative (i.e., is heading downhill) when $t = 5$.

When $t = 5$, $y = f(t) = 10 + 500e^{-.2(5)}$. Thus, when $t = 5$, we have

$$.2(10 - y) = .2(10 - (10 + 500e^{-.2(5)})).$$

Using a graphing calculator to evaluate this, we find that

$$.2(10-y) \approx -36.78794412$$

when $t = 5$. To evaluate $f'(5)$ we first find the formula for $f'(t)$. A straightforward application of the chain rule gives

$$f'(t) = -100e^{-.2t}.$$

Thus $f'(5) = -100e^{-.2(5)}$. Evaluating this with a graphing calculator, we find that

$$f'(5) \approx -36.78794412$$

Thus we have confirmed (to eight decimal places, for $t = 5$) that $y = 10 + 500e^{-.2t}$ is a solution to the differential equation $y' = .2(10-y)$.

Alternatively, set $Y_1 = f(t)$, and then use nDeriv in the MATH menu as shown below.

10.2 Separation of Variables

This is probably the most difficult section in Chapter 10 because it requires substantial integration and algebraic skills. Each example illustrates one or two situations you will encounter in the exercises. After you study Examples 1, 2, and 3, you may work on Exercises 1–6, 13–15, and 17, respectively.

Remember to use the examples properly. Study them carefully and then try to work the exercises *without referring to the examples*. If you really get stuck, take a peek at the appropriate example, but do not study the entire solution. As a last resort, peek at a solution in this *Manual* if a solution is available. You simply must train yourself to work independently. If you yield to the temptation to "copy" our solutions, you will not survive an examination on this material.

Exercises 33–38 are here to help you prepare for Section 10.6. You will be rewarded later if you spend some time on these exercises now.

1. Separate the variables and integrate both sides with respect to *t*.

$$y^2 \frac{dy}{dt} = 5 - t,$$

$$\int y^2 \frac{dy}{dt} dt = \int (5-t)dt,$$

$$\int y^2 dy = \int (5-t)dt.$$

One antiderivative of y^2 is $\frac{1}{3}y^3$. One antiderivative of $5-t$ is $5t - \frac{1}{2}t^2$. Since antiderivatives of the same function differ by a constant,

$$\frac{1}{3}y^3 = 5t - \frac{1}{2}t^2 + C_1,$$

for some constant C_1. Then

$$y^3 = 3\left(5t - \frac{1}{2}t^2 + C_1\right).$$

Since C_1 is arbitrary, so is $3C_1$, and it is easier to write C in place of $3C_1$. Therefore,

$$y^3 = 15t - \frac{3}{2}t^2 + C, \quad \text{and} \quad y = \left(15t - \frac{3}{2}t^2 + C\right)^{1/3}.$$

7. $y' = \left(\frac{t}{y}\right)^2 e^{t^3} \Rightarrow \frac{dy}{dt} = \frac{t^2}{y^2}e^{t^3}$ By separating the variables,

$$y^2 \frac{dy}{dt} = t^2 e^{t^3}$$
$$\int y^2 \frac{dy}{dt}\,dt = \int t^2 e^{t^3}\,dt$$
$$\int y^2 dy = \int t^2 e^{t^3}\,dt.$$

One antiderivative of y^2 is $\frac{1}{3}y^3$, and similarly an antiderivative of $t^2 e^{t^3}$ is $\frac{1}{3}e^{t^3}$. (This is found by a substitution, $u = t^3$.) So

$$\frac{1}{3}y^3 = \frac{1}{3}e^{t^3} + C.$$

Hence, $y^3 = e^{t^3} + C$ (another constant C), and finally, $y = (e^{t^3} + C)^{1/3}$.

13. Write y' as $\frac{dy}{dt}$, integrate both sides of the equation with respect to t, and "cancel" the dt's on the left side of the equation.

$$y'e^y = te^{t^2} \Rightarrow e^y \frac{dy}{dt} = te^{t^2}$$
$$\int e^y dy = \int te^{t^2}\,dt.$$

One antiderivative of e^y is e^y. To find an antiderivative of te^{t^2}, let $u = t^2$. Then $\frac{du}{dt} = 2t$ and $du = 2tdt$.

$$te^{t^2}\,dt = \frac{1}{2}e^{t^2}(2t\,dt) = e^{t^2}\,dt = \frac{1}{2}e^u du \Rightarrow \int te^{t^2}\,dt = \frac{1}{2}\int e^u du = \frac{1}{2}e^u + C = \frac{1}{2}e^{t^2} + C$$

Thus,

$$e^y + C_1 = \frac{1}{2}e^{t^2} + C_2 \Rightarrow e^y = \frac{1}{2}e^{t^2} + C$$

To solve for y, take the natural logarithm of both sides:

$$\ln(e^y) = \ln\left(\frac{1}{2}e^{t^2} + C\right).$$

Simplify to arrive at the solution:

$$y = \ln\left(\frac{1}{2}e^{t^2} + C\right).$$

Note: To check that this is indeed the correct solution, differentiate to find y' and plug the formulas for y and y' into the equation $y'e^y = te^t$.

19. $y' = 2te^{-2y} - e^{-2y}, y(0) = 3.$

The equation can be rewritten as

$$\frac{dy}{dt} = 2te^{-2y} - e^{-2y} = e^{-2y}(2t-1).$$

Applying the method of separation of variables,

$$\frac{1}{e^{-2y}}\frac{dy}{dt} = 2t-1, \qquad \left(\frac{1}{e^{-2y}} = e^{2y}\right),$$

$$\int e^{2y}\frac{dy}{dt}\,dt = \int (2t-1)\,dt,$$

$$\int e^{2y}dy = \int (2t-1)\,dt,$$

$$\frac{1}{2}e^{2y} = t^2 - t + C.$$

Hence, $e^{2y} = 2t^2 - 2t + C$. (This C is actually twice the old C.) Take the logarithm of both sides.

$$\ln e^{2y} = \ln(2t^2 - 2t + C)$$

$$2y = \ln(2t^2 - 2t + C).$$

Therefore,

$$y = \frac{1}{2}\ln(2t^2 - 2t + C).$$

Next, choose C to make $y(0) = 3$. Set $t = 0$ and $y = 3$:

$$3 = \frac{1}{2}\ln(2\cdot 0^2 - 2(0) + C) = \frac{1}{2}\ln C,$$

$$6 = \ln C,$$

$$e^6 = C.$$

So finally,

$$y = \frac{1}{2}\ln(2t^2 - 2t + e^6).$$

25. $\frac{dy}{dt} = \frac{t+1}{ty}, t > 0, y(1) = -3.$ Separating the variables:

$$y\frac{dy}{dt} = \frac{t+1}{t} = 1 + \frac{1}{t},$$

$$\int y\frac{dy}{dt}\,dt = \int\left(1+\frac{1}{t}\right)dt,$$

$$\int y\,dt = \int\left(1+\frac{1}{t}\right)dt.$$

Therefore, $\frac{y^2}{2} = t + \ln|t| + C.$

Since $t>0$, rewrite $\ln|t|$ as $\ln t$, so

$$\frac{y^2}{2}=t+\ln t+C,$$
$$y^2=2t+2\ln t+C.$$

(C is now two times the value of the old C.)

$$y=\pm\sqrt{2t+2\ln t+C}.$$

Since $y(1)=-3$, and since a square root cannot be negative, use the "minus" form:

$$y=-\sqrt{2t+2\ln t+C},$$
$$-3=-\sqrt{2(1)+2\ln(1)+C},$$
$$-3=-\sqrt{2+0+C}, \qquad \text{(square both sides)}$$
$$9=2+C, \quad \text{or} \quad C=7.$$

Therefore, $$y=-\sqrt{2t+2\ln t+7}.$$

31. The constant function $y=0$ is a solution since this makes both sides of the equation equal zero. If $y\neq 0$ you may divide by y to obtain

$$\frac{1}{y}\frac{dy}{dp}=-\frac{1}{2}\cdot\frac{1}{p+3}$$
$$\int\frac{1}{y}\frac{dy}{dp}dp=\int-\frac{1}{2}\cdot\frac{1}{p+3}dp$$
$$\int\frac{1}{y}dy=-\frac{1}{2}\int\frac{1}{p+3}dp$$
$$\ln|y|=-\frac{1}{2}\ln|p+3|+C \qquad (C \text{ a constant})$$
$$\ln|y|=\ln|p+3|^{-1/2}+C$$
$$|y|=e^{\ln|p+3|^{-1/2}+C}=|p+3|^{-1/2}\cdot e^C$$
$$y=\pm e^C|p+3|^{-1/2}$$

The general solution (including the constant solution) has the form

$$y=A\,|\,p+3\,|^{-1/2}, \qquad A \text{ any constant.}$$

Note, however, that both the price p and the sales volume y should be positive quantities. So the only solutions that make some sense economically have the form

$$y=A(p+3)^{-1/2}, \quad A>0.$$

37. $\dfrac{dy}{dt} = -ay\ln\dfrac{y}{b}$. Note that y cannot be zero, because the logarithm would not be defined. However, $\ln\dfrac{y}{b}$ is zero when $y = b$. So, the only constant solution is $y = b$. If $y \neq b$, you may separate the variables:

$$\frac{1}{y\ln\frac{y}{b}}\cdot\frac{dy}{dt} = -a,$$

$$\int\frac{1}{y\ln\frac{y}{b}}\cdot\frac{dy}{dt}\,dt = \int -a\,dt,$$

$$\int\frac{1}{y\ln\frac{y}{b}}\,dy = \int -a\,dt = -at + C.$$

To evaluate the left-hand side, let

$$u = \ln\frac{y}{b} \Rightarrow du = \frac{1}{\left(\frac{y}{b}\right)}\cdot\frac{1}{b}dy = \frac{1}{y}dy.$$

Then

$$\int\frac{1}{\ln\frac{y}{b}}\cdot\frac{1}{y}dy = \int\frac{1}{u}du = \ln|u| + C,$$

$$= \ln\left|\ln\frac{y}{b}\right| + C.$$

Therefore, setting the two sides equal,

$$\ln\left|\ln\frac{y}{b}\right| = -at + C \qquad \text{(different } C\text{)}.$$

Next, take the exponential of both sides to produce:

$$\left|\ln\frac{y}{b}\right| = e^{-at+C} = e^{C}\cdot e^{-at} = Ce^{-at}\,(c = e^{C}) \qquad \text{(different } C \text{ again)}$$

$$\left|\ln\frac{y}{b}\right| = Ce^{-at} \qquad \text{(where } C \text{ is positive)}$$

$$\ln\frac{y}{b} = \pm Ce^{-at}$$

Write $\pm C$ as simply C, where now C can be positive or negative. The constant solution $y = b$ corresponds to $C = 0$. So the general solution, with an arbitrary C, is

$$\frac{y}{b} = e^{Ce^{-at}}, \qquad \text{and} \qquad y = be^{Ce^{-at}}.$$

10.3 First-Order Linear Differential Equations

Before you start this section, review the product rule for differentiation and the chain rule. For instance, make sure you understand the following identity: If y and $A(t)$ are functions of t, then

$$\frac{d}{dt}\left[e^{A(t)}y\right] = e^{A(t)}y' + ye^{A(t)}A'(t).$$

In this section, $A(t)$ stands for an antiderivative of a function $a(t)$. So that $A'(t) = a(t)$ and the preceding identity can be written as

$$\frac{d}{dt}\left[e^{A(t)}y\right] = e^{A(t)}y' + ye^{A(t)}a(t).$$

Exercises 1–6 are warm-up exercises. They do not require the full power of the method of this section. You should do as many of them as you can, in order to prepare for the remaining exercises in this section.

1. In this exercise, we are not asked to solve the equation $y' - 2y = t$. All we need is the integrating factor. To get this integrating factor you must first put the equation in standard form $y' + a(t)y = b(t)$. The given equation is already in standard form, with $a(t) = -2$. The integrating factor is obtained from $a(t)$ in two steps:

Step 1: Find an antiderivative $A(t)$ of $a(t)$. In this case,

$$A(t) = \int a(t)\,dt = \int -2\,dt = -2t.$$

You have to keep in mind that all we need is one antiderivative, so do not include an arbitrary constant when evaluating $A(t)$. That is, do not write $A(t) = -2t + C$. Just set $C = 0$ and take $A(t) = -2t$.

Step 2: Form the integrating factor $e^{A(t)}$. This is a straightforward step, since you have $A(t) = -2t$, then the integrating factor is e^{-2t}, and you are done.

7. We will solve the equation by following the step-by-step method described in the text.

Step 1: Put the equation in standard form $(y' + a(t)y = b(t))$. The equation is already in standard form $y' + y = 1$ where $a(t) = 1$.

Step 2: Find an integrating factor. This step is like Exercise 1. First find an antiderivative of $a(t)$.

$$A(t) = \int 1\,dt = t.$$

Note how we did not include an arbitrary constant. Now form the integrating factor $e^{A(t)} = e^t$.

Step 3: Multiply the equation through by the integrating factor and then simplify. (This is a crucial step in working the product rule backward, as you will now see.)

$$e^t(y' + y) = e^t \cdot 1$$

$$\overbrace{e^t y' + e^t y}^{\frac{d}{dt}[e^t y]} = e^t$$

$$\frac{d}{dt}[e^t y] = e^t.$$

We used the product rule in expressing $e^t y' + e^t y = \frac{d}{dt}[e^t y]$.

Step 4: Solve the differential equation. Integrate both sides of the last equation and remember that the integral cancels the derivative:

$$\frac{d}{dt}[e^t y] = e^t \quad \Rightarrow \quad e^t y = \int e^t\, dt = e^t + C.$$

(Here it is important to include the arbitrary constant when evaluating the integral, since we want all possible answers and not just one.) Solve for y by multiplying by e^{-t}:

$$e^{-t}e^t y = e^{-t}(e^t + C) \quad \Rightarrow \quad y = 1 + Ce^{-t}.$$

13. Follow the steps of Exercise 7.

Step 1: Put the equation in standard form $(y' + a(t)y = b(t))$. The equation is already in standard form

$$y' + \frac{1}{10+t}y = 0$$

where $a(t) = \dfrac{1}{10+t}$.

Step 2: Find an integrating factor.

$$A(t) = \int \frac{1}{10+t}\, dt = \ln|10+t| = \ln(10+t)$$

because $t > 0$ so $10 + t > 0$. The integrating factor is

$$e^{A(t)} = e^{\ln(10+t)} = 10 + t.$$

Step 3: Multiply the equation through by the integrating factor and then simplify.

$$(10+t)\left(y' + \frac{1}{10+t}y\right) = (10+t)\cdot 0$$

$$\overbrace{(10+t)y' + y}^{\frac{d}{dt}[(10+t)y]} = 0$$

$$\frac{d}{dt}[(10+t)y] = 0.$$

Step 4: Integrate both sides:

$$\frac{d}{dt}[(10+t)y] = 0 \quad \Rightarrow \quad (10+t)y = C,$$

because 0 is the derivative of constant functions. Finally, solve for y by multiplying by $\dfrac{1}{10+t}$:

$$(10+t)y = C \quad \Rightarrow \quad y = \frac{C}{10+t}.$$

Note that the choice $C = 0$ yields the function $y = 0$, which is clearly a solution of the differential equation $y' + \dfrac{1}{10+t}y = 0$.

19. Follow the step-by-step method of integrating factors.

Step 1: Put the equation in standard form $(y' + a(t)y = b(t))$. The equation is already in standard form

$$y' + y = 2 - e^t, \quad \text{where } a(t) = 1.$$

Step 2: Find an integrating factor.

$$A(t) = \int 1\, dt = t.$$

The integrating factor is $e^{A(t)} = e^t$.

Step 3: Multiply the equation through by the integrating factor and then simplify.

$$e^t(y' + y) = e^t \cdot (2 - e^t)$$

$$\overbrace{e^t y' + e^t y}^{\frac{d}{dt}[e^t y]} = 2e^t - e^{2t}$$

$$\frac{d}{dt}[e^t y] = 2e^t - e^{2t}.$$

Step 4: Integrate both sides:

$$e^t y = \int (2e^t - e^{2t})dt = 2e^t - \frac{1}{2}e^{2t} + C.$$

Finally, solve for y:

$$y = e^{-t}\left(2e^t - \frac{1}{2}e^{2t} + C\right) = 2 - \frac{1}{2}e^t + Ce^{-t}.$$

25. Solving an initial value problem is really like solving two separate problems. First, you must solve the equation

$$y' + y = e^t.$$

Your solution will contain an arbitrary constant C. To each value of C corresponds one solution of the differential equation. Next, use the initial condition $y(0) = 1$ to determine the unique solution of the initial value problem.

To solve the equation, use an integrating factor. The equation is in standard form with $a(t) = 1$. So, $A(t) = t$ and the integrating factor is e^t. Multiply the equation by the integrating factor and simplify to get

$$\frac{d}{dt}[e^t y] = e^t \cdot e^{2t} \quad \Rightarrow \quad \frac{d}{dt}[e^t y] = e^{3t}.$$

Integrate both sides and then solve for y:

$$e^t y = \int e^{3t} dt = \frac{1}{3}e^{3t} + C \quad \Rightarrow \quad y = \frac{1}{3}e^{2t} + Ce^{-t}.$$

This is the general solution of the differential equation. It consists of infinitely many functions; one for each value of C. Next, we choose the constant C in order to satisfy the initial condition $y(0) = -1$:

$$-1 = y(0) \quad \Rightarrow \quad -1 = \frac{1}{3}e^{2\cdot 0} + Ce^{-0} = \frac{1}{3} + C \quad \Rightarrow \quad C = -\frac{4}{3}.$$

Thus the (unique) solution of the initial value problem is $y = \frac{1}{3}e^{2t} - \frac{4}{3}e^{-t}$.

10.4 Applications of First-Order Linear Differential Equations

There is no doubt that word problems are the most challenging problems in calculus since there is no step-by-step method to do these problems. You have to read each question very carefully and then try to identify the kind of problem. Once you have identified the problem, you can look back at some of the examples that we have solved in the text and model your answer after its solution. The goal is to reach a point where you can do the exercises without looking at the solved examples in your text.

1. (a) Since this question is about rate of change, we answer it by looking at the differential equation. Let $P(t)$ denote the amount of money in the account at time t. We know from Example 1 that $P(t)$ satisfies the differential equation

$$y' - .06y = 2400$$

The rate of change of $P(t)$ is $P'(t)$. From the differential equation

$$P'(t) = .06P(t) + 2400. \qquad (1)$$

When the amount in the bank, $P(t)$, is $30,000, the rate of change is

$$P' = .06(30{,}000) + 2400 = 4200 \text{ dollars per year.}$$

Note that we did not need to know t in order to answer this question.

(b) Here we are told that $P(t)$ is growing at a rate twice as fast as the rate of the annual contributions. That is the rate of growth is $2 \times 2400 = 4800$ dollars per year. Plugging this value of P' into the equation (1) and solving for P, we find

$$4800 = .06P(t) + 2400 \quad \Rightarrow \quad P(t) = \frac{4800 - 2400}{.06} = 40{,}000 \text{ dollars.}$$

Thus $40,000 was in the account when the account was growing at the rate of $4800 per year. Here again note that we answered the question without knowing the value of t.

(c) In this part, we are given that $P(t) = \$40{,}000$ and we are asked to find t. From Example 1, we have

$$P(t) = -40{,}000 + 41{,}000e^{.06t}.$$

Plugging $P(t) = 40{,}000$ and solving for t, we find

$$\begin{aligned} 40{,}000 &= -40{,}000 + 41{,}000e^{.06t} \\ 80{,}000 &= 41{,}000e^{.06t} \\ \frac{80{,}000}{41{,}000} &= e^{.06t}\ \frac{80}{41} = e^{.06t} \\ \ln\left(\frac{80}{41}\right) &= .06t \\ t &= \frac{1}{.06}\ln\left(\frac{80}{41}\right) \approx 11.1 \text{ years.} \end{aligned}$$

Thus it takes approximately 11.1 years or 11 years and 2 months for the account to reach $40,000 dollars.

7. This is a problem about paying back a loan. The closest example that we have to direct us through the solution is Example 2 in the text. You should read it very carefully and understand it thoroughly before you attempt to solve this exercise.

Let $P(t)$ denote the amount that the person owes at time t (in years from the time the loan was taken). We are asked to determine the value of k, the rate of annual payments, if the loan is to be paid in full in exactly 10 years; that is, if $P(10) = 0$. To answer this question, we must find a formula for $P(t)$ first.

Two influences act on $P(t)$: The interest that is added, and the payments that are subtracted (remember that a payment subtracts from the amount owed). Thus, the differential equation satisfied by $P(t)$ is

$$y' = .075y - k, \quad \text{or} \quad y' - .075y = -k,$$

where the last equation is in standard form. We are also given important information about $P(t)$, namely, that the initial amount owed was 10,000. Hence $y(0) = 100{,}000$. So $P(t)$ is the solution of the initial value problem

$$y' - .075y = -k \quad y(0) = 100{,}000.$$

We solve this problem using the integrating factor method from the previous section:

$$a(t) = -.075$$

$$A(t) = \int a(t)\,dt = -.075t$$

$$\text{Integrating factor} = e^{A(t)} = e^{-.075t}$$

Multiplying the equation by the integrating factor and simplifying, we obtain

$$e^{-.075t}(y' - .075y) = -ke^{-.075t}$$

$$\frac{d}{dt}[e^{-.075t}y] = -ke^{-.075t}.$$

Integrating with respect to t and solving for y, we find

$$e^{-.075t}y = \int -ke^{-.075t}\,dt = \frac{k}{.075}e^{-.075t} + C$$

$$y = e^{.075t}\left(\frac{k}{.075}e^{-.075t} + C\right)$$

$$y = \frac{k}{.075} + Ce^{.075t}.$$

Note that the solution contains an arbitrary constant C (as it should) and it also depends on k the rate of annual payments. The value of C will be determined from the initial condition: $P(0) = 100{,}000$. Plugging $t = 0$ into the formula for y, we find

$$y(0) = 100{,}000 = \frac{k}{.075} + Ce^{.075 \cdot 0} = \frac{k}{.075} + C$$

$$C = 100{,}000 - \frac{k}{.075}.$$

Plugging this value of C into the formula for y, we obtain

$$P(t) = \frac{k}{.075} + \overbrace{(100{,}000) - \tfrac{k}{.075}}^{=C}e^{.075t}.$$

This is the formula of $P(t)$ in terms of the annual payments k.

Now if the loan is to be paid in full in 10 years, this implies that $P(10) = 0$. Hence,

$$0 = P(10) = \frac{k}{.075} + \left(100{,}000 - \frac{k}{.075}\right)e^{.075\cdot 10}$$

$$-\frac{k}{.075} = \left(100{,}000 - \frac{k}{.075}\right)e^{.75}$$

$$\frac{k}{.075}e^{.75} - \frac{k}{.075} = 100{,}000e^{.75}$$

$$k\left(\frac{e^{.75}}{.075} - \frac{1}{.075}\right) = 100{,}000e^{.75}$$

$$k = \frac{100{,}000e^{.75}}{\frac{e^{.75}}{.075} - \frac{1}{.075}} \approx 14{,}214.4 \text{ dollars per year.}$$

Thus in order for the \$100,000 loan to be paid in full in 10 years, the rate of payments should be about \$14,214.40 per year (or \$1184.50 per month).

13. We have to solve the initial value problem

$$y' = .1(10 - y), \quad y(0) = 350.$$

Put the equation in standard form and solve it using the step by step method of integrating factor:

$$y' + .1y = 1 \qquad \text{(Equation in standard form)}$$

$$a(t) = .1, \quad A(t) = .1t$$

$$\text{Integrating factor} = e^{A(t)} = e^{.1t}$$

$$e^{.1t}(y' + .1y) = e^{.1t} \qquad \text{(Multiply by the integrating factor.)}$$

$$\frac{d}{dt}[e^{.1t}y] = e^{.1t} \qquad \text{(Simplify the equation.)}$$

$$e^{.1t}y = \int e^{.1t}dt = \frac{1}{.1}e^{.1t} + C \qquad \text{(Integrate.)}$$

$$y = 10 + Ce^{-.1t} \qquad \text{(Multiply by } e^{-.1t}.\text{)}$$

To satisfy the initial condition $y(0) = 350$, we must have

$$350 = y(0) = 10 + Ce^{-.1\cdot 0} = 10 + C \text{ or } C = 340.$$

Thus the temperature of the rod at any time t is given by $f(t) = 10 + 340e^{-.1t}$.

19. (a) In this problem, the concentration of creatinine in the dialysate solution at any time t (in hours) is denoted by $f(t)$ and measured by grams per liter. The concentration $f(t)$ satisfies the differential equation.

$$y' = k(110 - y).$$

We are told that when the concentration was $f(t) = 75$ grams per liter, it was rising at the rate of 10 grams per liter per hour. So, when $y = 75$, $y' = 10$. Plugging these values into the differential equation and solving for k, we find

$$10 = k(110 - 75) \Rightarrow k = \frac{10}{35} = \frac{2}{7} \approx .286.$$

(b) Using the value of k that we just found, we obtain the differential equation satisfied by $f(t)$

$$y' = \frac{2}{7}(110 - y).$$

This equation expresses the rate of change of $f(t)$ as a function of $f(t)$. We also know that initially the concentration in the dialysate is 0. Hence $y(0) = 0$. The rate of change of the concentration after 4 hours of dialysis is given in part (a). It is 10 grams per liter per hour. To obtain the rate of change of the concentration at the beginning of the session, we set $t = 0$ in the differential equation and use $y(0) = 0$:

$$y'(0) = \frac{2}{7}(110 - y(0)) = \frac{2}{7}110 \approx 31.43 \text{ grams per liter per hour.}$$

Thus at the beginning of the session, the creatinine substance was filtering into the dialysate solution at the rate of 31.43 grams per liter per hour, which is over three times the rate after four hours of dialysis. So if we replace the solution after four hours by a fresh solution, we would triple the rate of creatinine clearance from the body.

25. **(a)** Let $f(t)$ (in milligrams) denote the amount of morphine in the body at time t in hours. The rate of change of $f(t)$ is affected by two influences: the rate at which the body removes the substance, and the rate at which morphine is injected in the body. The first rate of change is given by $-.35f(t)$, since the body removes the substance at a rate proportional to the amount of the substance present at time t, with constant of proportionality $k = .35$. The second rate of change is given to us: It is t milligrams per hour. Thus $f(t)$ satisfies the differential equation

$$y' = -.35y + t.$$

(b) Since the body was free of morphine at the beginning of the infusion, we have the initial condition $y(0) = 0$. Putting the equation in standard form, we obtain the initial value problem:

$$y' + .35y = t \quad y(0) = 0.$$

To determine $f(t)$, we must solve this initial value problem. We use an integrating factor:

$$a(t) = .35, \quad A(t) = .35t$$

$$\text{Integrating factor } = e^{A(t)} = e^{.35t}$$

$$e^{.35t}(y' + .35y) = e^{.35t}t \qquad \text{(Multiply by the integrating factor.)}$$

$$\frac{d}{dt}[e^{.35t}y] = e^{.35t}t \qquad \text{(Simplify the equation).}$$

$$e^{.35t}y = \int e^{.35t}t\,dt = \frac{1}{(.35)^2}e^{.35t}(.35t - 1) + C \qquad \text{(Integrate.)}$$

In evaluating the last integral, you can use integration by parts or appeal to (1) in Section 10.4, with $a = 1, b = 0, c = .35$. Multiplying both sides by $e^{-.35t}$, we obtain

$$y = \frac{1}{(.35)^2}(.35t - 1) + Ce^{-.35t}$$

In order to satisfy the initial condition, we must have

$$0 = y(0) = \frac{1}{(.35)^2}(-1) + C$$

$$C = \frac{1}{(.35)^2} = \frac{1}{.1225}$$

Hence the amount of morphine in the body at time t is

$$f(t) = \frac{1}{(.35)^2}(.35t - 1) + \frac{1}{(.35)^2}e^{-.35t}$$
$$= \frac{1}{(.35)^2}(.35t - 1 + e^{-.35t}).$$

After 8 hours, there were $f(8) = \frac{1}{(.35)^2}\left(.35(8) - 1 + e^{-.35(8)}\right) \approx 15.2$ mg of morphine in the body.

10.5 Graphing Solutions of Differential Equations

We hope you get to study this section! Some instructors omit it because they think it is too hard for first-year calculus students. (No other book at this level attempts to teach the material.) But we have found that our students do as well on the qualitative theory as on almost any material in the second half of the text. The secret is to spend about one week on this section.

One of the difficulties here is learning to sketch yz-graphs where the y-axis is horizontal. The difficulty is real though it is mainly psychological. To help you get used to yz-graphs, we have included them for Exercises 7–20. Study them there so you will be able to produce similar graphs in Exercises 21–37.

1. In graphing the autonomous equations described in Exercises 1–6, you may find it helpful to fill in the blanks in the following sentences with either *increases* or *decreases*:

As t increases, y __________.
As y increases, $z = y'$ __________.

In sketching the yz-graph for Exercise 1, a key point to keep in mind is that $z = y'$, i.e., z represents the derivative (or slope) of y. Thus the statement

"the slope of y is always positive,"

a statement about the ty-graph, can be translated to

$$\text{"}z \text{ is always positive,"} \qquad (1)$$

a statement about the yz-graph. Similarly, the statement

"as t increases, the slope of y becomes less positive"

can be translated to

"as t increases, z decreases."

Since y increases as t increases (be careful here), we can thus conclude:

$$\text{"as } y \text{ increases, } z \text{ decreases."} \qquad (2)$$

Fortified with the information in (1) and (2), we are now ready to sketch the yz-graph. Begin with the point $(y, z) = (1, z(1))$ corresponding to $t = 0$. From (1), we know that the value of $z(1)$ is positive. (See Figure 1(a)). As t moves from 0 to 4, the value of y will increase, and thus our graph will move to the right from the starting point $(1, z(1))$. From (2), we know that the graph will move downward as y increases, so we can indicate the initial direction of the graph with an arrow as in Figure 1(b).

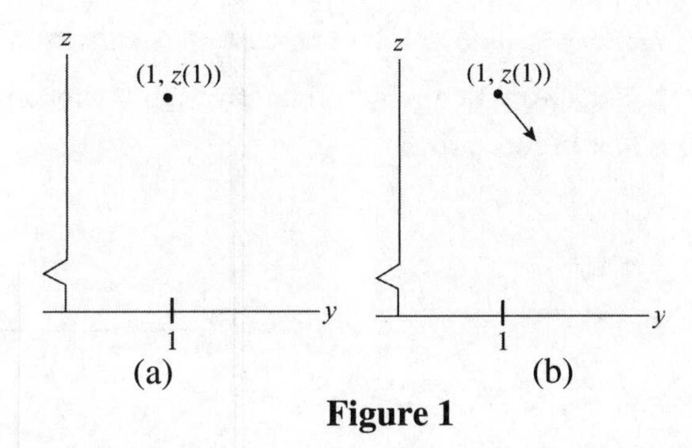

(a) (b)

Figure 1

From (1) we know that the graph will not descend below $y = 0$ in the domain (1, $y(3)$). A possible *yz*-graph is sketched in Figure 2. Note that we do not necessarily know that the graph is linear, however.

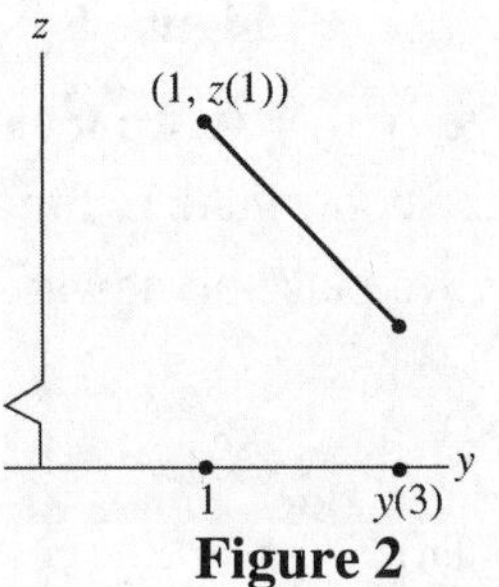

Figure 2

This is a graph that shows the behavior of y' (remember, $z = y'$). We can use this graph to sketch the particular solution of the equation. Start with the fact that (0, 1) is on the graph of the particular solution (Figure 3).

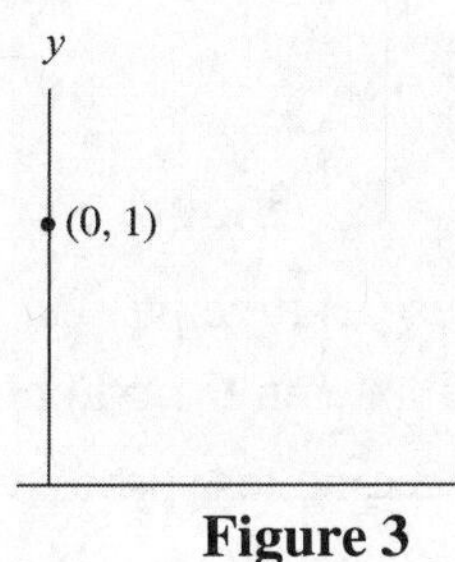

Figure 3

The domain is $0 \le t \le 3$. What happens as *t* moves from 0 towards 3? Since y' is positive, we know that *y* increases as *t* increases. Denote this with an upward arrow, as in Figure 4(a). Next, we need to determine whether the curve will be concave up or concave down. We can answer this by answering the question: "what happens to y' as *y* increases?" Referring to Figure 2, we see that y' decreases as *y* increases. Thus the slope of *y* is decreasing, so the curve is concave down, as in Figure 4(b).

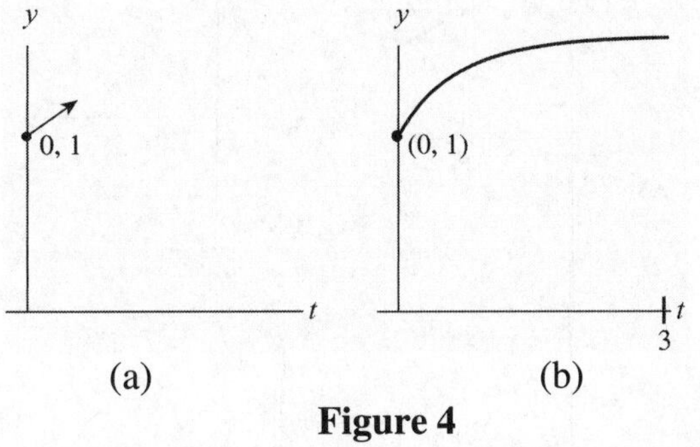

(a) (b)

Figure 4

Since y' is always positive, the autonomous equation $y' = g(y)$ does not have any constant solutions.

7. Given $y' = 3 - \frac{1}{2}y$, let $g(y) = 3 - \frac{1}{2}$. The graph of $z = g(y)$ is shown in Figure 5(a). Set $3 - \frac{1}{2} = 0$ and find $y = 6$. Thus $g(y)$ has a zero when $y = 6$. Therefore the constant function $y = 6$ is a solution of the differential equation. See Figure 5(b).

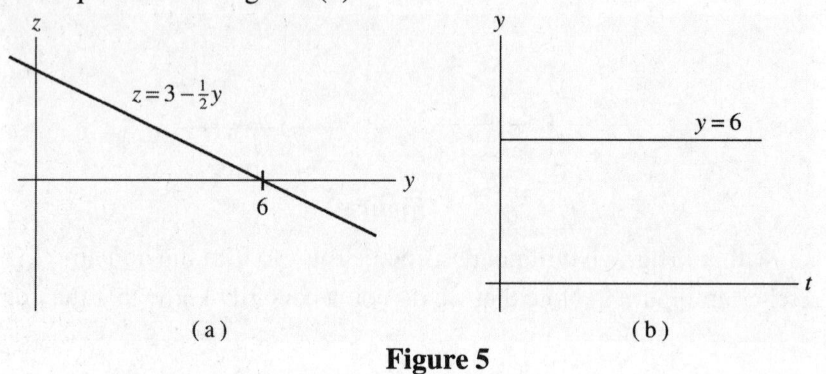

Figure 5

To sketch the solution corresponding to $y(0) = 4$, we locate this initial value on the y-axes in Figure 6(a) and Figure 6(b), and note that the z-coordinate in Figure 6(a) is positive when $y = 4$. That is, the derivative is positive when $y = 4$. So, we place an upward arrow at the initial point in Figure 6(b).

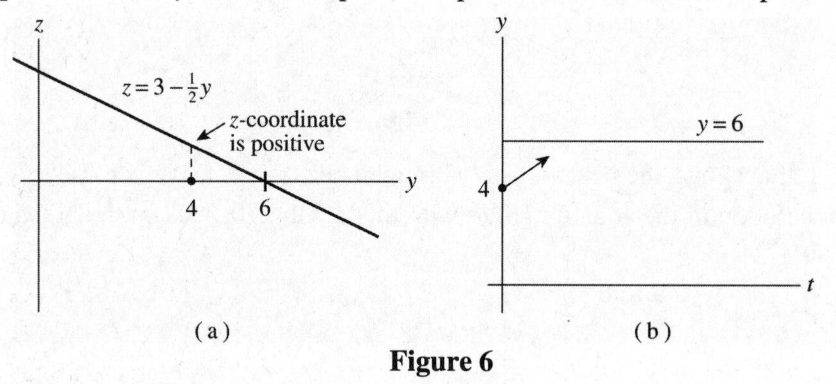

Figure 6

From Figure 2b, the y-values will increase as time passes, so y will move to the right on the yz-graph. See Figure 3(a). As a result, the z-coordinate on the graph of $z = 3 - \frac{1}{2}$ will become less positive. That is, the slope of the solution curve will become less positive. Thus the solution curve is concave down, as in Figure 3(b).

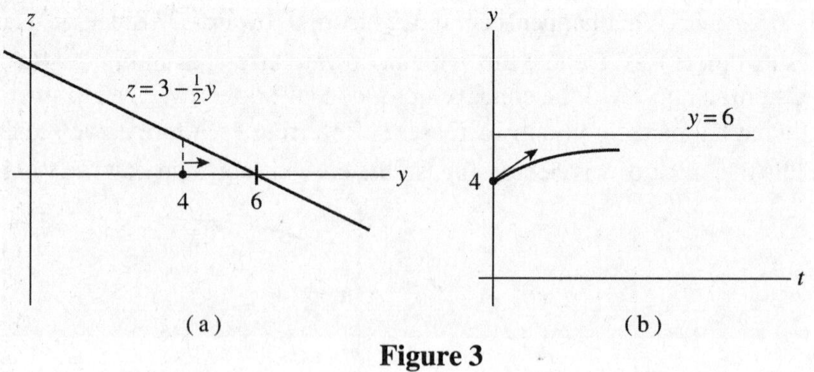

Figure 3

To obtain the graph of the solution satisfying $y(0)=8$, begin by making the sketch in Figure 4. First, note that the z-coordinate in Figure 4a is negative when $y=8$. (That is, the derivative is negative when $y=8$.) So place a downward arrow at $y=8$ in Figure 4(b).

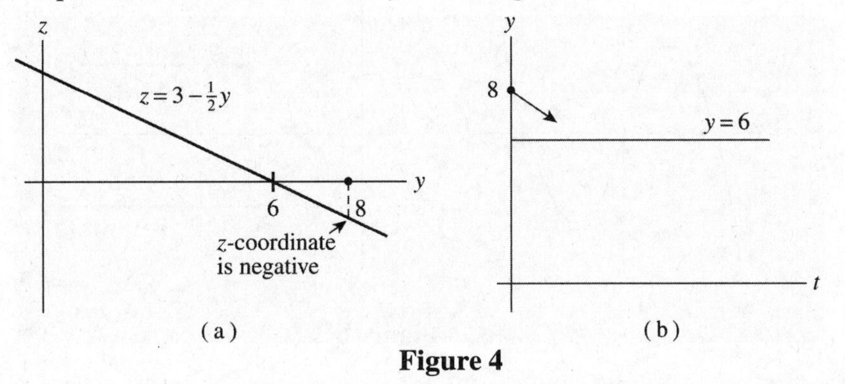

Figure 4

The downward arrow in Figure 4(b) shows that the y-values will decrease as time passes. This means that y will move to the *left* on the yz-graph. See Figure 5(a). As a result, the z-coordinates on the yz-graph will become less negative and the *slopes* on the ty-graph will become less negative. (Reread the preceding explanation because this is a key idea.)

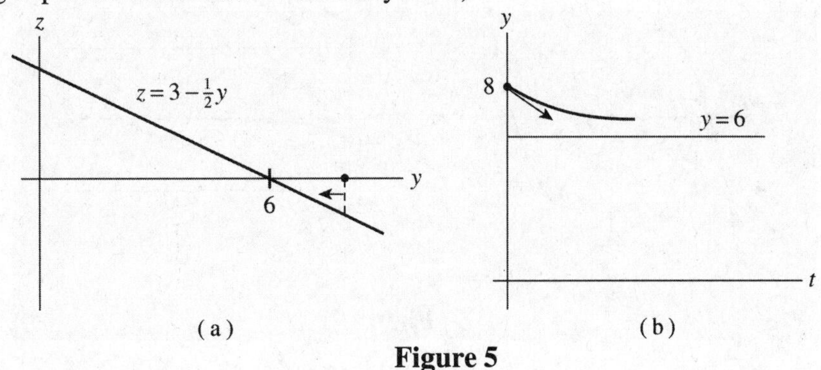

Figure 5

The solution curve in Figure 5(b) is concave up because as time passes the negative slopes on the curve become less negative.

13. First, sketch the graph of $z=y^3-9y$. Find the zeros of $g(y)=y^3-9y$ by setting $g(y)=0$ and solving for y.

$$y^3-9y=0,$$
$$y(y^2-9)=0,$$
$$y(y+3)(y-3)=0.$$

The graph of $z=y^3-9y$ crosses the y-axis at $y=0$, $y=\pm 3$. Next, to find where the graph has a relative maximum and relative minimum, set $\frac{dz}{dy}=0$ and solve for y.

$$\frac{d}{dy}(y^3-9y)=3y^2-9=0.$$

Thus $3y^2 = 9$, $y^2 = 3$, and $y = \pm\sqrt{3}$. See Figure 6(a). The constant solutions are shown in Figure 6(b).

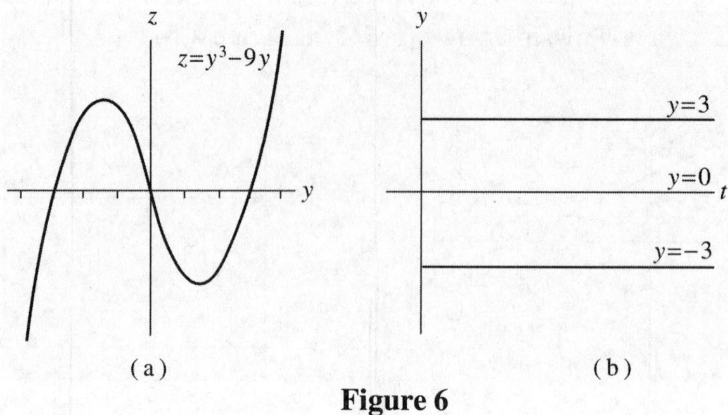

Figure 6

Figure 7 shows the sketches needed to obtain the graphs of the solutions such that $y(0) = -4$, $y(0) = -1$, and $y(0) = 4$. Note that the solution where $y(0) = -1$ cannot cross the constant solution $y = 0$.

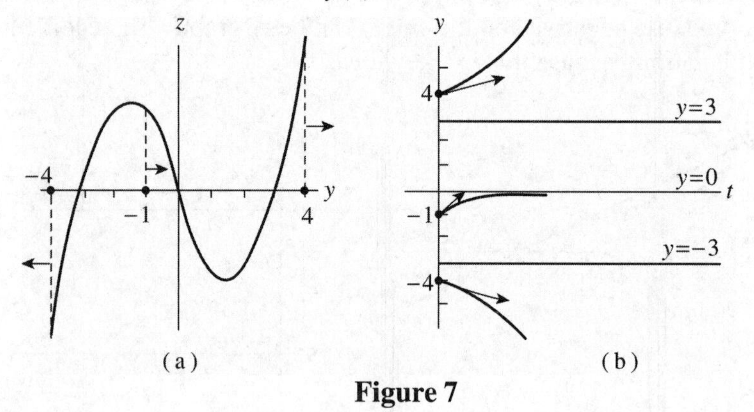

Figure 7

Figure 8 shows the work for the solution with $y(0) = 2$. Since the initial z-coordinate in Figure 8(a)a is negative, the y-values of the solution curve will decrease and y will move to the *left* on the yz-graph. At first, the z-coordinate will become more negative (i.e., the slope of the solution curve will become more negative). Then, as y moves to the left past $y = \sqrt{3}$, the z-coordinates on the yz-graph will become less negative (i.e., the slope of the solution curve will become less negative). Thus the solution curve will have an inflection point at $y = \sqrt{3}$, as in Figure 8(b).

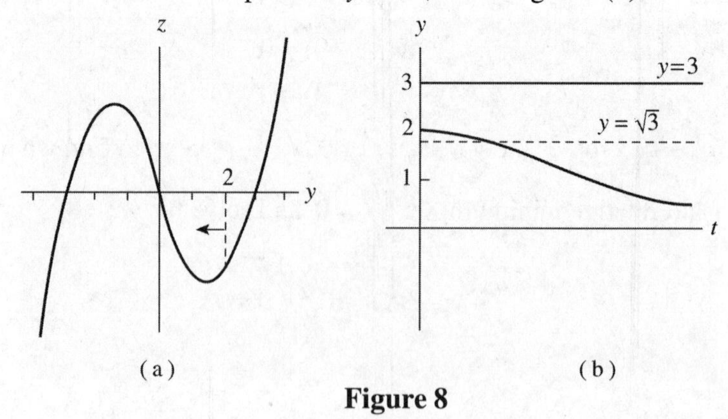

Figure 8

19. The graph of $z = g(y)$ shown in Figure 9(a) has zeros at $y = 1$ and $y = 6$. So, the constant solutions of $y' = g(y)$ are $y = 1$ and $y = 6$. The relative maximum of $g(y)$ occurs when $y = 4$, so a solution curve will have an inflection point whenever it crosses the dashed line shown in Figure 9(b). The inflection point at $y = 2$ on the graph of $g(y)$ has no influence on the general shape of solution curves, so you should ignore it.

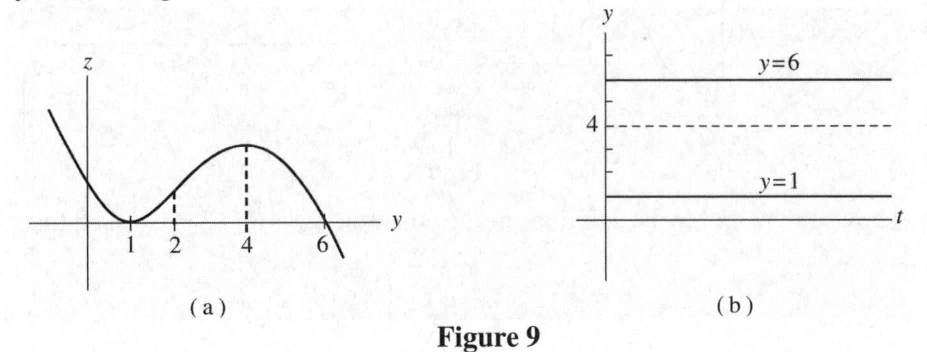

Figure 9

Figure 10 shows the work for the solution of $y' = g(y)$ such that $y(0) = 0$, $y(0) = 1.2$, $y(0) = 5$, and $y(0) = 7$.

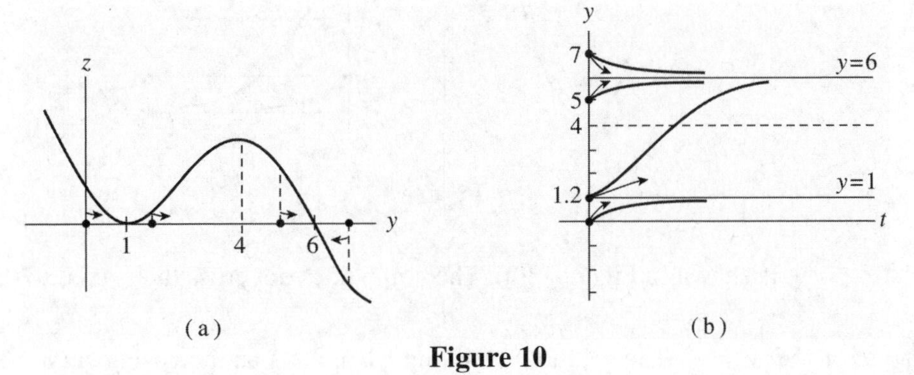

Figure 10

Warning: Although the function $g(y)$ in Exercise 19 has a minimum at $y = 1$, the minimum value is zero. Notice that Rule 7 on page 493 applies only to *nonzero* relative maximum or minimum points.

25. The graph of $z = y^2 - 3y - 4$ is a parabola. It opens upward because the coefficient of y^2 is positive. To find where the graph crosses the y-axis, solve

$$y^2 - 3y - 4 = 0,$$
$$(y+1)(y-4) = 0,$$
$$y = -1, \quad \text{or} \quad y = 4.$$

Set $\frac{dz}{dy} = 0$ and find that $2y - 3 = 0$ and $y = \frac{3}{2} = 1.5$. Thus the parabola has a minimum at $y = 1.5$.

This is enough information to produce the initial sketch in Figure 11(b).

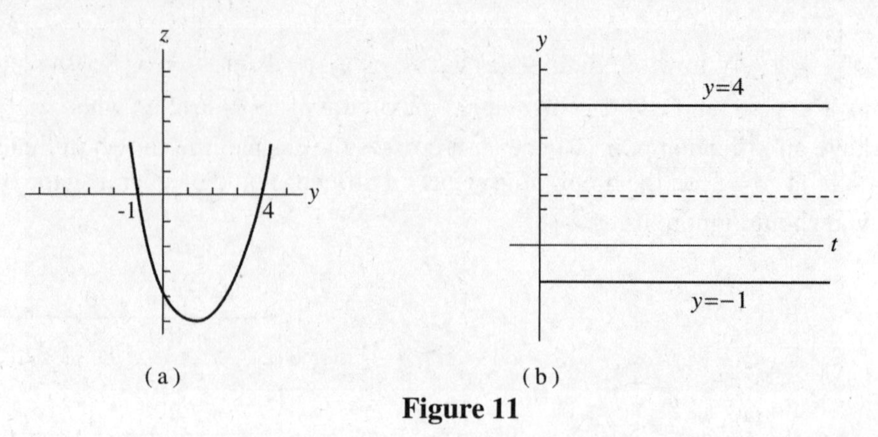

Figure 11

Figure 12 shows the work for the solutions of the equation $y' = y^2 - 3y - 4$ such that $y(0) = 0$ and $y(0) = 3$.

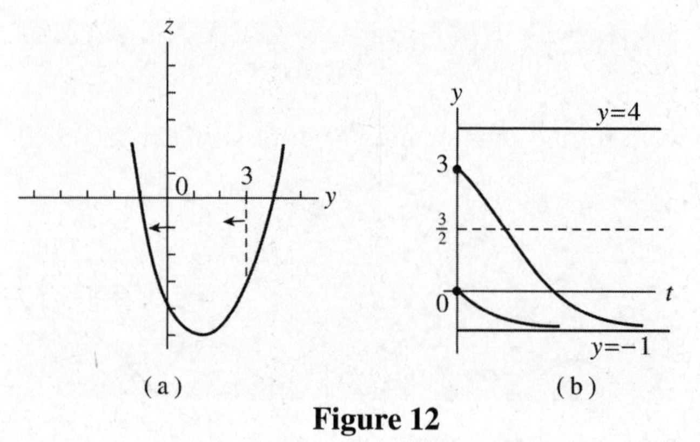

Figure 12

31. The graph of $z = \dfrac{1}{y}$ is shown in Figure 13(a). The graph does not cross the y-axis, so there are no constant solutions of $y' = \dfrac{1}{y}$. The solution satisfying $y(0) = 1$ is an increasing curve because the z-coordinate of the initial point on the yz-graph is positive. However, as y increases on the yz-graph, the z-coordinates become less positive; i.e., the slopes of the solution curve become less positive. A similar solution holds for the solution satisfying $y(0) = -1$, where the slopes are negative and become less negative. See Figure 13(b).

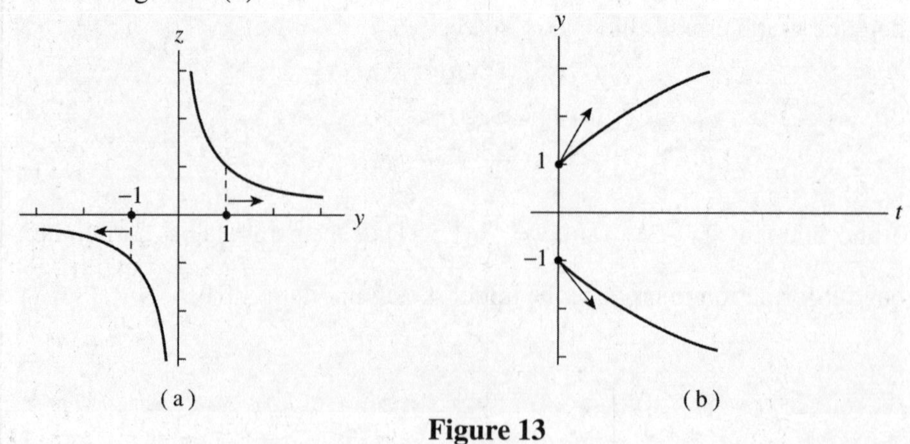

Figure 13

37. Let $f(t)$ be the height of the sunflower at time t. Then $f'(t)$ is the rate at which the sunflower is growing at time t. You are told that this rate is proportional to the product of its height and the difference between its height at maturity and its current height. If you let H be the sunflower's height at maturity, then

$$f'(t) = kf(t)[H - f(t)] \tag{1}$$

or equivalently,

$$y' = ky(H - y). \tag{2}$$

Before you can sketch a solution, you must determine whether k is positive or negative. Clearly, $f(t)$ is increasing, so $f'(t)$ is positive. Also, $f(t)$ is less than H, so the factor $H - f(t)$ is positive. From this it follows that k must be positive.

From (2) you must sketch

$$z = ky(H - y) = kHy - ky^2.$$

The graph of this equation is a parabola which opens down, with zeros at 0 and H, and maximum at $H/2$. See Figure 14(a).

The initial height of the sunflower is $y(0)$, which is slightly greater than 0. Since z is positive here, this solution is increasing as it leaves the initial point. As y moves to the right, the z-coordinate becomes more positive until y reaches $H/2$. After that, the z-coordinate becomes less positive, and hence the curve has an inflection point at $H/2$, and is concave down after that point. Also, it is clear from equation (1) that the constant functions $y = 0$ and $y = H$ are solutions. See Figure 14(b).

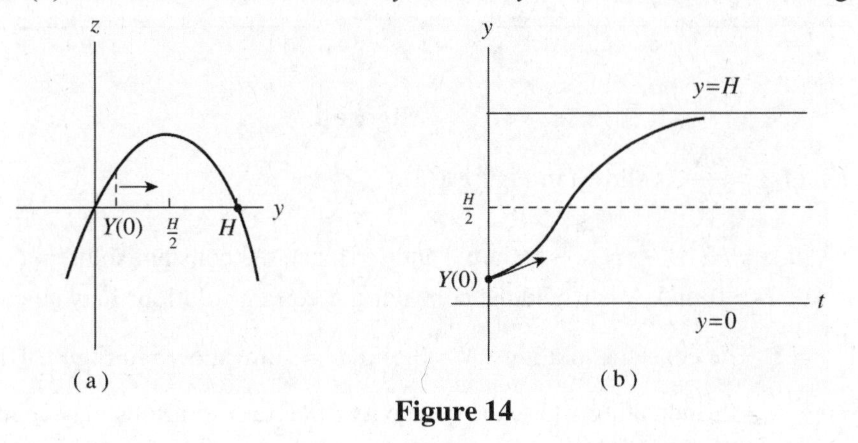

Figure 14

Helpful Hint: Exercise 37 is a key exercise. Work on it now! You will see more exercises like this in Section 10.6.

10.6 Applications of Differential Equations

Here we are! If you have been faithful in your work on the "word problems" in Sections 10.1–10.5, you should be ready for Section 10.6. The exercises fall into two categories:

1. Problems involving one constant of proportionality, and
2. "One-compartment" problems.

The first category is analyzed carefully in the first two pages of the section. Most of Section 5.4 concerned applications of the first category of problems. One of the most important applications there involved logistic growth, and this subject is discussed again in Section 10.6.

One-compartment problems are very common in applications. Multi-compartment problems also arise. You can read about them in *Mathematical Techniques for Physiology and Medicine*, by William Simon (New York: Academic Press, 1972). This book was at one time the text for a course at the Rochester School of Medicine and Dentistry.

The material on population genetics was suggested to us by one of our students. While in our class, she was taking a course on population genetics where she saw a qualitative analysis of certain differential equations. Only the *yz*-graphs were shown in her textbook, *Population Genetics*, by C. C. Li (Chicago: University of Chicago Press, 1995). Our student used her knowledge from our calculus course to provide the *ty*-graphs of the solution curves. Now you, too, can have a glimpse into this fascinating topic.

1. **(a)** $\frac{dN}{dt} = N(1-N),\ N(0) = .75$

The carrying capacity is $K = 1$ and the intrinsic rate is $r = 1$.

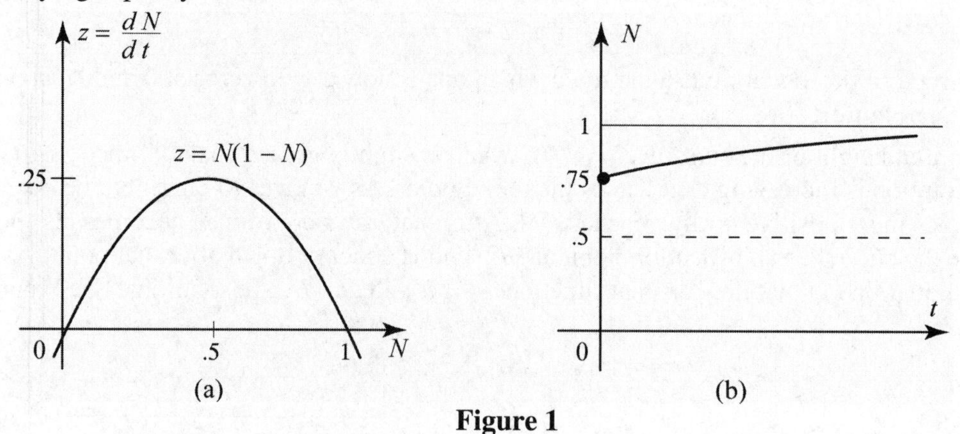

Figure 1

(b) The graph of $z = \frac{dN}{dt}$ is shown in Figure 1(a).

(c) The zeros of $z = N(1-N) = N - N^2$ are 0 and 1. Hence the constant solutions to the differential equation are $N = 0$ and $N = 1$, and the concavity of certain solutions may change at $N = \frac{0+1}{2} = .5$. The constant solutions $N = 0$ and $N = 1$ are shown in Figure 1(b), along with the dashed line $N = .5$ indicating where the concavity of certain solutions may change.

(d) When $N = .75$, $\frac{dN}{dt} = (.75)(1-.75)$ is positive, the solution curve corresponding to the initial condition $N(0) = .75$ is always increasing. The curve will not pass through the line $N = 1$, and the curve's concavity will not change since it will not pass through $N = .5$. Thus we sketch a curve starting at the point (0, .75) that is increasing, concave down, and approaches the horizontal asymptote $N = 1$. See Figure 1(b).

7. Let $y = f(t)$ be the percentage of the population having the information at time t. Then $0 \le f(t) \le 100$, and $100 - f(t)$ is the percentage that does not have the information. The problem says that

$$\begin{bmatrix}\text{the rate of} \\ \text{spread of} \\ \text{information}\end{bmatrix} \begin{bmatrix}\text{is} \\ \text{proportional} \\ \text{to}\end{bmatrix} \begin{bmatrix}\text{precentage} \\ \text{not having} \\ \text{information}\end{bmatrix}.$$

Thus,

$$f'(t) = k \cdot [100 - f(t)] \qquad (1)$$

for some constant k. Equivalently

$$y' = k(100 - y). \qquad (2)$$

Before you can sketch a solution, you must determine whether k is positive or negative. Clearly, $f(t)$ is increasing, so $f'(t)$ is positive. Also, $f(t)$ is not more than 100%, so $100 - f(t)$ is also positive. From this it follows that k is positive.
From (2) you must sketch

$$\begin{aligned} z &= k(100 - y) \\ &= 100k - ky, \end{aligned}$$

where k is positive. The graph of this equation is a straight line with negative slope. Also, when $y = 100$ we have $z = 0$. See Figure 2(a) below. For the solution where $y(0) = 1$, use the method in Section 10.4 and obtain the curve in Figure 2(b).

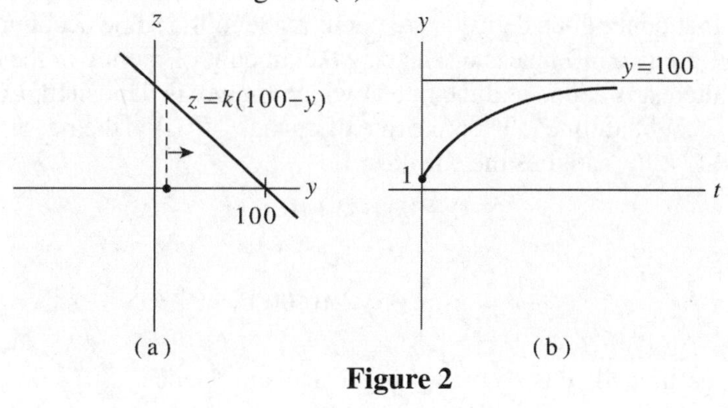

Figure 2

13. Let $y = f(t)$ be the amount of substance A at time t. Then $f'(t)$ is the rate at which substance A is converted into substance B. You are told that this rate is proportional to the square of the amount of A. Thus,

$$f'(t) = k[f(t)]^2 \qquad (1)$$

where k is a constant of proportionality. Equivalently,

$$y' = ky^2. \qquad (2)$$

Before you can sketch a solution, you must determine whether k is positive or negative. Clearly, $f(t)$ is decreasing, so $f'(t)$ is negative. Since $f^2(t)$ is never negative, k must be negative.

From (2) you must sketch,

$$z = ky^2,$$

where k is negative, which determines a parabola that opens down. See Figure 3(a) below. To sketch a solution where $y(0)$ is some positive number (representing the initial amount of substance A), use the method in Section 10.4 and obtain the curve in Figure 3(b).

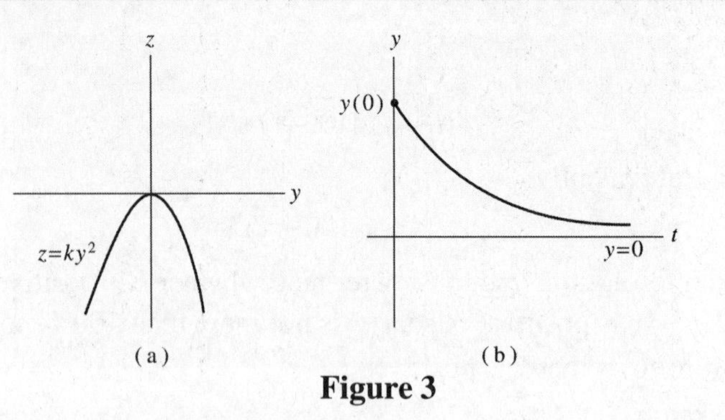

Figure 3

19. (a) At first, ignore the deposits to the account. In Section 5.2, we showed that if no deposits or withdrawals are made, then $f(t)$ satisfies the equation

$$y' = .05y.$$

That is, the account grows at a rate proportional to the amount in the account. We conclude that interest is being added to the account at a rate proportional to the amount in the account.

Now suppose that continuous deposits are being made to the same account at the rate of \$10,000 per year. There are two influences on the way the amount of money in the account changes—the rate at which interest is added and the rate at which money is deposited. Let $f(t)$ be the amount of money in the account at time t. Then the rate of change of $f(t)$ is the net effect of these two influences. That is, $f(t)$ satisfies the equation

$$f'(t) = .05f(t) + 10{,}000,$$

or equivalently,

$$y' = .05y + 10{,}000. \tag{1}$$

At time $t = 0$, assume there is no money in the account. Hence $y(0) = 0$.

(b) We note that $y = \frac{-10{,}000}{.05} = -200{,}000$ is a constant solution, since it makes both sides of equation (1) equal zero. Assuming that $y \neq -200{,}000$, use separation of variables:

$$y' = .05(y + 200{,}000),$$

$$\int \frac{1}{(y + 200{,}000)} \frac{dy}{dt} dt = \int .05 dt,$$

$$\int \frac{1}{(y + 200{,}000)} dy = \int .05 dt,$$

$$\ln|y + 200{,}000| = .05t + C, \qquad C \text{ a constant},$$

$$|y + 200{,}000| = e^{.05t+C} = e^C \cdot e^{.05t}.$$

If $y \neq -200{,}000$, then $y + 200{,}000$ is either positive or negative for all t. Thus,

$$y = \pm e^C \cdot e^{.05t} - 200{,}000.$$

The general solution (including the constant solution) has the form

$$y = Ae^{.05t} - 200{,}000, \qquad A \text{ a constant}.$$

Use the fact that $y(0) = 0$ to obtain $A = 200{,}000$. Hence,

$$y = 200{,}000e^{.05t} - 200{,}000 = 200{,}000(e^{.05t} - 1).$$

The amount in the account after 5 years is given by

$$y(t) = 200{,}000(e^{.25} - 1) \approx \$56{,}806$$

25. First sketch

$$z = -.0001q^2(1-q).$$

Clearly $z = 0$ when $q = 0$ or $q = 1$. Note that $z = .0001q^3 - .0001q^2$, which shows that the graph is a cubic curve. To find relative extreme points set $\dfrac{dz}{dq} = 0$.

$$.0003q^2 - .0002q = 0,$$
$$.0001q(3q-2) = 0.$$

So $q = 0$, or $3q - 2 = 0$ and $q = 2/3$. Since $\dfrac{d^2z}{dq^2} = .0006q - .0002 = .0002(3q-1)$, you can see that $\dfrac{d^2z}{dq^2}$ is negative at $q = 0$ and positive at $q = \dfrac{2}{3}$.

Here is a rough sketch of the graph.

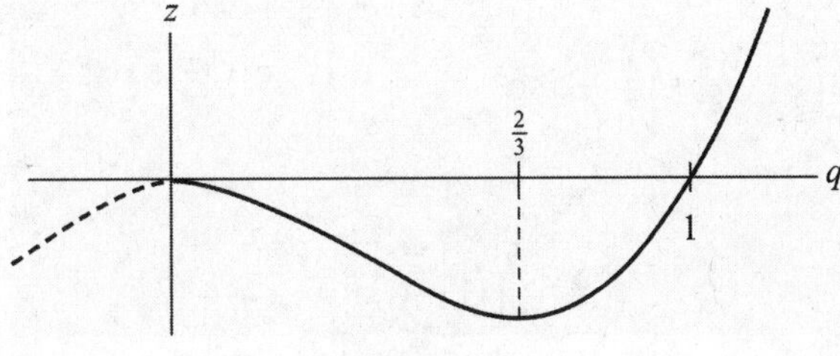

Figure 4

You are asked to find a solution when $q(0)$ slightly less than 1, so you should assume $q(0) > 2/3$. Using the method of Section 10.4, we obtain

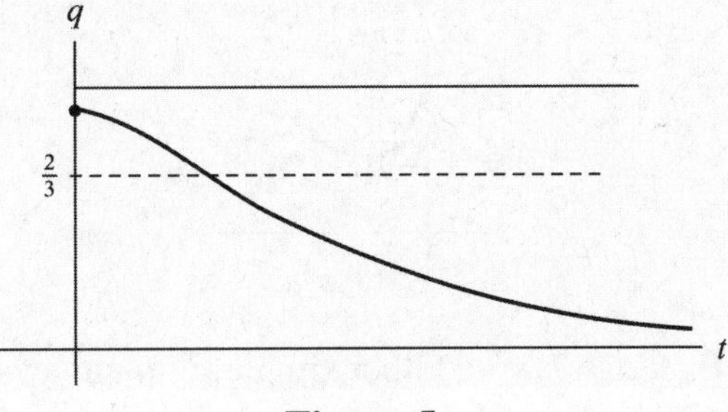

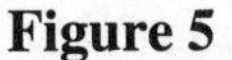
Figure 5

10.7 Numerical Solution of Differential Equations

1. If $f(t)$ is a solution of $y' = ty - 5$, then

$$f'(t) = tf(t) - 5,$$

for all t in the domain of $f(t)$. The graph of $f(t)$ passing through (2, 4) means that $f(2) = 4$. Set $t = 2$ in the equation above, and find that

$$f'(2) = 2f(2) - 5,$$
$$= 2 \cdot 4 - 5 = 3.$$

So, the slope of the graph is 3 at $t = 2$.

Helpful Hint: Make sure you understand the solutions to Exercises 1–4. They make good test questions and they prepare you for both Euler's method and the theory in Section 10.7.

7. Here $g(t, y) = 2t - y + 1$, $a = 0$, $b = 2$, $y(0) = 5$, and $h = \dfrac{2-0}{4} = \dfrac{1}{2}$. Starting with $(t_0, y_0) = (0, 5)$, compute $g(0, 5) = 2(0) - 5 + 1 = -4$. Thus,

$$t_1 = \frac{1}{2}, \quad y_1 = 5 + (-4)\cdot\frac{1}{2} = 3.$$

Next, $g\left(\frac{1}{2}, 3\right) = 2\left(\frac{1}{2}\right) - 3 + 1 = -1$, so

$$t_2 = 1, \quad y_2 = 3 + (-1)\frac{1}{2} = \frac{5}{2}.$$

Next $g\left(1, \frac{5}{2}\right) = 2(1) - \frac{5}{2} + 1 = \frac{1}{2}$, so

$$t_3 = \frac{3}{2}, \quad y_3 = \frac{5}{2} + \left(\frac{1}{2}\right)\frac{1}{2} = \frac{11}{4}.$$

And finally, $g\left(\frac{3}{2}, \frac{11}{4}\right) = 2\left(\frac{3}{2}\right) - \frac{11}{4} + 1 = \frac{5}{4}$, so

$$t_4 = 2,$$

$$y_4 = \frac{11}{4} + \left(\frac{5}{4}\right)\frac{1}{2} = \frac{27}{8}.$$

Thus the approximation to the solution $f(t)$ is given by the polygonal path shown in the answer section of the text. The last point $\left(2, \frac{27}{8}\right)$ is close to the graph of $f(t)$ at $t = 2$, so $f(2) \approx \frac{27}{8}$.

Helpful Hint: The fractions in Exercise 7 were rather simple. In many cases decimals are easier to use, particularly when h is smaller than $\frac{1}{4}$. Of course, a calculator becomes indispensable in such cases.

13. $y' = .5(1 - y)(4 - y)$

To generate the following graphs, set the TI83/TI84 in sequence mode and then use the following in the sequence **Y=** editor.

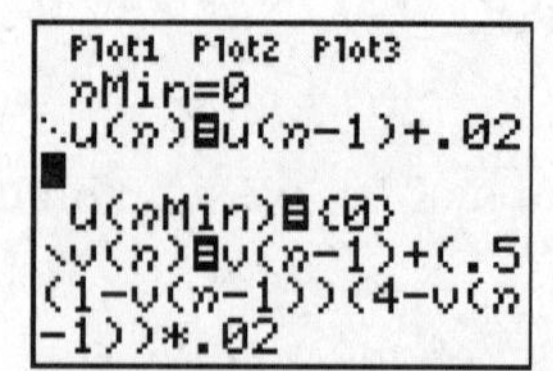

Set **vMin** equal to the value of $y(0)$.

a. $y(0) = -1$; solution is type C: increasing, concave down, and asymptotic to the line $y = 1$.

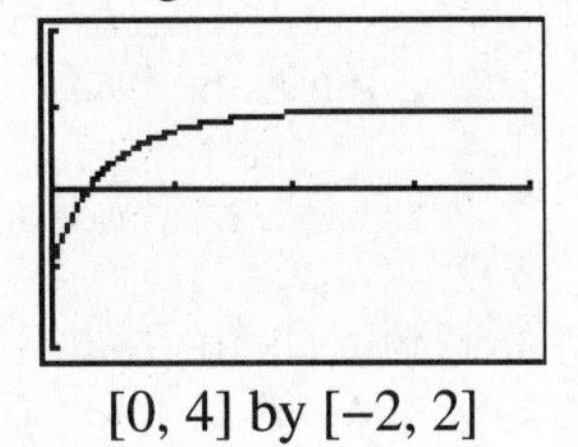

[0, 4] by [−2, 2]

b. $y(0) = 1$; solution is type A: constant solution.

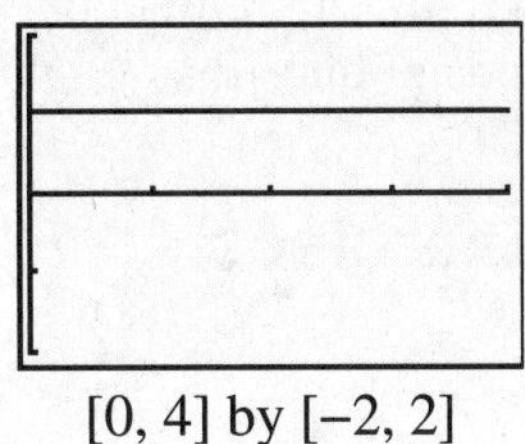

[0, 4] by [−2, 2]

c. $y(0) = 2$; solution is type E: decreasing, concave up, and asymptotic to the line $y = 1$.

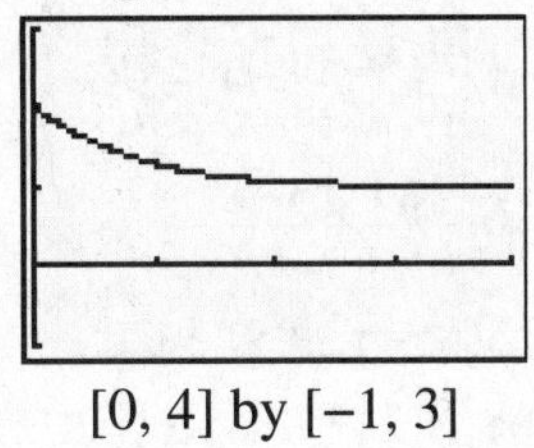

[0, 4] by [−1, 3]

d. $y(0) = 3.9$; solution is type B: decreasing, has an inflection point, and asymptotic to the line $y = 1$.

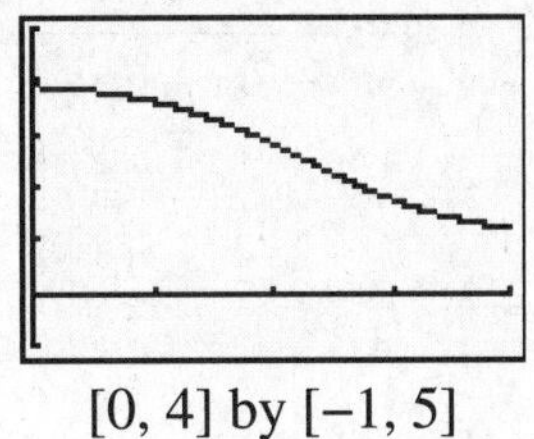

[0, 4] by [−1, 5]

e. $y(0) = 4.1$; solution is type D: concave up and increasing indefinitely.

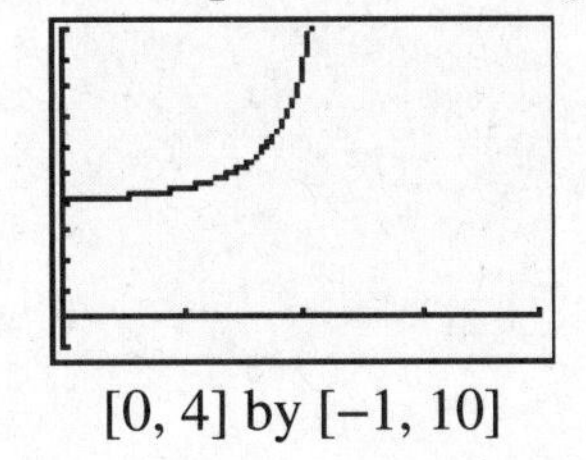

[0, 4] by [−1, 10]

Review of Chapter 10

Except for one or two differential equations in Section 10.1, all of the equations have been *first-order* equations that involve only y', y, and perhaps some functions of t. We have three methods for analyzing these equations: (1) separation of variables to find explicit solutions; (2) Euler's method to find an approximate solution; and (3) qualitative analysis of the equation. The chart below lists the methods that may be applied to first-order differential equations.

Form of the Equation	Methods of Solution
$y' = g(t, y)$	Euler's method
$y' = p(t)q(y)$	Euler's method, separation of variables
$y' = g(y)$	Euler's method, separation of variables, qualitative analysis

On an exam, your instructor might ask you to use more than one method on the same differential equation. This chart will help you anticipate what kinds of questions might be asked.

Chapter 10 Supplementary Exercises

1. Use separation of variables to obtain

$$\int y^2 \frac{dy}{dt}\,dt = \int (4t^3 - 3t^2 + 2)dt,$$

$$\int y^2 dy = \int (4t^3 - 3t^2 + 2)dt,$$

$$\frac{1}{3}y^3 = t^4 - t^3 + 2t + C_1,$$

$$y^3 = 3(t^4 - t^3 + 2t + C_1),$$

$$y^3 = 3(t^4 - t^3 + 2t) + C, \quad C \text{ a constant.}$$

Hence,

$$y = \sqrt[3]{3t^4 - 3t^3 + 6t + C}.$$

7. Use separation of variables to obtain

$$yy' = 6t^2 - t,$$

$$\int y\frac{dy}{dt}\,dt = \int (6t^2 - t)\,dt,$$

$$\int y\,dy = \int (6t^2 - t)\,dt,$$

$$\frac{1}{2}y^2 = 2t^3 - \frac{1}{2}t^2 + C_1,$$

$$y^2 = 2\left(2t^3 - \frac{1}{2}t^2\right) + C, \quad C \text{ a constant.}$$

Hence,

$$y = \pm\sqrt{4t^3 - t^2 + C}.$$

Since $y(0) = 7$, select the positive square root. (If $y(0)$ were negative here, you would use the negative square root.) Since $y(0) = \sqrt{C} = 7$, we have $C = 49$. Therefore,

$$y = \sqrt{4t^3 - t^2 + 49}.$$

13. Suppose $f(t)$ is a solution of $y' = (2 - y)e^{-y}$ and t_0 is a number such that $f(t_0) = 3$. If $y = f(t_0) = 3$, then

$$\begin{aligned} y' &= (2 - y)e^{-y} \\ &= (2 - 3)e^{-3} \\ &= -e^{-3} = \frac{-1}{e^3}. \end{aligned}$$

This is a negative number, so we conclude that y' is negative when $y = 3$. Thus $f(t)$ is decreasing when $t = t_0$, i.e., $f(t)$ is decreasing when $f(t) = 3$.

19. From $y' = \ln y$ we obtain the function $z = \ln y$, whose graph is shown in Figure 1. Thus $g(y)$ has a zero at $y = 1$. Therefore, the constant function $y = 1$ is a solution of the differential equation.

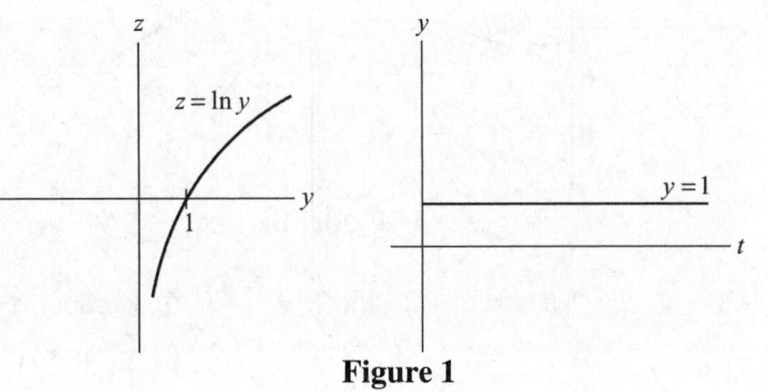

Figure 1

To sketch the solution corresponding to $y(0) = 2$, observe that when $y = 2$, $g(y)$ is positive. Hence the solution is increasing as it leaves the initial point. As y moves to the right, the z-coordinate gets more positive, and hence the solution is concave up, and will continue to increase without bound. See Figure 2.

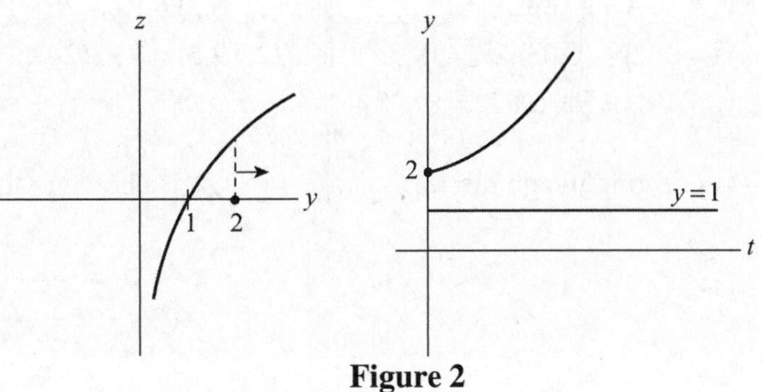

Figure 2

25. **(a)** Let $N = f(t)$ be the city's population at time t. Then $f'(t)$ is the rate of change of the population of the city at time t. Since the birth rate is 3.5%, and the death rate is 2%, and 3000 people leave the city each year,

$$\begin{aligned} f'(t) &= .035f(t) - .02f(t) - 3000 \\ &= .015f(t) - 3000, \end{aligned}$$

or equivalently,

$$N' = .015N - 3000.$$

(b) We seek a constant function N which satisfies the differential equation

$$N' = .015N - 3000$$
$$= .015(N - 200{,}000).$$

Clearly, the constant function $N = 200{,}000$ satisfies the equation since it makes both sides equal to zero. However, it is unlikely that a city could maintain this constant population in practice, for the graph indicates that the constant solution is unstable.

Note that $z = .015N - 3000$ is the graph of a line with positive slope, and with a zero at $N = 200{,}000$. Using the method of Section 10.4, we obtain

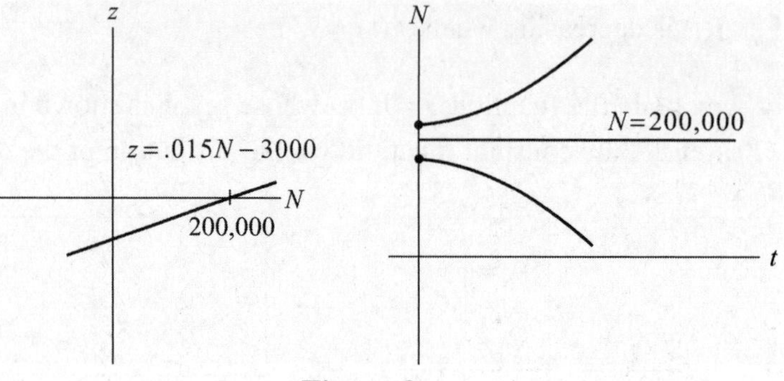

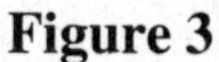

Figure 3

31. Here $g(t, y) = .1y(20 - y)$, $a = 0$, $b = 3$, $y_0 = 2$, and $h = \frac{3-0}{6} = .5$. Compute:

$$\begin{aligned}
&t_0 = 0, && y_0 = 2, && g(0, 2) = .2(20 - 2) = 3.6,\\
&t_1 = .5, && y_1 = 2 + (3.6)(.5) = 3.8, && g(1/2, 3.8) = .38(20 - 3.8) = 6.16,\\
&t_2 = 1, && y_2 = 3.8 + (6.16)(.5) = 6.9, && g(1, 6.9) = .69(20 - 6.9) = 9.04,\\
&t_3 = 1.5, && y_3 = 6.9 + (9.04)(.5) = 11.4, && g(3/2, 11.4) = 1.14(20 - 11.4) = 9.8,\\
&t_4 = 2, && y_4 = 11.4 + (9.8)(.5) = 16.3, && g(2, 16.3) = 1.63(20 - 16.3) = 6.03,\\
&t_5 = 2.5, && y_5 = 16.3 + (6.03)(.5) = 19.3, && g(5/2, 19.3) = 1.93(20 - 19.3) = 1.35,\\
&t_6 = 3, && y_6 = 19.3 + (1.35)(.5) = 19.98.
\end{aligned}$$

The polygonal path connecting the points $(t_0, y_0), \ldots, (t_6, y_6)$ is shown in the answer section of the text.

Chapter 11
Taylor Polynomials and Infinite Series

11.1 Taylor Polynomials

Taylor polynomials are often used in applications to approximate more complex functions, and they have an important connection with the Taylor series, to be discussed in Section 11.5.

The phrase "at $x = 0$" in "Taylor polynomial of $f(x)$ at $x = 0$" indicates that the coefficients of the polynomial are computed by evaluating $f(x)$ and its derivatives at $x = 0$. The phrase "at $x = 0$" does *not* mean that the x in the polynomial must be zero. However, in an application, the values of x are usually taken to be close to zero. In such cases, the remainder theorem is useful for estimating how close the values of a Taylor polynomial are to the values of the function $f(x)$.

1.
$$\begin{aligned} f(x) &= \sin x, & f(0) &= 0, \\ f'(x) &= \cos x, & f'(0) &= 1, \\ f''(x) &= -\sin x, & f''(0) &= 0, \\ f^{(3)}(x) &= -\cos x, & f^{(3)}(0) &= -1. \end{aligned}$$

Therefore, the third Taylor polynomial at $x = 0$ is

$$\begin{aligned} p_3(x) &= 0 + \frac{1}{1!}x + \frac{0}{2!}x^2 + \frac{-1}{3!}x^3 \\ &= x - \frac{1}{6}x^3. \end{aligned}$$

7.
$$\begin{aligned} f(x) &= xe^{3x}, & f(0) &= 0, \\ f'(x) &= \underbrace{3xe^{3x} + e^{3x}}_{\text{from product rule}}, & f'(0) &= 1, \\ f''(x) &= 9xe^{3x} + 3e^{3x} + 3e^{3x}, & f''(0) &= 6, \\ f^{(3)}(x) &= \underbrace{27xe^{3x} + 9e^{3x}}_{\text{from product rule}} + 9e^{3x} + 9e^{3x} & f^{(3)}(0) &= 27. \end{aligned}$$

Therefore,

$$\begin{aligned} p_3(x) &= 0 + \frac{1}{1!}x + \frac{6}{2!}x^2 + \frac{27}{3!}x^3 \\ &= x + 3x^2 + \frac{9}{2}x^3. \end{aligned}$$

13. Since

$$f(x) = f'(x) = f''(x) = \ldots = f^{(n)}(x) = e^x,$$
$$f(0) = f'(0) = f''(0) = \ldots = f^{(n)}(0) = e^0 = 1.$$

Therefore, the nth Taylor polynomial for $f(x) = e^x$ at $x = 0$ is:

$$\begin{aligned} p_n(x) &= 1 + \frac{1}{1!}x + \frac{1}{2!}x^2 + \frac{1}{3!}x^3 + \ldots + \frac{1}{n!}x^n \\ &= 1 + x + \frac{1}{2}x^2 + \frac{1}{3!}x^3 + \ldots + \frac{1}{n!}x^n. \end{aligned}$$

Helpful Hint: If your class plans to cover Section 11.5, then you should memorize the formulas for the Taylor polynomials of e^x and $\frac{1}{1-x}$, as described in Examples 2 and 3.

19.

$$\begin{aligned} f(x) &= \cos x, & f(\pi) &= -1, \\ f'(x) &= -\sin x, & f'(\pi) &= 0, \\ f''(x) &= -\cos x, & f''(\pi) &= 1, \\ f^{(3)}(x) &= \sin x, & f^{(3)}(\pi) &= 0, \\ f^{(4)}(x) &= \cos x, & f^{(4)}(\pi) &= -1. \end{aligned}$$

Thus the third Taylor polynomial at $x = \pi$ is

$$\begin{aligned} p_3(x) &= -1 + \frac{0}{1!}(x-\pi) + \frac{1}{2!}(x-\pi)^2 + \frac{0}{3!}(x-\pi)^3 \\ &= -1 + \frac{1}{2}(x-\pi)^2, \end{aligned}$$

and the fourth Taylor polynomial is

$$\begin{aligned} p_4(x) &= -1 + \frac{0}{1!}(x-\pi) + \frac{1}{2!}(x-\pi)^2 + \frac{0}{3!}(x-\pi)^3 + \frac{-1}{4!}(x-\pi)^4 \\ &= -1 + \frac{1}{2}(x-\pi)^2 - \frac{1}{24}(x-\pi)^4. \end{aligned}$$

Notice that $p_3(x)$ is a polynomial of degree 2, not degree 3, because $f^{(3)}(\pi) = 0$. See Practice Problem 1.

25. Be sure to try the Practice Problems before you attempt this problem. You should realize from Practice Problem 2 that if $f(x)$ is a polynomial of degree 3, then the third Taylor polynomial of $f(x)$ at $x = 0$ is $f(x)$ itself. This is true because $f(x)$ is a polynomial that agrees with $f(x)$ and all of its derivatives at $x = 3$.

Now write the abstract *formula* for the third Taylor polynomial of $f(x)$, set that equal to $f(x)$, and equate corresponding coefficients. From this you should be able to discover what the values of $f''(0)$ and $f'''(0)$ must be. Try to do this before reading the rest of this solution.

The third Taylor polynomial of $f(x)$ is

$$p_3(x) = f(0) + \frac{f'(0)}{1}x + \frac{f''(0)}{2!}x^2 + \frac{f'''(0)}{3!}x^3$$

and

$$f(x) = 3 + 4x + \left(-\frac{5}{2!}\right)x^2 + \left(\frac{7}{3!}\right)x^3.$$

Since $p_3(x)$ equals $f(x)$ for all x, the polynomials must have the same coefficients. That is, $f(0) = 3$, $f'(0) = 4$, $\frac{f''(0)}{2!} = -\frac{5}{2!}$ and $\frac{f'''(0)}{3!} = \frac{7}{3!}$. Accordingly, $f''(0) = -5$ and $f'''(0) = 7$.

31. From Example 3, you should know that the fourth Taylor polynomial of $Y_1 = 1/(1-X)$ is the polynomial given by

$$\mathbf{Y_2 = 1 + X + X\wedge 2 + X\wedge 3 + X\wedge 4.}$$

When you graph Y_1 and Y_2 in the window [−1, 1] by [−1, 5], you will probably find that their graphs appear identical for x between about −.66 and .53, or perhaps .55. So b is about .55.

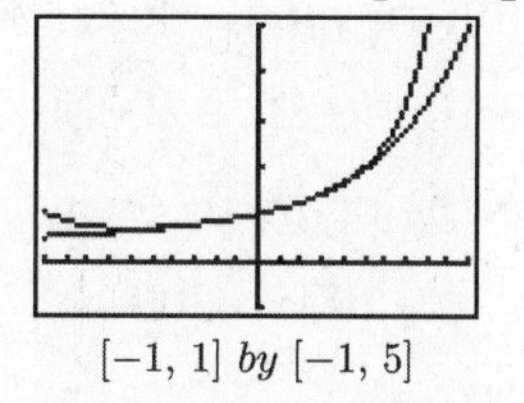

[−1, 1] *by* [−1, 5]

You can use **TRACE** to display the y-values of points on the two curves with the same x-coordinates. At $x = .5106383$, for instance, $Y_1 - Y_2 \approx 2.04348 - 1.97253 = .07095$. Another method for comparing values of Y_1 and Y_2 is shown in Figure 5 on page 519 in the text.

11.2 The Newton-Raphson Algorithm

A calculator is indispensable for this section. In order to memorize the Newton-Raphson algorithm, it is important to work several problems by hand using a calculator. If you have access to an IBM-compatible computer, you should experiment with Visual Calculus, which will show you geometrically how the algorithm works.

1. $\sqrt{5}$ is a zero of the function $x^2 - 5$. Note that $\sqrt{5}$ lies between 2 and 3 because $2^2 < 5 < 3^2$. For an initial approximation, take $x_0 = 2$. (The value $x_0 = 3$ would work just as well.) Since $f'(x) = 2x$, we have

$$x_1 = x_0 - \frac{x_0^2 - 5}{2x_0} = 2 - \frac{(2)^2 - 5}{2(2)} = 2 - \frac{-1}{4} = \frac{9}{4} = 2.25$$

$$x_2 = 2.25 - \frac{(2.25)^2 - 5}{2(2.25)} = 2.25 - \frac{.0625}{4.5} = 2.2361$$

$$x_3 = 2.2361 - \frac{(2.2361)^2 - 5}{2(2.2361)} = 2.2361 - \frac{.00014}{4.4722} = 2.23607.$$

7. Here $f(x) = \sin x + x^2 - 1$ and $x_0 = 0$. First, compute $f'(x) = \cos x + 2x$, and then use the Newton-Raphson algorithm:

$$x_1 = x_0 - \frac{\sin x_0 + x_0^2 - 1}{\cos x_0 + 2x_0} = 0 - \frac{\sin 0 + 0^2 - 1}{\cos 0 + 2(0)} = 0 - \frac{-1}{1} = 1$$

$$x_2 = 1 - \frac{\sin 1 + 1^2 - 1}{\cos(1) + 2(1)} = 1 - \frac{.84147}{2.54030} = .66875$$

$$x_3 = .66875 - \frac{.62001 + .44723 - 1}{.78460 + 1.3375} = .66875 - .03168 = .63707.$$

13. Let i be the monthly rate of interest. The present value of an amount A to be received in k months is $A(1+i)^{-k}$. Therefore, you must solve the following equation for i:

$$\begin{bmatrix}\text{amount of initial}\\ \text{investment}\end{bmatrix} = \begin{bmatrix}\text{sum of present}\\ \text{values of returns}\end{bmatrix}$$

$$500 = 100(1+i)^{-1} + 200(1+i)^{-2} + 300(1+i)^{-3}.$$

Multiply both sides by $(1+i)^3$, take all terms to the left, and obtain

$$500(1+i)^3 - 100(1+i)^2 - 200(1+i) - 300 = 0.$$

Let $x = 1 + i$, and solve the resulting equation by the Newton-Raphson algorithm with $x_0 = 1.1$.

$$f(x) = 500x^3 - 100x^2 - 200x - 300 = 0,$$

$$f'(x) = 1500x^2 - 200x - 200.$$

$$x_1 = 1.1 - \frac{500(1.1)^3 - 100(1.1)^2 - 200(1.1) - 300}{1500(1.1)^2 - 200(1.1) - 200}$$

$$\approx 1.1 - .018 = 1.082$$

$$x_2 = 1.082 - \frac{500(1.082)^3 - 100(1.082)^2 - 200(1.082) - 300}{1500(1.082)^2 - 200(1.082) - 200}$$

$$\approx 1.082 - (-.00008) \approx 1.082.$$

Therefore, the solution is $x = 1.082$. Hence, $i = .082$ and the investment had an internal rate of return of 8.2% per month.

19. The slope of the tangent line is 4, so $f'(3) = 4$. Also, $f(3) - 4(3) + 5 = 17$. If the initial guess is 3, then the Newton-Raphson algorithm produces:

$$x_1 = 3 - \frac{f(3)}{f'(3)} = 3 - \frac{17}{4} = \frac{-5}{4}.$$

25. When $f(x) = x^{1/3}$, and $f'(x) = \frac{1}{3}x^{-2/3}$, the Newton-Raphson formula is

$$x_1 = x_0 - \frac{f(x_0)}{f'(x_0)} = x_0 - \frac{x_0^{1/3}}{\left(\frac{1}{3}\right)x_0^{-2/3}}$$

$$= x_0 - 3x_0 = -2x_0.$$

Similarly, $x_2 = -2x_1, x_3 = -2x_2, \ldots$, and $x_{n+1} = -2x_n$. If $x_0 = 1$, then the Newton-Raphson algorithm produces the sequence 1, –2, 4, –8, and so on. This sequence does *not* converge to the actual root, $r = 0$. Can you draw a picture, based on Figure 10(a) in the text, to show what is happening?

11.3 Infinite Series

Most students who take this calculus course will see an application of infinite series in some other course. More often than not, the application will involve a geometric series.

Make sure that you completely master the material in this section. It will help you to understand the concepts in the sections to follow.

1. $1+\frac{1}{6}+\frac{1}{6^2}+\frac{1}{6^3}+\frac{1}{6^4}+\cdots=\left(\frac{1}{6}\right)^0+\left(\frac{1}{6}\right)^1+\left(\frac{1}{6}\right)^2+\left(\frac{1}{6}\right)^3+\left(\frac{1}{6}\right)^4+\cdots.$

Thus $a = 1$, $r=\frac{1}{6}$, and the series converges to

$$\frac{a}{1-r}=\frac{1}{1-\frac{1}{6}}=\frac{1}{\frac{5}{6}}=\frac{6}{5}.$$

7. Find r by dividing any term by the preceding term. So,

$$r=\frac{\frac{1}{5^4}}{\frac{1}{5}}=\frac{5}{5^4}=\frac{1}{5^3}=\frac{1}{125}.$$

The first term of the series is $a=\frac{1}{5}$, so the sum of the series is

$$\frac{\frac{1}{5}}{1-\frac{1}{125}}=\frac{1}{5}\cdot\frac{125}{124}=\frac{25}{124}.$$

13. Find r by dividing any term by the preceding term. So,

$$r=\frac{4}{5}.$$

The first term of the series is $a = 5$, so the sum of the series is

$$\frac{5}{1-\frac{4}{5}}=5\cdot 5=25.$$

19. The technique here is to find a rational number that represents $.011\overline{011}$, and then to add 4 to this number. First, write

$$\begin{aligned}.011\overline{011}&=.011+.000011+.000000011+\ldots\\&=\frac{11}{1000}+\frac{11}{1000^2}+\frac{11}{1000^3}+\cdots,\end{aligned}$$

which is a geometric series with $a=\frac{11}{1000}$ and $r=\frac{1}{1000}$. The sum of the series is:

$$\frac{\frac{11}{1000}}{1-\frac{1}{1000}}=\frac{11}{1000}\cdot\frac{1000}{999}=\frac{11}{999}.$$

Then, add 4 to the sum.

$$4.011\overline{011}=4+\frac{11}{999}=\frac{3996}{999}+\frac{11}{999}=\frac{4007}{999}.$$

25. **(a)** Construct a sum of the present value of all future payments. That is, express capital value as

$$100+100(1.01)^{-1}+100(1.01)^{-2}+\cdots=\sum_{k=0}^{\infty}100(1.01)^{-k}.$$

(b) This is a geometric series with $a = 100$ and $r = (1.01)^{-1} = \frac{1}{1.01}$, (so $|r|<1$). So the sum of the series is

$$\frac{100}{1-\frac{1}{1.01}}=100\frac{1.01}{.01}=10{,}100.$$

Thus the capital value of the perpetuity is $10,100.

31. At the end of the first "day" (a 24-hour period), 25% of the original dose of M mg has been eliminated, and only $.75M$ remains. Immediately after the next dose, the body contains $M + .75M$ mg of the drug. At the end of the second day, only 75% of that $M + .75M$ remains, which may be written as $.75(M + .75M)$, or $.75M + (.75)^2M$. When a dose of M mg is given again, the new amount in the body is

$$M+.75M+(.75)^2M \qquad \text{(after the third dose).}$$

After n full days, just after the next dose is given, the amount in the body is

$$M+.75M+(.75)^2M+\cdots+(.75)^nM \text{ milligrams.}$$

This is a partial sum of a geometric series whose initial term is M and whose ratio is $r = .75$. When n is large, the value of this partial sum will be very close to the sum of the infinite series

$$\sum_{k=0}^{\infty}M(.75)^k=\frac{M}{1-.75}=\frac{M}{.25}=4M.$$

To make this amount equal to 20 mg, M must be 5 mg. This is the "maintenance dose."

37. $\sum_{j=1}^{\infty}5^{-2j}=\frac{1}{5^2}+\frac{1}{5^4}+\frac{1}{5^6}+\frac{1}{5^8}+\cdots.$

This is a geometric series with $a = 1/5^2$ and $r = 1/5^2$. So the sum of the series is

$$\frac{\frac{1}{25}}{1-\frac{1}{25}}=\frac{1}{25}\cdot\frac{25}{24}=\frac{1}{24}.$$

43. The first entry in the **sum(seq(. . .))** command displays the general term, 6(.5)^X, from which you can see that $a = 6$ and the ration is $r = .5$. The remaining entries in the **sum(seq(. . . X, 0, 10, 1))** command show that the "counter" X, begins at 0 and ends at 10, with increment (step size) 1. (On the TI-83 and TI-86, the value for the increment may be omitted.) Thus the calculator is computing the partial sum

$$6+6(.5)+6(.5)^2+\cdots+6(.5)^{10}.$$

The exact sum of the corresponding geometric series is $\dfrac{a}{1-r}=\dfrac{6}{1-.5}=\dfrac{6}{.5}=12.$

49. Unfortunately, the T1-82 **sum(seq(. . .))** command is limited to 99 terms, so let's begin there. You should find that

$$\textbf{sum(seq(X^{\wedge}–2, X, 1, 99, 1))} \approx \textbf{1.634884}$$

(to six decimal places). Also, $\pi^2/6 = 1.644934$, which is a difference of about .01. With 99 replaced by 999, the computation is much slower, but the result 1.6439934 differs from $\pi^2/6$ only by about .001.

The result on a TI-84 is shown below.

```
sum(seq(1/(X²),X
,1,999,1)
        1.643933567
π²/6
        1.644934067
■
```

11.4 Series with Positive Terms

The integral test provides a nice link between infinite series and improper integrals. Your understanding of both topics should be strengthened by studying them together. It may be helpful to review Section 9.6 before you begin the exercises in Section 11.4.

Problems involving the comparison test can be very difficult. Knowing what series to use for comparison is a skill that requires a lot of experience—more than we expect you to obtain in this course. We have made Exercises 21–26 more reasonable by suggesting one or two series for comparison.

1. To apply the integral test, compute

$$\int_1^\infty \frac{3}{\sqrt{x}}dx = \int_1^\infty 3x^{-1/2}dx = \lim_{b\to\infty}\int_1^b 3x^{-1/2}dx$$

$$= \lim_{b\to\infty}\left(6x^{1/2}\Big|_1^b\right) = \lim_{b\to\infty}(6b^{1/2} - 6) = \infty.$$

The series

$$\sum_{k=1}^{\infty}\frac{3}{\sqrt{k}}$$

is divergent, because the corresponding improper integral is divergent.

7. Study the improper integral

$$\int_2^\infty \frac{x}{(x^2+1)^{3/2}}dx.$$

Begin by finding an antiderivative of $x(x^2+1)^{-3/2}$ using the method of substitution. Set $u = x^2 + 1$ and $du = 2x\,dx$. Then

$$\int \frac{x}{(x^2+1)^{3/2}}dx = \frac{1}{2}\int (x^2+1)^{-3/2}\cdot 2x\,dx = \frac{1}{2}\int u^{-3/2}du$$

$$= \frac{1}{2}(-2u^{-1/2}) = -(x^2+1)^{-1/2} + C.$$

Next,

$$\int_2^b \frac{x}{(x^2+1)^{3/2}}\,dx = -(x^2+1)^{-1/2}\Big|_2^b = -\frac{1}{\sqrt{b^2+1}} - \left(-\frac{1}{\sqrt{2^2+1}}\right).$$

Finally,

$$\int_2^\infty \frac{x}{(x^2+1)^{3/2}}\,dx = \lim_{b\to\infty}\left(\frac{1}{\sqrt{5}} - \frac{1}{\sqrt{b^2+1}}\right) = \frac{1}{\sqrt{5}}.$$

Since the improper integral is convergent, the corresponding infinite series is also convergent.

Warning: The sum of the series in Exercise 7 is *not* $\frac{1}{\sqrt{5}}$. The integral test only shows that the series converges.

13. Evaluate

$$\int_1^\infty xe^{-x^2}\,dx.$$

To find an antiderivative for xe^{-x^2}, either guess e^{-x^2} and adjust the guess to obtain $-\frac{1}{2}e^{-x^2}$ or use substitution with $u=-x^2$, $du=-2x\,dx$, and compute

$$\int xe^{-x^2}\,dx = -\frac{1}{2}\int -2xe^{-x^2}\,dx = -\frac{1}{2}\int e^u\,du = -\frac{1}{2}e^u + C = -\frac{1}{2}e^{-x^2} + C.$$

Next, compute

$$\begin{aligned}\int_1^\infty xe^{-x^2}\,dx = \lim_{b\to\infty}\int_1^b xe^{-x^2}\,dx \lim_{b\to\infty}\left(-\frac{1}{2}e^{-x^2}\Big|_1^b\right)\\ = \lim_{b\to\infty}\left(-\frac{1}{2}e^{-b^2} - \left(-\frac{1}{2}e^{-1}\right)\right)\\ = \lim_{b\to\infty}\left(-\frac{1}{2}e^{-b^2} + \frac{1}{2}e^{-1}\right) = \frac{1}{2}e^{-1}.\end{aligned}$$

Since the improper integral is convergent, so is the infinite series $\sum_{k=1}^{\infty} ke^{-k^2}$.

19. Evaluate the integral

$$\int_1^\infty \frac{x}{e^x}\,dx = \int_1^\infty xe^{-x}\,dx$$

using integration by parts. Set

$$\begin{aligned} f(x) &= x, \quad g(x) = e^{-x},\\ f'(x) &= 1, \quad G(x) = -e^{-x}.\end{aligned}$$

Then

$$\int \frac{x}{e^x}dx = -xe^{-x} - \int -e^{-x}dx = -xe^{-x} + \int e^{-x}dx$$
$$= -xe^{-x} - e^{-x} + C$$

and

$$\int_1^b \frac{x}{e^x}dx = (-xe^{-x} - e^{-x})\Big|_1^b$$
$$= (-be^{-b} - e^{-b}) - (-e^{-1} - e^{-1})$$
$$= 2e^{-1} - be^{-b} - e^{-b}.$$

For the final step, use the fact (given in the text) that $\lim\limits_{b\to\infty} be^{-b} = 0.$

$$\int_1^\infty \frac{x}{e^x}dx = \lim_{b\to\infty}(2e^{-1} - be^{-b} - e^{-b}) = 2e^{-1} - 0 - 0.$$

Since the improper integral is convergent, so is the infinite series $\sum\limits_{k=1}^{\infty} \frac{k}{e^k}$.

25. First compare

$$\sum_{k=1}^{\infty} \frac{1}{5^k}\cos^2\left(\frac{k\pi}{4}\right) = \frac{1}{5}\cos^2\left(\frac{\pi}{4}\right) + \frac{1}{25}\cos^2\left(\frac{\pi}{2}\right) + \frac{1}{125}\cos^2\left(\frac{3\pi}{4}\right) + \cdots$$

with

$$\sum_{k=1}^{\infty} \cos^2\left(\frac{k\pi}{4}\right) = \cos^2\left(\frac{\pi}{4}\right) + \cos^2\left(\frac{\pi}{2}\right) + \cos^2\left(\frac{3\pi}{4}\right) + \cdots.$$

Note that each term of the first series is less than the corresponding term of the second series. Since $\left|\cos\frac{k\pi}{4}\right| = \frac{1}{\sqrt{2}}$ when k is odd, and $\left|\cos\frac{k\pi}{4}\right| = 0$ or 1 (alternately) when k is even, the second series has the form

$$\frac{1}{2} + 0 + \frac{1}{2} + 1 + \frac{1}{2} + 0 + \frac{1}{2} + 1 + \cdots.$$

Since these terms keep repeating, the series diverges. We have learned nothing about the convergence of the first series. Next, compare the first series with

$$\sum_{k=1}^{\infty} \frac{1}{5^k} = \frac{1}{5} + \frac{1}{5^2} + \frac{1}{5^3} + \frac{1}{5^4} + \cdots,$$

which is a convergent geometric series because $|r| = \frac{1}{5} < 1$. Now, since

$$\frac{1}{5^k}\cos^2\left(\frac{k\pi}{4}\right) \le \frac{1}{5^k} \quad \text{for all values of } k,$$

each term of $\sum 5^{-k}$ is greater than the corresponding term of the first series shown above. Since $\sum 5^{-k}$ converges, so does the first series.

31. $$\sum_{k=0}^{\infty}\frac{8^k+9^k}{10^k}=\frac{8^0+9^0}{10^0}+\frac{8^1+9^1}{10^1}+\frac{8^2+9^2}{10^2}+\cdots$$

$$=\left(\frac{8^0}{10^0}+\frac{9^0}{10^0}\right)+\left(\frac{8^1}{10^1}+\frac{9^1}{10^1}\right)+\left(\frac{8^2}{10^2}+\frac{9^2}{10^2}\right)+\cdots$$

$$=\sum_{k=0}^{\infty}\left(\frac{8^k}{10^k}+\frac{9^k}{10^k}\right)$$

$$=\sum_{k=0}^{\infty}\frac{8^k}{10^k}+\sum_{k=0}^{\infty}\frac{9^k}{10^k} \quad \text{by Exercise 29,}$$

$$=\sum_{k=0}^{\infty}\left(\frac{8}{10}\right)^k+\sum_{k=0}^{\infty}\left(\frac{9}{10}\right)^k.$$

Each term in this sum is a convergent geometric series. The first has $a=1$ and $r=.8$, and its sum is $\frac{a}{1-r}=\frac{1}{1-.8}=5$. The second has $a=1$ and $r=.9$, and its sum is $\frac{1}{1-.9}=10$. By Exercise 29, the original series is convergent, and its sum is $5+10=15$.

11.5 Taylor Series

One of the main ideas in this section is that a Taylor series is a function. Be sure to read the text on Pages 546 and 548–549. You need to know the general formula for a Taylor series at $x=0$ and the specific formulas for e^x and $1/(1-x)$. Check with your instructor to see if you also need to memorize the Taylor series at $x=0$ for $\cos x$ and $\sin x$. Study Examples 3, 5, and 6 carefully to get ideas for the solutions of Exercises 5–27.

1. Compute:

$$f(x)=\frac{1}{2x+3}=(2x+3)^{-1}, \qquad f(0)=\frac{1}{3},$$

$$f'(x)=-2(2x-3)^{-2}, \qquad f'(0)=\frac{-2}{9}=-\frac{2}{9},$$

$$f''(x)=8(2x+3)^{-3}, \qquad f''(0)=\frac{8}{27},$$

$$f^{(3)}(x)=-48(2x+3)^{-4}, \qquad f^{(3)}(0)=\frac{-48}{81}=-\frac{16}{27},$$

$$f^{(4)}(x)=384(2x+3)^{-5}, \qquad f^{(4)}(0)=\frac{384}{243}=\frac{128}{81}.$$

Then write

$$f(x)=\frac{1}{3}+\frac{-\frac{2}{9}}{1!}x+\frac{\frac{8}{27}}{2!}x^2+\frac{-\frac{16}{27}}{3!}x^3+\frac{\frac{128}{81}}{4!}x^4+\cdots$$
$$=\frac{1}{3}-\frac{2}{9}x+\frac{4}{27}x^2-\frac{8}{81}x^3+\frac{16}{243}x^4+\cdots$$
$$=\frac{1}{3}-\frac{2}{9}x+\frac{2^2}{3^3}x^2-\frac{2^3}{3^4}x^3+\frac{2^2}{3^5}x^4+\cdots$$
$$=\frac{1}{3}-\frac{2}{9}x+\frac{4}{27}x^2-\frac{2^3}{3^4}x^3+\frac{2^4}{3^5}x^4-\cdots.$$

7. In the Taylor series at $x=0$ for $\frac{1}{1-x}$, replace x by $-x^2$ to obtain

$$\frac{1}{1-(-x^2)}=\frac{1}{1+x^2}$$
$$=1-x^2+x^4-x^6+\cdots.$$

13. In the Taylor series at $x=0$ for e^x, replace x by $-x$ to obtain

$$e^{-x}=1+(-x)+\frac{1}{2!}(-x)^2+\frac{1}{3!}(-x)^3+\frac{1}{4!}(-x)^4-\cdots.$$

Hence,

$$1-e^{-x}=1-\left(1-x+\frac{1}{2!}x^2-\frac{1}{3!}x^3+\frac{1}{4!}x^4-\cdots\right)$$
$$=x-\frac{1}{2!}x^2+\frac{1}{3!}x^3-\frac{1}{4!}x^4+\cdots.$$

19. In the Taylor series at $x=0$ for $\cos x$, replace x by $3x$ to obtain

$$\cos 3x=1-\frac{9}{2!}x^2+\frac{81}{4!}x^4-\frac{729}{6!}x^6+\cdots.$$

Now differentiate both sides and divide by -3 to obtain

$$-3\sin 3x=-\frac{18}{2!}x+\frac{4(81)}{4!}x^3-\frac{6(729)}{6!}x^5+\cdots.$$
$$\sin 3x=\frac{6}{2!}x-\frac{4(27)}{4!}x^3+\frac{6(243)}{6!}x^5-\cdots$$
$$=3x-\frac{27}{3!}x^3+\frac{243}{5!}x^5-\cdots$$
$$=3x-\frac{3^3}{3!}x^3+\frac{3^5}{5!}x^5-\cdots.$$

25. In the Taylor series expansion at $x=0$ for $\dfrac{1}{\sqrt{1+x}}$, replace x by $-x$ to obtain

$$\frac{1}{\sqrt{1-x}}=1+\frac{1}{2}x+\frac{1\cdot 3}{2\cdot 4}x^2+\frac{1\cdot 3\cdot 5}{2\cdot 4\cdot 6}x^3.$$

31. If you are reading this before you work the exercise, stop and study Practice Problem 4. Then try the exercise again before reading on with the solution. We observe that the term in the series containing x^5 is

$$\frac{f^{(5)}(0)}{5!}x^5=\tfrac{2}{5}x^5.$$

So

$$\frac{f^{(5)}(0)}{5!}=\frac{2}{5}, \quad \text{and} \quad f^{(5)}(0)=(5!)\frac{2}{5}=(4!)2=24\cdot 2=48.$$

37. In the Taylor expansion at $x=0$ for $\frac{1}{1-x}$ replace x by $-x^3$ to obtain

$$\frac{1}{1+x^3}=1-x^3+x^6-x^9+\cdots.$$

Then integrate both sides, term by term:

$$\int\frac{1}{1+x^3}dx=\int(1-x^3+x^6-x^9+\cdots)\,dx$$

$$=\left[x-\frac{1}{4}x^4+\frac{1}{7}x^7-\frac{1}{10}x^{10}+\cdots\right]+C.$$

43. Since $e^x=1+x+\frac{1}{2}x^2+\frac{1}{6}x^3+\cdots$, for all x, we have $e^x>\frac{1}{6}x^3$ when $x>0$. Then idea now is to find a function which approaches 0 as $x\to\infty$, with the property that for every value of x greater than 0, this function is greater than the function x^2e^{-x}. From the inequality above, we have

$$\frac{1}{e^x}<\frac{6}{x^3}$$

and so

$$x^2e^{-x}<x^2\cdot\frac{6}{x^3}=\frac{6}{x}.$$

As $x\to\infty$, $\dfrac{6}{x}$ approaches 0. Since $\dfrac{6}{x}>x^2e^{-x}$ for all $x>0$, it follows that x^2e^{-x} must approach 0 as $x\to\infty$.

Chapter 11: Supplementary Exercises

1. $f(x) = x(x+1)^{3/2}$, $\quad f(0) = 0$,

$f'(x) = \frac{3}{2}x(x+1)^{1/2} + (x+1)^{3/2}$, $\quad f'(0) = 1$,

$$f''(x) = \frac{3}{4}x(x+1)^{-1/2} + \frac{3}{2}(x+1)^{1/2} + \frac{3}{2}(x+1)^{1/2}$$

$$= \frac{3}{4}x(x+1)^{-1/2} + 3(x+1)^{1/2}. \qquad f''(0) = 3.$$

Hence,

$$p_2(x) = 0 + \frac{1}{1!}x + \frac{3}{2!}x^2 = x + \frac{3}{2}x^2.$$

7. $f(t) = -\ln(\cos 2t)$, $\quad f(0) = -\ln(1) = 0$,

$f'(t) = \frac{2\sin 2t}{\cos 2t}$, $\quad f'(0) = 0$,

$$f''(t) = \frac{(\cos 2t)(4\cos 2t) + (2\sin 2t)(2\sin 2t)}{\cos^2 2t}$$

$$= \frac{4\cos^2 2t + 4\sin^2 2t}{\cos^2 2t} = 4 + \frac{4\sin^2 2t}{\cos^2 2t}, \qquad f''(0) = 4.$$

Hence,

$$p_2(t) = 0 + \frac{0}{1!}t + \frac{4}{2!}t^2 = 2t^2.$$

The area under the graph of $y = f(t)$ is approximately equal to the area under the graph of $y = p_2(t)$. Thus,

$$[\text{Area}] \approx \int_0^{1/2} 2t^2 dt = \frac{2}{3}t^3\Big|_0^{1/2} = \frac{2}{3}\cdot\frac{1}{8} = \frac{1}{12}.$$

13. The series is a geometric series with $a = 1$ and $r = -\frac{3}{4}$. Since $|-\frac{3}{4}| < 1$, the series is convergent and converges to $\frac{1}{1+\frac{3}{4}} = \frac{4}{7}$.

19. Recall that

$$e^x = 1 + \frac{1}{1!}x + \frac{1}{2!}x^2 + \frac{1}{3!}x^3 + \frac{1}{4!}x^4 + \cdots.$$

Therefore

$$1 + 2 + \frac{2^2}{2!} + \frac{2^3}{3!} + \frac{2^4}{4!} + \cdots$$

$$= 1 + \frac{1}{1!}(2) + \frac{1}{2!}(2)^2 + \frac{1}{3!}(2)^3 + \frac{1}{4!}(2)^4 + \cdots = e^2.$$

25. Use the integral test, and consider $\int_1^\infty \frac{\ln x}{x}dx$. First, let $u = \ln x$ and $du = \frac{1}{x}dx$. Then

$$\int \frac{\ln x}{x}dx = \int u\,du = \frac{1}{2}u^2 + C = \frac{1}{2}(\ln x)^2 + C.$$

Next,

$$\int_1^b \frac{\ln x}{x}dx = \frac{1}{2}(\ln x)^2\Big|_1^b = \frac{1}{2}(\ln b)^2 = 0.$$

This quantity grows arbitrarily large as $b \to \infty$, so the improper integral above is divergent. By the integral test, the infinite series

$$\sum_{k=1}^{\infty} \frac{\ln k}{k}$$

is also divergent.

31.
$$\begin{aligned}\frac{1}{(1-3x)^2} &= \frac{1}{3}\frac{d}{dx}\left[\frac{1}{1-3x}\right]\\ &= \frac{1}{3}\frac{d}{dx}[1+3x+3^2x^2+3^3x^3+\cdots]\\ &= 1+6x+27x^2+4\cdot 3^3x^3+\cdots.\end{aligned}$$

37. **(a)** The fifth Taylor polynomial consists of all terms in the sixth Taylor polynomial with degree ≤ 5. Therefore, $p_5(x) = x^2$.

(b) Since the coefficient of x^3 in $p_6(x)$ is 0, $\frac{f^{(3)}(0)}{3!} = 0$. Therefore, $f^{(3)}(0) = 0$.

(c) $\int_0^1 \sin x^2 dx \approx \int_0^1\left(x^2 - \frac{1}{6}x^6\right)dx = \left(\frac{x^3}{3} - \frac{1}{42}x^7\right)\Big|_0^1 = \frac{1}{3} - \frac{1}{42}$

$\approx .3095$ (exact value: .3103).

43. $\sum_{k=1}^{\infty} 10{,}000e^{-.08k} = \sum_{k=1}^{\infty} 10{,}000(e^{-.08})^k = \frac{10{,}000(e^{-.08})}{1-e^{-.08}} \approx \$120{,}066.66.$

Chapter 12
Probability and Calculus

12.1 Discrete Random Variables

Although this section is provided mainly as a background for the sections that follow, there are interesting and important problems here. The main concepts are: relative frequency table, probability, expected value, variance, and (discrete) random variable. The probability density histograms on page 560 will be used in Section 12.2 to motivate the definition of a probability density function.

1. There are two possible outcomes, 0 and 1, with probabilities $\frac{1}{5}$ and $\frac{4}{5}$, respectively. Thus

$$E(X) = 0\cdot\frac{1}{5} + 1\cdot\frac{4}{5} = \frac{4}{5}.$$

$$\text{Var}(X) = \left(0 - \frac{4}{5}\right)^2 \cdot \frac{1}{5} + \left(1 - \frac{4}{5}\right)^2 \cdot \frac{4}{5} = .128 + .032 = .16.$$

Standard deviation of $X = \sqrt{.16} = .4.$

7. **(a)** Recall that the area of a circle of radius r is given by $[\text{Area}] = \pi r^2$. Thus the area of a circle of radius 1 is π, and the area of a circle of radius $\frac{1}{2}$ is $\pi\left(\frac{1}{2}\right)^2 = \frac{1}{4}\pi$. To find the percentage of points lying in the circle of radius $\frac{1}{2}$, set up the ratio

$$\frac{[\text{Area of circle of radius } \frac{1}{2}]}{[\text{Area of circle of radius } 1]} = \frac{\frac{1}{4}\pi}{\pi} = \frac{1}{4} = .25 = 25\%.$$

Thus 25% of the points lie within $\frac{1}{2}$ unit of the center.

(b) The area of a circle of radius c is: $[\text{Area}] = \pi c^2$. To find the percentage of points lying within this circle set up the ratio

$$\frac{[\text{Area of circle of radius } c]}{[\text{Area of circle of radius } 1]} = \frac{\pi c^2}{\pi} = c^2 = 100c^2\%.$$

12.2 Continuous Random Variables

We recommend that you study the theoretical parts of Sections 6.1 and 6.3 (particularly the Fundamental Theorem of Calculus) as background for the discussion here of the cumulative distribution function $F(x)$ and the probability density function $f(x)$. Given either function, you should be able to find the other with no difficulty. Observe that $F'(x) = f(x)$ and that $F(x)$ is the unique antiderivative of $f(x)$ for which $F(A) = 0$, when X is a random variable on $A \le x \le B$. You should also be able to use either $F(x)$ or $f(x)$ to compute various probabilities of the form $\Pr(a \le X \le b)$.

Pay attention to Example 6 because it provides a nice connection with earlier material on improper integrals and it introduces ideas that are needed in Section 12.4. For these reasons, questions similar to Example 6 often appear on exams.

1. Clearly, $f(x) \ge 0$, since $\frac{1}{18}x \ge 0$ when $0 \le x \le 6$. Thus Property I is satisfied. For Property II, check that

$$\int_0^6 f(x)dx = \int_0^6 \frac{1}{18}xdx = \frac{1}{36}x^2\Big|_0^6 = \frac{1}{36}(6)^2 - 0 = 1.$$

Thus Property II is also satisfied, and $f(x)$ is indeed a probability density function.

7. In order to satisfy Property I, k must be nonnegative, for if $k < 0$ and $1 \le x \le 3$, then $f(x) = kx < 0$. For Property II, set

$$\int_1^3 kxdx = 1,$$

and above for k.

$$\int_1^3 kxdx = \frac{k}{2}x^2\Big|_1^3 = \frac{k}{2}(9) - \frac{k}{2} = 4k = 1.$$

Therefore the value $k = \frac{1}{4}$ satisfies both Properties I and II, and $f(x) = \frac{1}{4}x, 1 \le x \le 3,$ is a probability density function.

13. The set of possible values of X is determined by the domain of the density function $f(x)$. In this problem the domain is $0 \le x \le 4$, so the values of X range between 0 and 4. Since X can't be less than 0, the probability that X is less than or equal to 1 is the same as the probability that X is between 0 and 1. Thus $\Pr(X \le 1)$ is an abbreviation for $\Pr(0 \le X \le 1)$. Similarly, since X cannot be larger than 4, $\Pr(3.5 \le X)$ is an abbreviation for $\Pr(3.5 \le X \le 4)$.

(a) $\Pr(X \le 1)$ is represented by the area under the graph of $f(x) = \frac{1}{8}x$ where $0 \le x \le 1$.

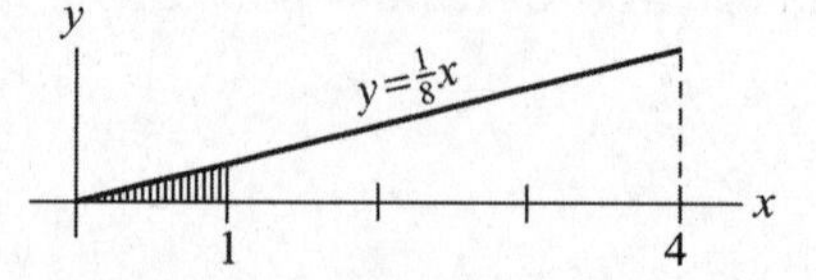

(b) $\Pr(2 \le X \le 2.5)$ is represented by the area under the graph of $f(x) = \frac{1}{8}x$ where $2 \le x \le 2.5$.

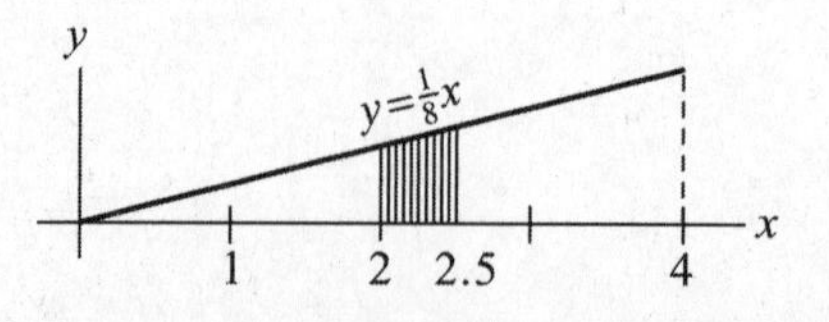

(c) $\Pr(3.5 \le X)$ is represented by the area under the graph of $f(x) = \frac{1}{8}x$ where $3.5 \le x \le 4$.

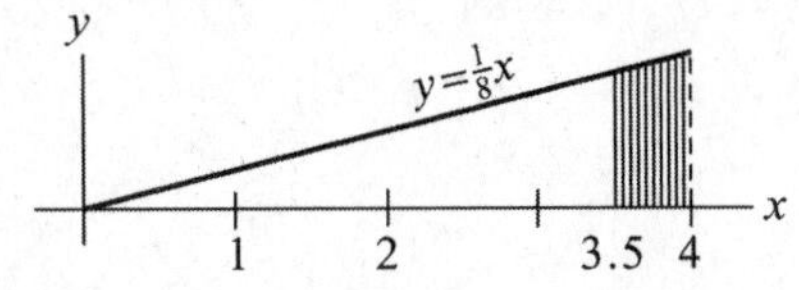

19. The probability that the lifetime X of a battery is *at least* 35 hours is given by $\Pr(35 \le X)$. However, we are told in this problem that the values of X range between 30 and 50. So $\Pr(35 \le X) = \Pr(35 \le X \le 50)$, and

$$\Pr(35 \le X) = \int_{35}^{50} \frac{1}{20}\,dx = \frac{1}{20}x\bigg|_{35}^{50} = \frac{50}{20} - \frac{35}{20} = \frac{15}{20} = \frac{3}{4}.$$

Thus the probability is $\frac{3}{4}$ that the battery will last longer than 35 hours.

25. **(a)** $\Pr(2 \le X \le 3) = \int_{1}^{3} \frac{1}{21}x^2\,dx = \frac{1}{63}x^3\bigg|_{2}^{3} = \frac{27}{63} - \frac{8}{63} = \frac{19}{63}.$

(b) Since $F(x)$ is an antiderivative of $f(x) = \frac{1}{21}x^2$,

$$F(x) = \int \frac{1}{21}x^2\,dx = \frac{1}{63}x^3 + C.$$

Since X has values greater than or equal to 1, $F(1) = 0$. Set $F(1) = \frac{1}{63} + C = 0$ to find that $C = -\frac{1}{63}$, so

$$F(x) = \frac{1}{63}x^3 - \frac{1}{63} = \frac{1}{63}(x^3 - 1).$$

(c) $\Pr(2 \le X \le 3) = F(3) - F(2)$

$$= \frac{1}{63}(3^3 - 1) - \frac{1}{63}(2^3 - 1) = \frac{26}{63} - \frac{7}{63} = \frac{19}{63}.$$

31. $\Pr(0 \le X \le 5) = \int_{0}^{5} 2ke^{-kx}\,dx = \frac{2k}{-k}e^{-kx}\bigg|_{0}^{5} = -2e^{-kx}\Big|_{0}^{5}.$

Since $k = (\ln 2)/10 \approx .0693$,

$$\Pr(0 \le X \le 5) = -2e^{-.0693x}\Big|_{0}^{5} = -2e^{(-.0693)(5)} - [-2e^{0}]$$
$$= -1.4142 + 2 = .5858.$$

Alternate calculation:

$$-2e^{-kx}\Big|_{0}^{5} = -2e^{-k(5)} - (-2e^{0}) = -2e^{-5(\ln 2)/10} + 2.$$

Observe that

$$-2e^{-5(\ln 2)/10} = -2e^{-(1/2)\ln 2} = -2e^{(\ln 2)^{-1/2}}$$
$$= -2 \cdot 2^{-1/2} = -2^{1/2} = -\sqrt{2}.$$

Thus

$$\Pr(0 \le X \le 5) = -\sqrt{2} + 2 \approx -1.4142 + 2 = .5858.$$

37. **(a)** Clearly, $f(x)=4x^{-5}\geq 0$ for all values of x greater than or equal to 1, so Property I is satisfied. For Property II, check that

$$\int_1^{\infty} 4x^{-5}dx = 1.$$

If $b \geq 1$, then

$$\int_1^b 4x^{-5}dx = -x^{-4}\Big|_1^b = -(b)^{-4} - [-(1)^{-4}] = 1 - \frac{1}{b^4}.$$

Since as $b \to \infty$, the number b^4 gets arbitrarily large and $\frac{1}{b^4}$ approaches 0. Thus

$$\int_1^{\infty} 4x^{-5}dx = \lim_{b\to\infty}\int_1^b 4x^{-5}dx = \lim_{b\to\infty}\left(1-\frac{1}{b^4}\right) = 1.$$

(b) Since $F(x)$ is an antiderivative of $f(x)=4x^{-5}$,

$$F(x) = \int 4x^{-5}dx = -x^{-4} + C.$$

Since X has values greater than or equal to 1, $F(1)=0$. Set $F(1)=-1+C=0$ to find that $C=1$, so

$$F(x) = -x^{-4} + 1 = 1 - x^{-4}.$$

(c) $\Pr(1\leq X\leq 2) = F(2) - F(1) = \frac{15}{16} - 0 = \frac{15}{16}$. Observe that $\Pr(1\leq X\leq 2) + \Pr(2\leq X) = 1$, since *all* values of X are greater than or equal to 1. Hence,

$$\Pr(2\leq X) = 1 - \Pr(1\leq X\leq 2) = 1 - \frac{15}{16} = \frac{1}{16}.$$

Helpful Hint: Decide now what you would do on an exam if you were given Exercise 37 with no mention of parts (a), (b), and $\Pr(1\leq x\leq 2)$. Certainly there would be no need to find the cumulative distribution function. Would you want to compute $\Pr(1\leq X\leq 2)$? Compute (a) and (b) below, and decide which approach is easier for you.

(a) $\Pr(2\leq X) = 1 - \Pr(1\leq X\leq 2) = 1 - \int_1^2 4x^{-5}dx.$

(b) $\Pr(2\leq X) = \int_2^{\infty} 4x^{-5}dx.$

12.3 Expected Value and Variance

The expected value of a random variable X is easier to understand and compute than the variance. However, both concepts are important in statistics. An explanation of how $E(X)$ and Var (X) are related to the graph of the probability density function for X will be given in Section 12.4, in the discussion of normal random variables. Check with your instructor to see if you need to know the definition of Var (X), or if the alternate formula for Var (X) will suffice. The alternate formula is easier to compute.

Calculations of $E(X)$ and Var (X) sometimes seem to require integration by parts, but this can be avoided when the probability density function is a polynomial. Observe how the integrals

$$\int_0^1 xf(x)dx \quad \text{and} \quad \int_0^1 x^2 f(x)dx$$

in Practice Problem 2 are simplified by expanding $f(x)$ so that $xf(x)$ and $x^2 f(x)$ are simple polynomials that are easily integrated.

1. Compute

$$E(X)=\int_0^6 x\cdot\frac{1}{18}x\,dx=\int_0^6\frac{1}{18}x^2dx=\frac{1}{54}x^3\Big|_0^6=4.$$

To compute Var (*X*) by the alternate formula, first compute

$$\int_0^6 x^2\cdot\frac{1}{18}x\,dx=\int_0^6\frac{1}{18}x^3dx=\frac{1}{72}x^4\Big|_0^6=\frac{1296}{72}=18.$$

Then use this value and *E*(*x*) to obtain

$$\text{Var}(X)=\int_0^6 x^2\cdot\frac{1}{18}x\,dx-E(X)^2=18-(4)^2=2.$$

7. Compute

$$E(X)=\int_0^1 x\cdot 12x(1-x)^2dx=\int_0^1 12x^2(1-2x+x^2)dx$$

$$=\int_0^1(12x^2-24x^3+12x^4)dx$$

$$=\left(4x^3-6x^4+\frac{12}{5}x^5\right)\Big|_0^1=\frac{2}{5}.$$

$$\text{Var}(X)=\int_0^1 x^2\cdot 12x(1-x)^2dx-E(X)^2$$

$$=\int_0^1(12x^3-24x^4+12x^5)dx-\left(\frac{2}{5}\right)^2$$

$$=\left(3x^4-\frac{24}{5}x^5+2x^6\right)\Big|_0^1-\frac{4}{25}$$

$$=\frac{1}{5}-\frac{4}{25}=\frac{1}{25}.$$

13. The average time is the expected value of *X*.

$$E(X)=\int_0^{12}\frac{1}{72}x^2dx=\frac{1}{216}x^3\Big|_0^{12}=\frac{1728}{216}=8.$$

Therefore, the average time spent reading the editorial page is 8 minutes.

19. Set $\int_0^M\frac{1}{18}x\,dx=\frac{1}{2}$ and solve for *M*.

$$\int_0^M\frac{1}{18}x\,dx=\frac{1}{36}x^2\Big|_0^M=\frac{1}{36}M^2=\frac{1}{2}.$$

So $M^2=18$, and $M=\sqrt{18}=3\sqrt{2}$. Note that $-3\sqrt{2}$ is not a solution since $0\le M\le 6$.

25. By definition $E(X)=\int_A^B xf(x)dx$. Integrate by parts, setting

$$h(x)=x,\quad f(x)\quad [= g(x) \text{ in integration by parts formula}]$$
$$h'(x)=1,\quad F(x)\quad [= \text{antiderivative of } f(x)]$$

where $f(x)$ is any probability density function on $A \le x \le B$ and $F(x)$ is the cumulative distribution function corresponding to $f(x)$. Then

$$\begin{aligned} E(X) &= \int_A^B xf(x)dx = xF(x)\Big|_A^B - \int_A^B 1\cdot F(x)dx \\ &= B\cdot F(B) - A\cdot F(A) - \int_A^B F(x)dx \\ &= B\cdot 1 - A\cdot 0 - \int_A^B F(x)dx \qquad [F(B)=1,\ F(A)=0] \\ &= B - \int_A^B F(x)dx, \end{aligned}$$

12.4 Exponential and Normal Random Variables

This section deserves your attention because exponential and normal random variables arise so often in applications. In a situation when an exponential density function ke^{-kx} is appropriate for a random variable X, the value of $E(X)$ is often estimated from experimental data, and then the constant k is obtained from the relation

$$E(X)=\frac{1}{k},\quad \text{or equivalently,}\quad k=\frac{1}{E(x)}.$$

You need to know this relation in order work Exercises 5–14.

The text's discussion of arbitrary normal random variables provides an opportunity to review changes of variable in a definite integral. Check with your instructor about how much detail you should show when you make a substitution $z=(x-\mu)/\sigma$.

1. Here $k=3$, so

$$E(X)=\frac{1}{3},\quad \text{and}\quad \text{Var}(X)=\frac{1}{9}.$$

7. The mean is given by $E(X)=\frac{1}{k}=3$. Hence $k=\frac{1}{3}$, and the probability density function is $f(x)=\frac{1}{3}e^{-(1/3)x}$. The probability that a customer is served in less than 2 minutes is given by

$$\begin{aligned} \Pr(0\le X\le 2) &= \int_0^2 \frac{1}{3}e^{-(1/3)x}dx = -e^{-(1/3)x}\Big|_0^2 \\ &= -e^{-2/3}-(-e^0) \\ &= 1-e^{-2/3}. \end{aligned}$$

13. The average life span (or mean) is given by $E(X) = \frac{1}{k} = 72$. Hence $k = \frac{1}{72}$, and the probability density function is $f(x) = \frac{1}{72}e^{-(1/72)x}$.

(a) The probability that a component lasts for more than 24 months is given by

$$\begin{aligned}\Pr(24 \le X) &= \int_{24}^{\infty} \frac{1}{72} e^{-(1/72)x} dx \\ &= \lim_{b\to\infty} \int_{24}^{b} \frac{1}{72} e^{-(1/72)x} dx \\ &= \lim_{b\to\infty} \left[-e^{-(1/72)x} \Big|_{24}^{b} \right] \\ &= \lim_{b\to\infty} \left[e^{-1/3} - e^{-b/72} \right]\end{aligned}$$

Note that as $b \to \infty$, the number $-\frac{b}{72}$ approaches $-\infty$, so $e^{-b/72}$ approaches 0. Therefore,

$$\Pr(24 \le X) = \lim_{b\to\infty} [e^{-1/3} - e^{-b/72}] = e^{-1/3}.$$

(b) To find the reliability function, repeat the calculations in (a) with the number 24 replaced by the variable t.

$$\begin{aligned}r(t) = \Pr(t \le X) &= \int_{t}^{\infty} \frac{1}{72} e^{-(1/72)x} dx \\ &= \lim_{b\to\infty} \int_{t}^{b} \frac{1}{72} e^{-(1/72)x} dx \\ &= \lim_{b\to\infty} \left[-e^{-(1/72)x} \Big|_{t}^{b} \right] \\ &= \lim_{b\to\infty} \left[e^{-t/72} - e^{-b/72} \right] \\ &= e^{-t/72}.\end{aligned}$$

19. To find a relative maximum, first set $f'(x) = 0$ and solve for x.

$$f'(x) = -xe^{-x^2/2} = 0.$$

Since $e^{-x^2/2}$ is strictly positive for all values of x, you can see that $f'(x) = 0$ if and only if $x = 0$. To check that $f(x)$ has a maximum at $x = 0$, check $f''(0)$.

$$\begin{aligned}f''(x) &= x^2 e^{-x^2/2} - e^{-x^2/2} \\ f''(0) &= 0 - 1 = -1 < 0.\end{aligned}$$

Thus $f(x)$ is indeed concave down at $x = 0$. Therefore $f(x)$ has a relative maximum at $x = 0$.

25. **(a)** We have $\mu = 6$ and $\sigma = \frac{1}{2}$. Note that $\dfrac{x-\mu}{\sigma} = \dfrac{x-6}{1/2} = 2x - 12$ and

$$\frac{1}{\sigma\sqrt{2\pi}} = \frac{1}{(1/2)\sqrt{2\pi}} = \frac{2}{\sqrt{2\pi}}.$$

Thus, the normal density function for X, the gestation period, is

$$f(x) = \frac{2}{\sqrt{2\pi}} e^{-(1/2)(2x-12)^2}$$

So,

$$\Pr(6 \le X \le 7) = \int_6^7 \frac{2}{\sqrt{2\pi}} e^{-(1/2)(2x-12)^2}\,dx.$$

Use the substitution $z = 2x-12$, $dz = 2dx$. If $x = 6$, then $z = 0$, and if $x = 7$, then $z = 2$. So

$$\int_6^7 \frac{2}{\sqrt{2\pi}} e^{-(1/2)(2x-12)^2}\,dx = \int_0^2 \frac{1}{\sqrt{2\pi}} e^{-(1/2)z^2}\,dz.$$

(Do not forget to change the limits of integration.) The value of this integral is the area under the standard normal curve form 0 to 2. Use Table 1 in the Appendix of the text to find

$$\Pr(6 \le X \le 7) = A(2) = .4772.$$

Therefore, about 47.72% of births occur after a gestation period of between 6 and 7 months.

(b) Proceeding as in part (a), use the same substitution to evaluate

$$\int_5^6 \frac{2}{\sqrt{2\pi}} e^{-(1/2)(2x-12)^2}\,dx,$$

noting that if $x = 5$, then $z = 2x-12 = -2$, and if $x = 6$, then $z = 0$. Therefore,

$$\Pr(5 \le X \le 6) = \int_{-2}^0 \frac{2}{\sqrt{2\pi}} e^{-(1/2)z^2}\,dz$$
$$= A(-2) = A(2) = .4772.$$

Therefore, about 47.72% of births occur after a gestation period of between 5 and 6 months.

Warning: You can use the TI-83 calculator to check your calculations of areas under normal curves, but do not use it to replace all of your written work. Many instructors use Section 12.4 to review the change of variable in a definite integral, and they may wish you to show these calculations and just use the area table $A(z)$ or the TI-83 to find areas under the *standard* normal curve. For example, the probability computed in exercise 25(a) is given by each of the commands

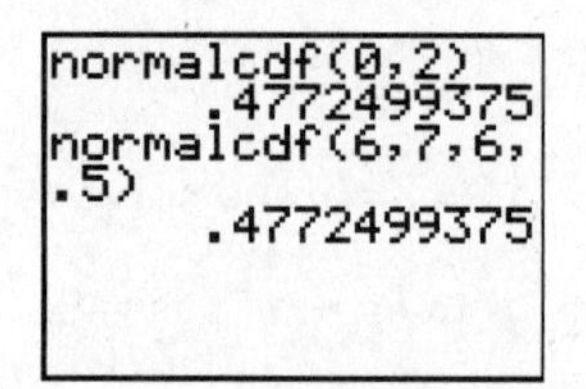

Your instructor may tell you to use only the first form of normalcdf on exams. To put normalcdf on the TI-83 home screen, press [2nd][DIST] 2. See page 582 in the text or a TI-83 manual for more information about the normal cumuluative distribution function (cdf).

31. The normal density function for this exercise is

$$f(x) = \frac{1}{(.8)\sqrt{2\pi}} e^{-(1/2)[(x-18.2)/.8]^2}.$$

Let X be the diameter of a bolt selected at random from the supply of bolts. Then

$$\Pr(20 \le X) = \int_{20}^{\infty} \frac{1}{(.8)\sqrt{2\pi}} e^{-(1/2)[(x-18.2)/.8]^2}\,dx.$$

Use the substitution $z = (x-18.2)/.8$, $dz = (1/.8)dx$, and note that if $x = 20$, then $z = 1.8/.8 = 2.25$, and if $x \to \infty$, then $z \to \infty$ too. Hence

$$\Pr(20 \le X) = \int_{2.25}^{\infty} \frac{1}{\sqrt{2\pi}} e^{-(1/2)z^2}\,dz.$$

The value of this integral is the area under the standard normal curve from 2.25 to ∞. If we consider areas under the standard normal curve, we see that

$$\begin{bmatrix}\text{area to}\\ \text{right of } 2.25\end{bmatrix}=\begin{bmatrix}\text{area to}\\ \text{right of } 0\end{bmatrix}-\begin{bmatrix}\text{area between } 0\\ \text{and } 2.25\end{bmatrix}$$

That is,

$$\begin{aligned}\Pr(20 \le X) &= .5 - A(2.25)\\ &= .5 - .4878\\ &= .0122.\end{aligned}$$

Therefore, about 1.22% of the bolts will be discarded.

37. Set $Y_1 = e\text{^}(-X\text{^}2/2)/\sqrt{(2\pi)}$. On a TI-83/84 calculator, you can set $Y_1 = \text{normalpdf(X)}$. The command is found in the DIST menu. To graph Y_1, use the window $[-5, 5]$ by $[-.2, .5]$, with $\text{Yscl} = .1$. For the definite integral of $x^2 f(x)$, use fnInt $(X\text{^}2 * Y_1, X, -8, 8)$.

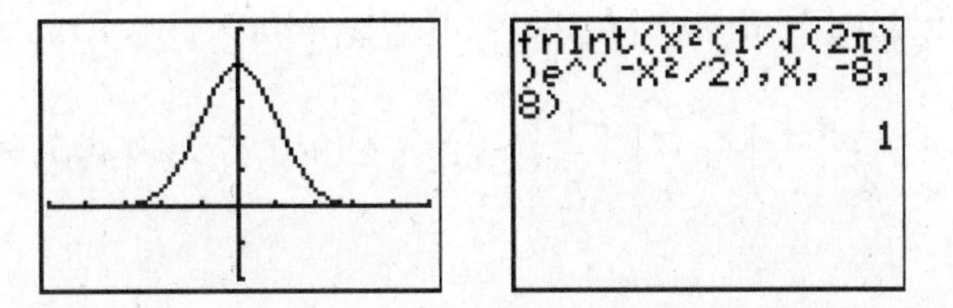

The fnInt command is on the MATH menu for the TI-82 and TI-83/84, and on the CALC menu for the TI-85 and TI-86. Repeat the command (use the [2nd] ENTRY key to save time), with ± 8 replaced by $\pm N$, for $N = 4$ and 6. Report your calculations, which should suggest that the improper integral of $x^2 f(x)$ from $-\infty$ to ∞ is 1. Since $E(Z) = 0$, this calculation shows that $\text{Var}(Z) = 1$ and hence the standard deviation of Z is 1, too.

12.5 Poisson and Geometric Random Variables

This last section of the text combines the probability theory in this chapter with the Taylor series for e^x (from Section 11.5) and the geometric series (Section 11.3). For convenience, state all probabilities to four decimal places.

1. From Example 1, $p_5 = .1008$. Since $p_n = (\lambda/n)\, p_{n-1}$,

$$\begin{aligned}p_6 &= \left(\frac{3}{6}\right) p_5 = 3 * .1008/6 = .0504\\ p_7 &= \left(\frac{3}{7}\right) p_6 = 3 * .0504/7 = .0216\\ p_8 &= \left(\frac{3}{8}\right) p_7 = 3 * 0.216/8 = .0081.\end{aligned}$$

7. Since the average of a Poisson random variable equals the parameter for the variable, $\lambda = 1.5$ for this problem.

(a) $\Pr(X = 0) = p_0 = e^{-\lambda} = e^{-1.5} = .2231.$

(b) First, compute $p_1 = (1.5/1)p_0 = .3347$, $p_2 = (1.5/2)p_3 = .2510$, and $p_3 = (1.5/3)p_2 = .1255$. Then

$$p_2 + p_3 = .2510 + .1255 = .3765.$$

(c) $\Pr(X \geq 4) = 1 - \Pr(X < 4)$

$$= 1 - (p_0 + p_1 + p_2 + p_3)$$
$$= 1 - (.2231 + .3347 + .2510 + .1255)$$
$$= 1 - .9343$$
$$= .0657.$$

13. The number of X of Red taxis that appear before the first Blue taxi appears is a geometric random variable, with "success" corresponding to the arrival of a Red taxi. At any given time, the probability of success is $p = 3/4$, because three out of every four cabs are Red.

(a) For $n \geq 1$, the probability of n successes (Red taxis) before the first failure (a Blue taxi) is

$$p_n = p^n(1-p) = \left(\frac{3}{4}\right)^n \left(\frac{1}{4}\right).$$

(b) The probability of observing at least three Red taxis in a row is 1 minus the probability of observing less than three Red taxis, namely, $1 - (p_0 + p_1 + p_2)$. This probability is

$$1 - \left(\frac{1}{4} + \left(\frac{3}{4}\right)\left(\frac{1}{4}\right) + \left(\frac{3}{4}\right)^2\left(\frac{1}{4}\right)\right) = 1 - (.25 + .1875 + .1406) \approx .4219.$$

(c) The average number of Red taxis observed in a row is given by $E(X) = p/(1-p) = \frac{3}{4} \div \frac{1}{4} = 3.$

19. It can be shown that the number of defective fuses found in a box selected at random is a Poisson random variable. Let $f(\lambda) = (\lambda^2/2)e^{-\lambda}$, the probability of selecting a box with two defective fuses. Then,

$$f'(\lambda) = \left(\frac{\lambda^2}{2}\right)\left(-e^{-\lambda}\right) + \lambda e^{-\lambda} \quad \text{(by the product rule)}$$
$$= \left(\lambda - \frac{\lambda^2}{2}\right)e^{-\lambda}$$
$$f''(\lambda) = \left(\lambda - \frac{\lambda^2}{2}\right)\left(-e^{-\lambda}\right) + (1-\lambda)e^{-\lambda}$$

A possible maximum occurs when

$$\left(\lambda - \frac{\lambda^2}{2}\right)e^{-\lambda} = 0, \quad \text{or} \quad \lambda\left(1 - \frac{\lambda}{2}\right)e^{-\lambda} = 0.$$

Since $e^{-\lambda} \neq 0$ and $\lambda \neq 0$ (because there are some defective fuses in the box sampled), we conclude that $\lambda = 2$. Checking $f''(2) = 0 + (1-2)e^{-2}$, we see that $f''(2)$ is negative, so $f(\lambda)$ has a maximum at $\lambda = 2$. Thus, given one sample which has two defective fuses, the maximum likelihood estimate of the Poisson parameter is $\lambda = 2$.

25. **(a)** The formula $p_n = (\lambda/n)p_{n-1}$ shows that $p_7 = (6.9/7)p_6 < p_6$. So the answer is "no," seven babies are less likely to be born than six babies.

(b) The answer requires computing $p_0 + \cdots + p_{15}$. Use the basic formula for p_n, and the **sum(seq** (. . .)) command, (discussed on pages 312–313 and 590 in the text):

$$\text{sum(seq}(\lambda^n/n! * p_0, n, 0, 15, 1))$$

Here $\lambda = 6.9$ and $p_0 = e^{-6.9}$. Note X is more convenient to use than n. The sum is

$$\text{sum(seq(6.9\^{}X/X!* e\^{} - 6.9,X,0,15,1))} \approx .997906$$

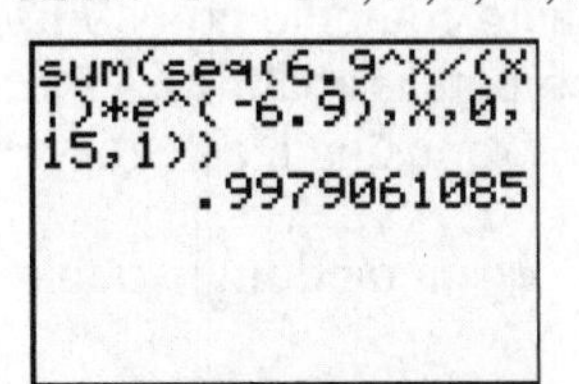

For the TI-83/84, you can use the command **poissonpdf**, which is entry "C" on the DISTR menu. Place this inside the sum(seq (. . .)) command, as follows:

$$\text{sum(seq(poissonpdf(6.9, X), X,0,15,1))}.$$

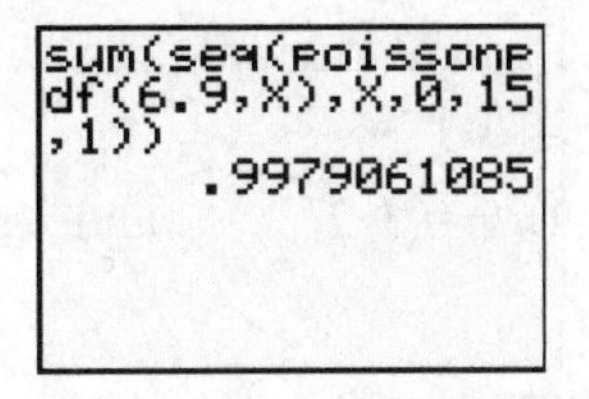

Warning: The TI-83/84 calculator has commands for both Poisson probabilities and geometric probabilities, on the DIST menu. However, the TI-83 definition of geometric probabilities is different from the one used in the text. ("Success" and "failure" are interchanged and the numbering of the probabilities $p_0, p_1, \ldots,$ is different.) You can safely use poissonpdf and poissoncdf to check your work, but ask your instructor about any limitations on using these functions on an exam. See the text, page 590, for details about the Poisson functions.

Review of Chapter 12

Work through the Review of the Fundamental Concepts at the end of the chapter and determine which concepts you must know and which terms and formulas you must memorize. Summarized below are the main skills you should have, listed by the various types of random variables.

(a) *Discrete random variable.* Compute $E(X)$, Var (X), and the standard deviation of X.

(b) *Continuous random variable on a finite interval.* Compute $E(X)$, Var (X); test if $f(x)$ is a probability density function; given one of $f(x)$ and $F(x)$, find the other; use $f(x)$ or $F(x)$ to compute probabilities. Remember that $\Pr(X \le b)$ stands for $\Pr(A \le X \le b)$, etc. Harder questions are found in exercises 32–36 of Section 12.2.

(c) *Continuous random variable on an infinite interval.* A typical probability density function is $f(x) = kx^{-(k+1)}$ for $x \ge 1$. Same skills as for (b), plus use the formula $\Pr(X \ge b) = 1 - \Pr(X \le b)$ to simplify calculations.

(d) *Exponential random variable.* The probability density function is $f(x) = ke^{-kx}$ for $x \geq 0$. Show that $f(x)$ *is* a probability density functions. Compute various probabilities and the cumulative distribution function. You probably will not have to compute $E(X)$ and Var (X), unless your instructor gives you the following limits to learn:

$$\lim_{b\to\infty} be^{-kb} = 0 \quad \text{and} \quad \lim_{b\to\infty} b^2e^{-kb} = 0 \quad \text{for } k > 0.$$

However, you should know that $E(X) = \frac{1}{k}$ and $\text{Var}(X) = \frac{1}{k^2}$.

(e) *Normal random variable.* Use a table to compute probabilities for a standard normal random variable, and make a change of variable to compute probabilities for an arbitrary normal random variable. Your instructor may set up guidelines for using a calculator to compute areas under normal curves.

(f) *Poisson random variable.* If X is a Poisson random variable with parameter λ, then you should know that $p_0 = e^{-\lambda}$,

$$p_n = \frac{\lambda^n}{n!}e^{-\lambda} (n = 1, 2, \ldots), \quad \text{and} \quad E(X) = \lambda.$$

(g) *Geometric random variable.* If X is a geometric random variable and the probability of success on one trial is p, then you should know that

$$p_n = p^n(1-p) \quad (n = 0, 1, 2, \ldots), \quad \text{and} \quad E(X) = p/(1-p).$$

Chapter 12: Supplementary Exercises

1. (a) $\Pr(X \leq 1) = \int_0^1 \frac{3}{8}x^2dx = \frac{1}{8}x^3\Big|_0^1 = \frac{1}{8} - 0 = .125.$

$$\Pr(1 \leq X \leq 1.5) = \int_1^{1.5} \frac{3}{8}x^2dx = \frac{1}{8}x^3\Big|_1^{1.5} = .4219 - .125 = .2969.$$

(b) $E(X) = \int_0^2 xf(x)dx = \int_0^2 x \cdot \frac{3}{8}x^2dx = \frac{3}{8} \cdot \frac{1}{4}x^4\Big|_0^2 = \frac{3}{2}.$

$$\text{Var}(X) = \int_0^2 x^2 f(x)dx - E(X)^2 = \int_0^2 x^2 \cdot \frac{3}{8}x^2dx - \left(\frac{3}{2}\right)^2$$

$$= \frac{3}{8}\int_0^2 x^4dx - \frac{9}{4} = \frac{3}{8} \cdot \frac{1}{5}x^5\Big|_0^2 - \frac{9}{4} = \frac{12}{5} - \frac{9}{4} = \frac{3}{20}.$$

7. (a) From the definition of expected value,

$$E(X) = 1(.599) + 11(.401)$$
$$= 5.01.$$

This can be interpreted as saying that if a large number of samples are tested in batches of ten samples per batch, then the average number of tests per batch will be close to 5.01.

(b) The 200 samples will be divided into 20 batches of ten samples. In part (a) we found that, on average, 5.01 tests must be run on each batch. So the laboratory should expect to run about $(20)(5.01) \approx 100$ tests altogether.

13. **(a)** Find the value of k that satisfied the equation

$$\int_5^{25} kxdx = 1.$$

So compute

$$\int_5^{25} kxdx = \frac{k}{2}x^2\Big|_5^{25} = \frac{k}{2}625 - \frac{k}{2}25 = 300k = 1.$$

Therefore $k = \frac{1}{300}$.

(b) $\Pr(20 \le X \le 25) = \int_{20}^{25} \frac{1}{300} xdx = \frac{1}{600}x^2\Big|_{20}^{25}$

$$= \frac{1}{600}625 - \frac{1}{600}400 = \frac{1}{600}(625 - 400) = .375.$$

(c) $E(X) = \int_5^{25} \frac{1}{300}x^2dx = \frac{1}{900}x^3\Big|_5^{25} = \frac{1}{900}(15{,}625 - 125) = 17.222.$

Therefore, the mean annual income is \$17,222.

19. To show that $\Pr(X \le 4) = 1 - e^{-4k}$, compute

$$\Pr(X \le 4) = \int_0^4 ke^{-kx}dx = -e^{-kx}\Big|_0^4 = -e^{-4k} - (-e^0) = 1 - e^{-4k}.$$

To make $\Pr(X \le 4) = .75$, the parameter k must satisfy

$$1 - e^{-4x} = .75$$
$$e^{-4x} = .25$$
$$-4x = \ln .25 \approx -1.386.$$

Thus

$$k = \frac{-1.386}{-4} \approx .35.$$

25. Recall that $\Pr(a \le Z)$ is the area under the standard normal curve to the right of a. Since $\Pr(a \le Z) = .4$, and $\Pr(0 \le Z) = .5$, the number a lies to the right of 0. That is, a is positive. Thus,

$$\begin{bmatrix}\text{area to}\\ \text{right of } a\end{bmatrix} = \begin{bmatrix}\text{area to}\\ \text{right of } 0\end{bmatrix} - \begin{bmatrix}\text{area between } 0\\ \text{and } a\end{bmatrix}$$
$$\Pr(a \le Z) = \Pr(0 \le Z) - \Pr(0 \le Z \le a)$$
$$.4 = .5 - \Pr(0 \le Z \le a).$$

So $\Pr(0 \le Z \le a) = A(a) = .1$. From the $A(z)$ table, look for the z that makes $A(z)$ as close to .1 as possible. Since $A(.25) = 0.987 \approx .1000$, we conclude that $a \approx .25$.

31. For $\lambda = 4$, $p_4 = \lambda^4 \frac{e^{-4}}{4!} = (4)^4 \frac{e^{-4}}{24} \approx .1954.$

VISUAL CALCULUS

The student oriented software package *Visual Calculus*, which runs on both Macintosh and PC computers, was developed to accompany the textbook. *Visual Calculus* enhances the visualization and therefore the understanding of the ideas behind the calculus. It is easy to use and focuses on the learning of calculus concepts. The software is accompanied by a 43-page manual in Adobe Reader format.

To obtain a CD containing *Visual Calculus*, send an email to David Schneider at dis@math.umd.edu. (You may freely make copies of the CD.) The software also is available for download from the website www.pearsoned.com/goldstein.

Visual Calculus contains two types of routines: Main Routines (contains the graphing and calculating routines) and Demonstration Routines (contains mainly visual demonstrations).

Main Routines

Analyze a Function and its Derivatives displays the graphs of one or more of f, f', f" and displays the values of these functions at any point. A Solve command finds zeros, relative extreme points, and inflection points. A Tangent Line command draws a tangent line to f(x) or f'(x) at the crosshairs.

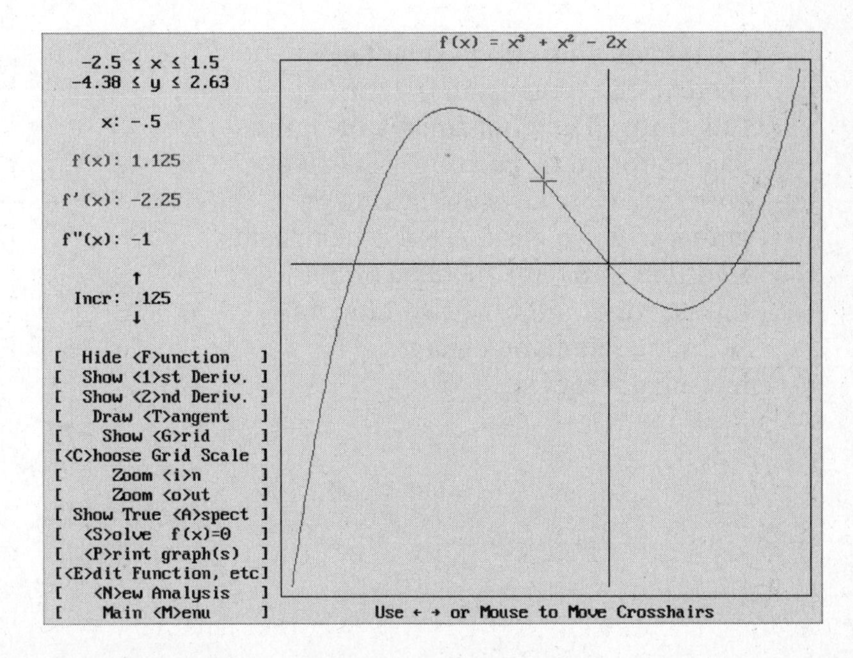

Examine Solutions of Differential Equations graphs the solutions of a differential equation of the form $y' = g(t,y)$ with an initial value and determines the coordinates of points on the graph. Graphs with different initial values can be superimposed. Successive Euler approximations may be displayed.

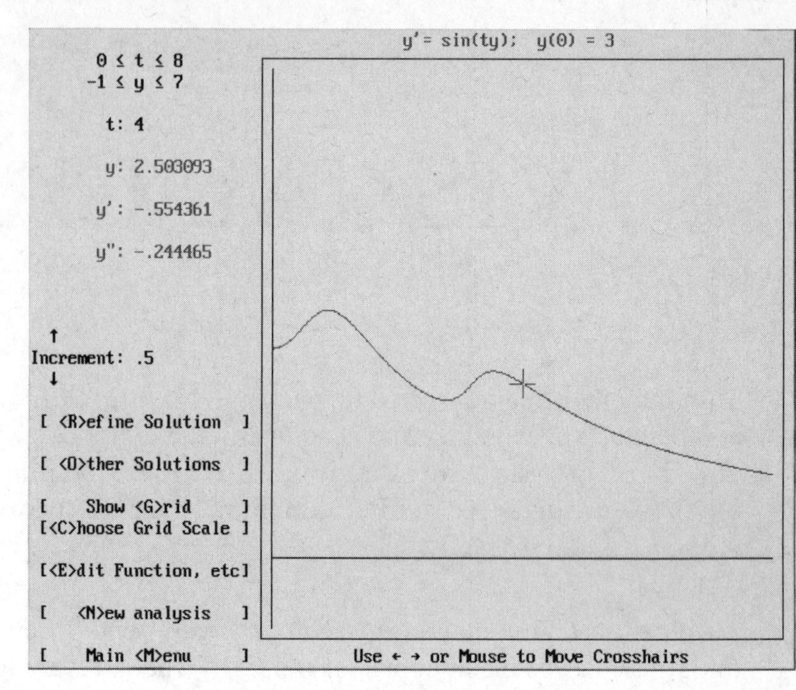

Function Evaluators (1 and 2 variables) These two routines display the value of a function and its derivatives. [1 variable: $f(x)$, $f'(x)$, and $f''(x)$ are displayed. 2 variables: $f(x,y)$, $\partial f/\partial x$, and $\partial f/\partial y$ are displayed.] A single push of a button increases the value of x (and/or y) by an amount input by the user and recomputes the function values.

Function Evaluator

f(x) = sqr(1 - x^2)

x = 1.1
Increment: .1

x	f(x)	f'(x)	f"(x)
0	1	0	-1
.1	.994987	-.1005	-1.01519
.2	.979796	-.20412	-1.06315
.3	.953939	-.31449	-1.15196
.4	.916515	-.43644	-1.29892
.5	.866025	-.57735	-1.5396
.6	.8	-.75	-1.95313
.7	.714143	-.9802	-2.74565
.8	.6	-1.33333	-4.62963
.9	.43589	-2.06474	-12.0745
1	0	Undefined	
1.1	Undefined		

Increment and Evaluate

New Problem

Main Menu

Graphs of Functions simultaneously sketches the graphs of 1, 2, or 3 functions. The coordinates of points can be found by moving crosshairs with the arrow keys or clicking on points with a mouse. A Solve command finds the zeros of a function or the intersection points of two functions. A Zoom command enlarges a rectangular region.

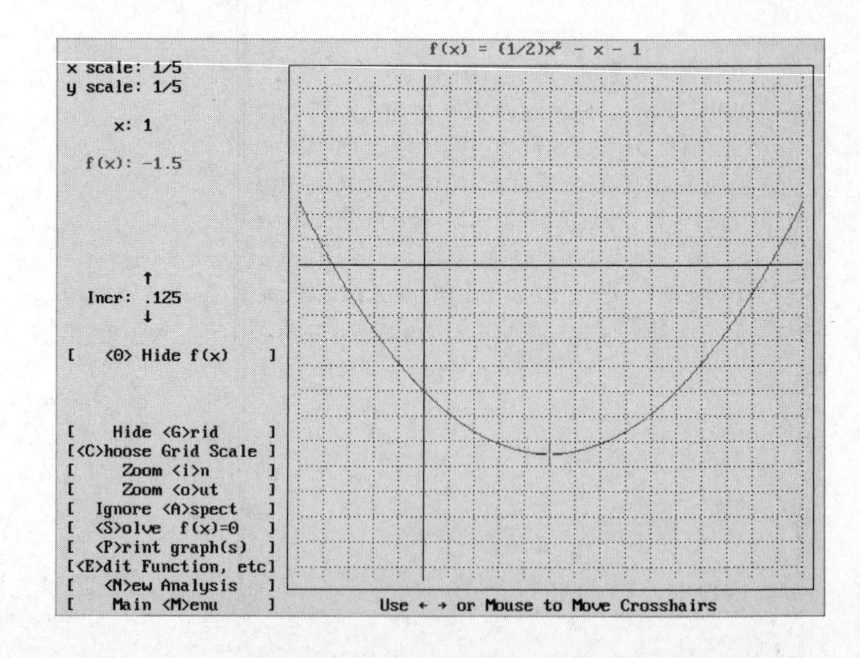

Integrals, Definite calculates the definite integral of a function. Also, illustrates five numerical methods.

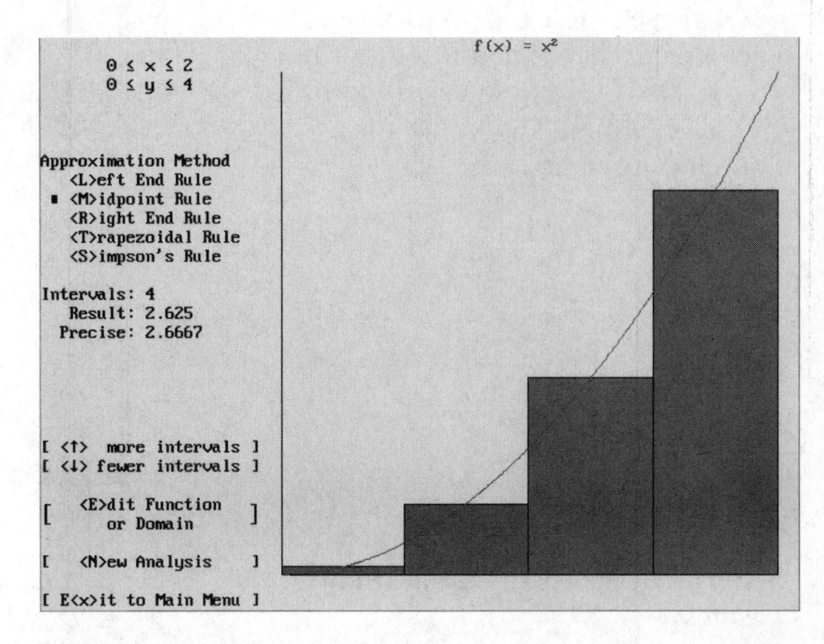

Least Squares calculates the line of best fit and also allows the user to move a line around the screen to find the best fit.

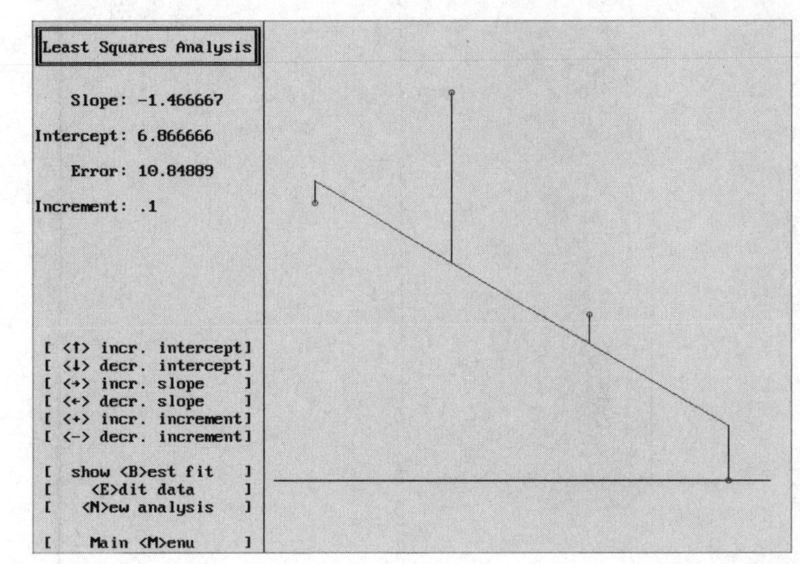

Newton's Method finds a zero of a function by successively drawing tangent lines.

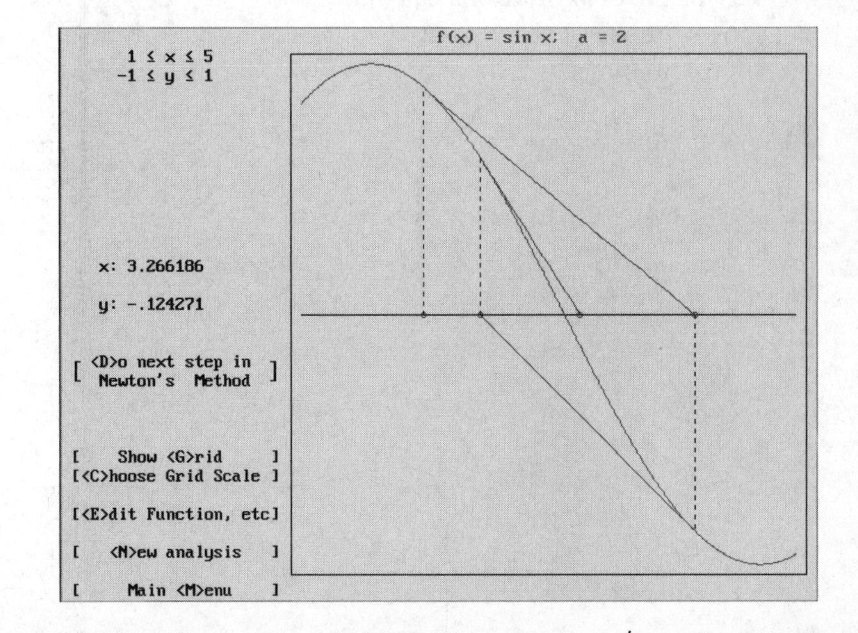

Observe Limit of f(x) as $x \to a$ evaluates f(a+h) and f(a−h) for h = .1, .01, .001, ... and guesses the limit. You can have any value for a including ∞ and -∞.

Limit of f(x) as x approaches a

f(x)=(x^2 - x - 6)/(x - 3) a = 3

Enter function or press F4 to choose from list

Clear f(x) | egin approaching a | Main Menu

h	f(a-h)	f(a+h)
.1	4.9	5.1
.01	4.99	5.01
.001	4.999	5.001
.0001	4.9999	5.0001
.00001	4.99999	5.00001

A good guess is that the limit is 5

Series evaluates finite and infinite sums.

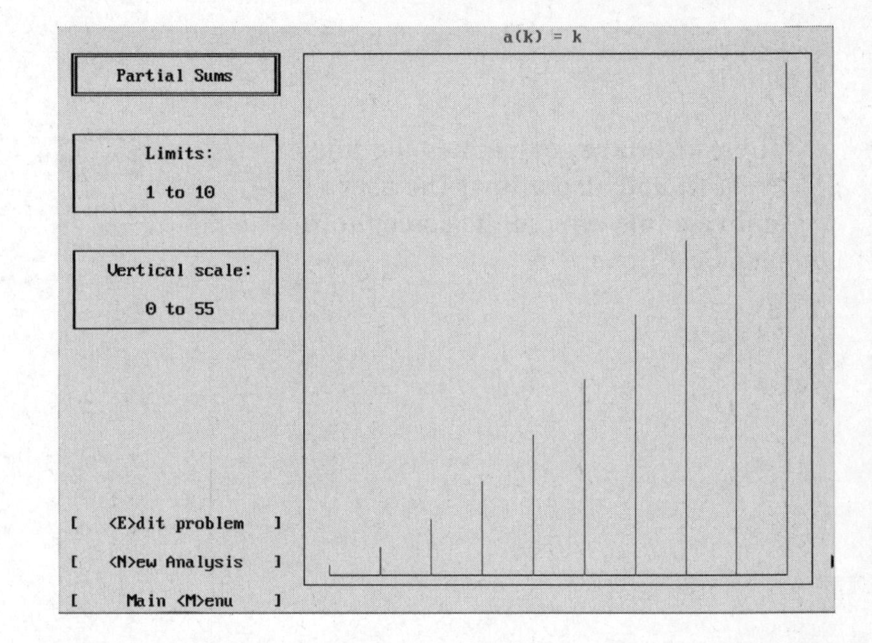

Taylor Approximations calculates and shows graphs of successive approximations.

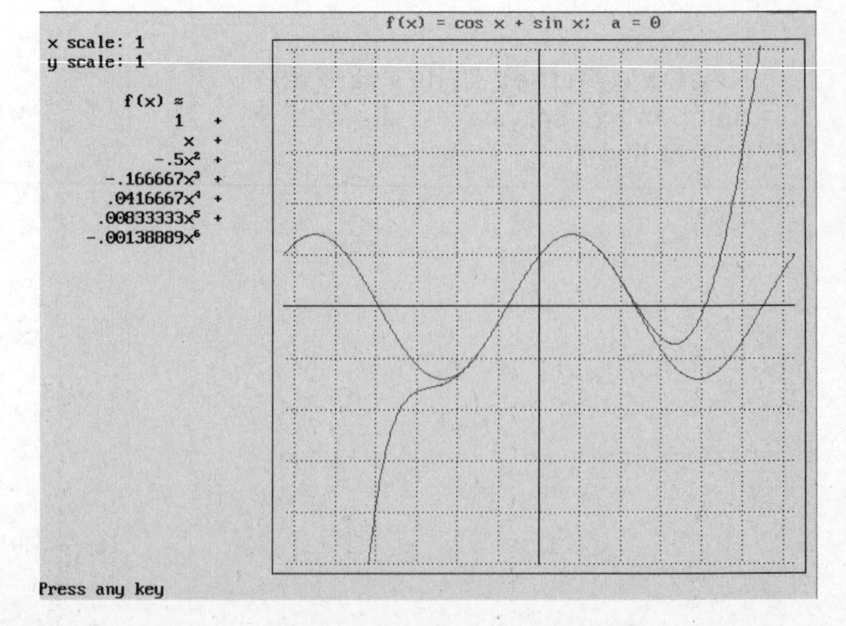

Demonstration Routines

Ball Thrown Straight Up shows the connection between the motion of a ball and the graph of its height function.

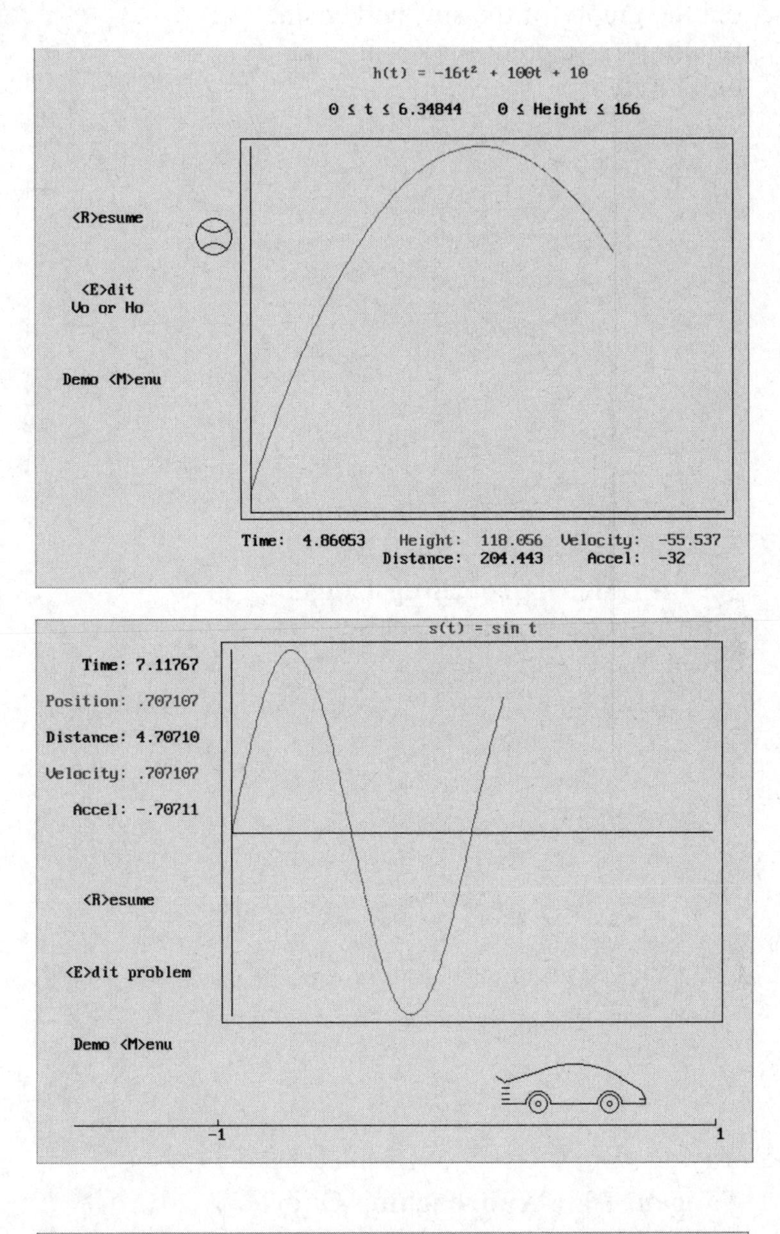

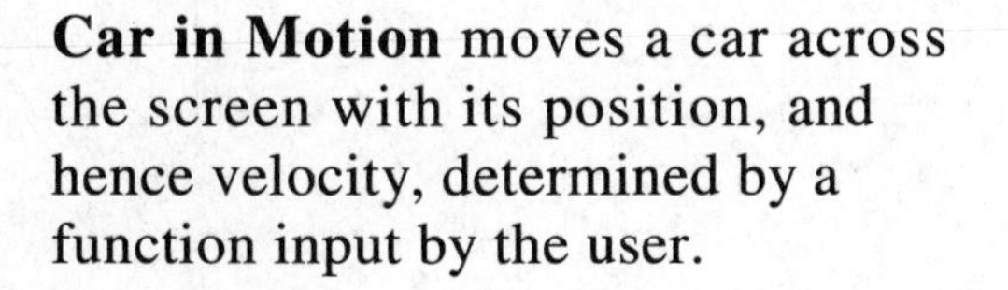

Car in Motion moves a car across the screen with its position, and hence velocity, determined by a function input by the user.

Garden Fencing Problem shows the connection between a function to be optimized and the corresponding physical problem.

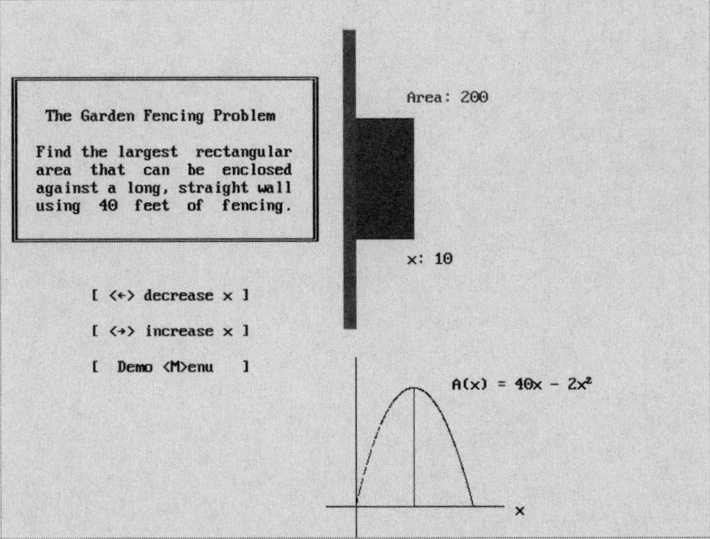

Plot Cosine and Sine Functions traces out the graphs of the sine and cosine functions as a point moves around the unit circle.

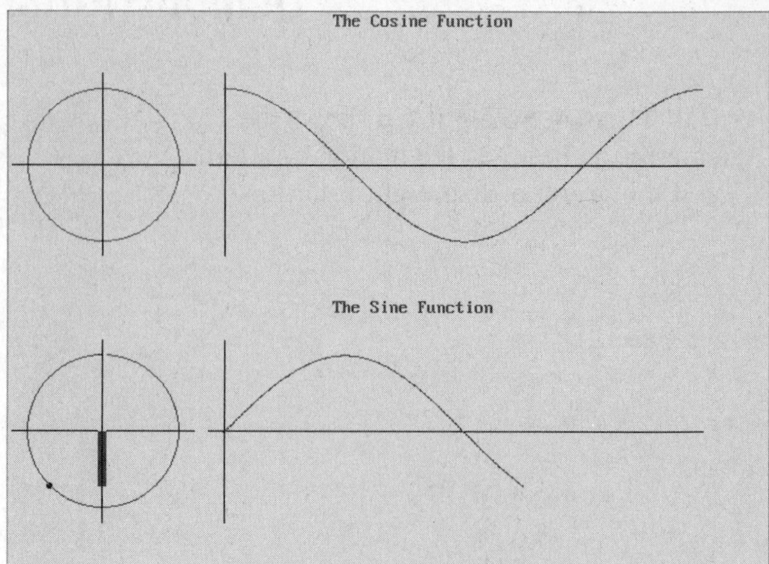

Secant Line Approaching Tangent Line moves a secant line closer and closer to a tangent line.

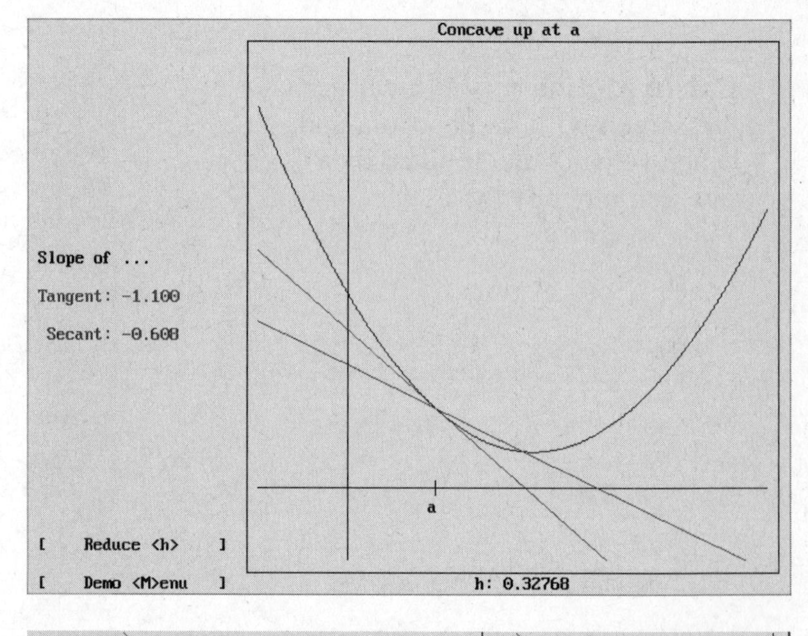

Tangent Line Approaching Curve illustrates the fact that locally a curve looks like its tangent line.

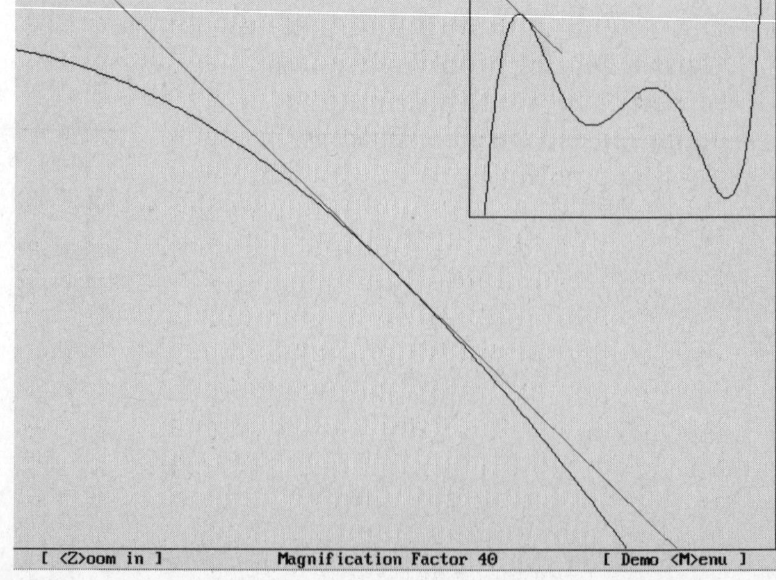